# Applied Fluid Mechanics

(English-Chinese bilingual version)

# 工程流体力学

（英汉双语版）

李　聪　毕德贵　杨一琼　高玉琼　张云澍　等编著

上海大学出版社

·上海·

图书在版编目(CIP)数据

工程流体力学:汉英对照 / 李聪等编著. -- 上海: 上海大学出版社,2024.10. -- ISBN 978-7-5671-4930-4

Ⅰ.TB126

中国国家版本馆 CIP 数据核字第 2024GD0145 号

责任编辑　王悦生
封面设计　柯国富
技术编辑　金　鑫　钱宇坤

### 工程流体力学(英汉双语版)

李　聪　毕德贵　杨一琼　高玉琼　张云澍　等编著
上海大学出版社出版发行
(上海市上大路 99 号　邮政编码 200444)
(https://www.shupress.cn　发行热线 021-66135112)
出版人　余　洋

\*

南京展望文化发展有限公司排版
上海颛辉印刷厂有限公司印刷　各地新华书店经销
开本 787mm×1092mm　1/16　印张 24.5　字数 627 千
2024 年 10 月第 1 版　2024 年 10 月第 1 次印刷
ISBN 978-7-5671-4930-4/TB·21　定价 88.00 元

版权所有　侵权必究
如发现本书有印装质量问题请与印刷厂质量科联系
联系电话: 021-57602918

# Preface

Fluid mechanics is a basic course for various engineering majors such as environmental engineering, energy and power in colleges and universities in China, and it is also a major professional course commonly opened in this discipline in the world. Because the background to build a world-class university with Chinese characteristics is China's higher education development strategy in the new era, an English course in engineering fluid mechanics is an inevitable requirement for the internationalization of engineering teaching in China. However, at present, the English version of fluid mechanics textbooks suitable for national conditions is relatively lacking. Moreover, the curriculum involves multidisciplinary interdisciplinarity; the existing textbooks have the characteristics of complex theoretical analysis and difficult to master; more concepts and equation derivations and easy to confuse; and high requirements for students' ability to analyze and deal with problems with advanced mathematical knowledge and comprehensive application of multi-knowledge, which affect the teaching effect of international students. Based on the above disadvantages, this book aims to revise and publish Einglish-Chinse bilingual engineering fluid mechanics textbooks suitable for environmental engineering majors, including the necessary basic theory of fluid motion, hydraulic characteristics, focusing on of fluid mechanics in civil engineering, hydraulic conservancy and other engineering, which are easier to understand and practical, so as to make up for the existing deficiencies.

This book is one of the first-class undergraduate textbooks of the University of Shanghai for Science and Technology. In the book, many examples and calculations of flow phenomena have been analyzed, and students and teachers can easily understand relevant knowledge. In Chapter 1, the introduction of the concept of fluid mechanics have been described and written by Li Cong, Sun Chunmeng and Yu Jianping; in Chapter 2, stress characteristics, differential equilibrium equations, and hydrostatic pressure calculations have been described and written by Yang Yiqiong and Jie Borui; in Chapter 3, the basic theory of fluid motion has been described and written by Bi Degui, Xu Qingshan and Wang Jingzheng; in Chapter 4, the fundamental equations of steady total flow have been described and written by Bi Degui, Zhou Yue and Wang Lihan; in Chapter 5, the analysis of similar theory and dimensional analysis have been described and written by Yang Yiqiong and Lin Huidong; in Chapter 6, the fluid resistance and energy loss have been described and written by Gao

Yuqiong; in Chapter 7, the principles and calculations of orifice outflow and nozzle flow have been described and written by Gao Yuqiong and Li Kexuan; in Chapter 8, the hydraulic characteristics, fluid motion in pressure pipe and calculations of the pressure pipelines have been described and written by Li Cong, Zhang Kai and Li Shanshan; in Chapter 9, the hydraulic characteristics and calculations of open channel flow have been described and written by Zhang Yunshu and Guo Jiaqi; in Chapter 10, the classification, basic equations and calculations of the weir flow have been described and written by Zhang Yunshu and Wang Yong.

# 前　　言

流体力学是我国高等院校工科环境工程、能源与动力等工科专业基础课,也是国际上本学科专业普遍开设的一门主要专业基础课。在建设有中国特色的世界一流大学是新时代我国高等教育发展战略的大背景下,开设工程流体力学全英文课程是我国工科专业教学国际化的必然要求。但目前适合我国教育国情的英文版流体力学教材相对缺乏,且由于课程涉及多学科交叉,现有教材存在理论分析复杂不易掌握、概念和方程推导较多且易混淆、对学生高等数学知识及多知识综合应用来分析和处理问题能力要求较高等特点,影响留学生教学效果。基于此,本书旨在修订出版适合于环境工程专业的英汉双语工程流体力学教材,除阐述必要的基础理论和计算之外,着重于流体力学在土木、水利等工程中的应用的计算案例,内容上更容易理解及实用,以弥补现有教材存在的不足。

本书是上海理工大学一流本科系列教材之一。在本书中,我们分析了许多流动现象的例子和计算,学生和教师可以很容易地理解相关的知识。第1章关于流体力学概念等内容由李聪、孙春萌和余建平撰写;第2章流体静力学包括应力特性、微分平衡方程和静水力等内容由杨一琼和介博锐撰写;第3章流体流动的基本原理由毕德贵、徐清山和王敬燊撰写;第4章稳态总流的基本方程由毕德贵、周悦和王立涵撰写;第5章相似理论和量纲分析由杨一琼和林辉东撰写;第6章流体阻力和能量损失由高玉琼和曾乾撰写;第7章中孔流出流和管嘴出流的原理及计算由高玉琼和李可玄撰写;第8章有压管道的水力特性和计算由李聪、章凯和李珊珊撰写;第9章明渠流的水力特性和计算由张云澍和郭佳琦撰写;第10章堰流的分类、基本公式和计算由张云澍和汪永撰写。

# Contents

**Chapter 1**  **Introduction** ·················································································· 1
    1.1  Summary ···························································································· 1
    1.2  Fluid and modeling ············································································ 5
    1.3  Major physical properties of fluid ···················································· 6
    1.4  Classification of fluid ········································································ 15
    1.5  Forces acting on the fluid ································································ 17

**Chapter 2**  **Hydrostatics** ···················································································· 20
    2.1  Stress characteristics in a stationary fluid ······································ 20
    2.2  Differential balance equation ·························································· 23
    2.3  Fluid hydrostatic pressure distribution under the action of gravity ········· 26
    2.4  Relative balance of liquid ································································ 32
    2.5  Total hydrostatic pressure of liquid acting on a horizontal surface ········ 35
    2.6  Total hydrostatic force on curved surface ······································ 42
    2.7  Buoyancy force and stability of snorkeling ···································· 50

**Chapter 3**  **The Basic Theory of Fluid Motion** ·················································· 56
    3.1  Description method of fluid motion ················································ 56
    3.2  Some basic concepts of fluid motion ·············································· 57
    3.3  Motion analysis of fluid parcel ························································ 62
    3.4  The basic equation for steady flow ·················································· 63
    3.5  Integration of Euler differential equation ······································ 68
    3.6  Steady plane potential flows ···························································· 69

**Chapter 4**  **Fundamental Equations of Steady Total Flow** ······························ 74
    4.1  Analysis method of total flow ·························································· 74
    4.2  Continuity equation of incompressible fluid ·································· 76
    4.3  The Bernoulli equation ···································································· 77
    4.4  Momentum equation ········································································ 83

**Chapter 5**  **Similarity Theory and Dimensional Analysis** ······························ 88
    5.1  Dimension ·························································································· 88
    5.2  Dimensional analysis ········································································ 89
    5.3  Similarity theory basis of fluid motion ·········································· 92
    5.4  Model experiment ············································································ 96

| Chapter 6 | **Fluid Resistance and Energy Loss** | 98 |
|---|---|---|
| 6.1 | Introduction | 98 |
| 6.2 | Laminar and turbulent flow | 102 |
| 6.3 | Laminar flow in a circular pipe | 106 |
| 6.4 | Turbulent flow | 109 |
| 6.5 | Linear head losses of turbulent flow in circular pipes | 113 |
| 6.6 | The linear head loss of the turbulent flow in a non circular tube | 121 |
| 6.7 | Local head loss | 123 |
| 6.8 | The concept of boundary layer | 126 |
| 6.9 | Drag force | 129 |
| Chapter 7 | **Orifice Outflow and Nozzle Flow** | 133 |
| 7.1 | Steady orifice outflow | 133 |
| 7.2 | Nozzle outflow | 137 |
| 7.3 | Unsteady outflow of orifice and nozzle | 139 |
| Chapter 8 | **Fluid Motion in Pressure Pipe** | 142 |
| 8.1 | Hydraulic calculation of equal diameter short pipe | 142 |
| 8.2 | Hydraulic calculation of long pipe | 146 |
| 8.3 | Hydraulic calculation basis of pipe networks | 153 |
| 8.4 | Water hammer in a pressure pipe | 160 |
| Chapter 9 | **Open channel flow** | 166 |
| 9.1 | Classification of open channel | 166 |
| 9.2 | Uniform open channel flow | 168 |
| 9.3 | Uniform flow in non-pressure circular conduits | 174 |
| 9.4 | Concepts of non-uniform flow in open channel | 176 |
| 9.5 | Gradually-varied flow water surface curve analysis in open channel | 185 |
| 9.6 | The calculation of open-channel gradually-varied flow water surface curve | 192 |
| Chapter 10 | **Weir Flow** | 196 |
| 10.1 | Definition and Classification of Weir Flow | 196 |
| 10.2 | Basic equation of weir flow | 199 |
| 10.3 | Sharp-crested weir | 200 |
| 10.4 | Ogee weir | 201 |
| 10.5 | Broad-crested weir | 203 |
| 10.6 | Hydraulic computation of flow cross small bridge | 205 |

# 目 录

**第1章 引言** ········································································· 209
   1.1 概论 ········································································· 209
   1.2 流体与建模 ································································ 212
   1.3 流体的主要物理性质 ··················································· 213
   1.4 流体的分类 ································································ 220
   1.5 作用在流体上的力 ······················································ 221

**第2章 流体静力学** ······························································· 224
   2.1 静止流体中的应力特性 ················································ 224
   2.2 微分平衡方程 ····························································· 226
   2.3 重力作用下的流体静压强分布 ······································ 228
   2.4 液体的相对平衡 ························································· 234
   2.5 作用在水平面上的液体总静压强 ·································· 236
   2.6 曲面上的总静压力 ······················································ 242
   2.7 浮力及浮潜体稳定 ······················································ 249

**第3章 流体动力学基础** ························································ 253
   3.1 描述流体运动的方法 ··················································· 253
   3.2 流体流动的一些基本概念 ············································ 254
   3.3 理想流体微团的运动 ··················································· 258
   3.4 恒定流的基本方程 ······················································ 259
   3.5 欧拉运动微分方程的积分 ············································ 264
   3.6 恒定平面势流 ····························································· 266

**第4章 稳态总流的基本方程** ················································· 270
   4.1 总流分析方法 ····························································· 270
   4.2 不可压缩流体的连续性方程 ········································ 272
   4.3 伯努利方程 ································································ 273
   4.4 动量方程 ··································································· 279

**第5章 相似原理和量纲分析** ················································· 283
   5.1 量纲 ··········································································· 283
   5.2 量纲分析法 ································································ 284
   5.3 流动相似理论基础 ······················································ 286
   5.4 模型实验 ··································································· 290

## 第 6 章 流体阻力和能量损失 ······ 292
- 6.1 简介 ······ 292
- 6.2 层流和湍流 ······ 296
- 6.3 圆管中的层流 ······ 298
- 6.4 湍流 ······ 301
- 6.5 圆管内湍流的沿程水头损失 ······ 304
- 6.6 非圆管中湍流的损失 ······ 310
- 6.7 局部水头损失 ······ 312
- 6.8 边界层的概念 ······ 315
- 6.9 阻力 ······ 317

## 第 7 章 孔口出流和管嘴出流 ······ 320
- 7.1 孔口恒定出流 ······ 320
- 7.2 管嘴出流 ······ 323
- 7.3 孔口和管嘴的非恒定出流 ······ 325

## 第 8 章 有压管道流动 ······ 327
- 8.1 等径短管的水力计算 ······ 327
- 8.2 长管的水力计算 ······ 330
- 8.3 管网水力计算的基础 ······ 336
- 8.4 有压管道中的水击 ······ 341

## 第 9 章 明渠流 ······ 346
- 9.1 明渠分类 ······ 346
- 9.2 明渠均匀流 ······ 347
- 9.3 无压圆管均匀流 ······ 352
- 9.4 明渠恒定非均匀流 ······ 354
- 9.5 明渠渐变流水面曲线分析 ······ 361
- 9.6 明渠渐变流水面曲线计算 ······ 366

## 第 10 章 堰流 ······ 370
- 10.1 堰流的定义和分类 ······ 370
- 10.2 堰流的基本公式 ······ 372
- 10.3 薄壁堰 ······ 373
- 10.4 实用堰 ······ 374
- 10.5 宽顶堰 ······ 375
- 10.6 小桥孔径水力计算 ······ 378

# Chapter 1　Introduction

## 1.1　Summary

### 1.1.1　Task of fluid mechanics

Fluid mechanics is an independent branch of mechanics that studies the mechanics of fluids (liquids, gases, and plasma) and the forces on them.

Fluid mechanics includes two branches. The first one deals with the fluids at rest, which embraces the study of the conditions under which fluids are at rest and in stable equilibrium. This branch is called fluid statics or hydrostatics. The second one deals with fluid flow—the natural science of fluids (liquids and gases) in motion. This branch is called fluid dynamics.

Fluid mechanics can be also divided into theoretical fluid mechanics and applied fluid mechanics (also called engineering fluid mechanics) according to the research contents. The first one focuses on theoretical research and mathematical reasoning. The second one focuses on how to solve practical engineering problems. This book belongs to the second one.

### 1.1.2　Development of fluid mechanics

#### 1. Major phases for fluid mechanics development

Fluid mechanics originates from a book *On Floating Bodies* written by Archimedes, a Greek scholar who lived in Sicily. He summarized mechanical properties of stationary liquid for the first time.

In 1687, Issac Newton discussed about fluid resistance and wave movement in the famous book *Philosophiae Naturalis Principia Mathematica*, indicating fluid mechanics stepped into main development periods.

Phase I: Energy estimation of liquid motion proposed by Daniel Bernoulli and analytic method of liquid movement proposed by Leonhard Euler laid a solid foundation for the studies of the law of liquid motion. As a result, a classical "Hydrodynamics" which belongs to mathematics was formed.

Phase II: Based on the classical "Hydrodynamics", Claude-Louis Navier and George Gabriel Stokes proposed a famous basic motion equation of viscous fluid, which laid a theoretical foundation for the long-term development of fluid mechanics. However, engineering problems cannot be solved due to the complexity of mathematical solution and the limitations of fluid model. Therefore, "experimental fluid mechanics" which used experimental methods to summarize empirical equation was formed. The development of initial "experimental fluid mechanics" was limited due to a lack of theoretical foundation.

Phase Ⅲ: The contents of classical hydrodynamics and experimental fluid mechanics had been continually renewed since the late 19th century. In the meantime, the combination of theoretical analytic methods and experimental analytic methods leads to the generation of modern fluid mechanics which emphasizes both theory and practice. Since 1960s, fluid mechanics has been applied increasingly widely due to the development and popularity of computer science.

### 2. Major events in the field of fluid mechanics

In 1738, Daniel Bernoulli, a Swiss mathematician and physicist proposed Bernoulli equation in his famous book *Hydrodynamica*.

In 1757, Leonhard Euler proposed an important set of equations for inviscid flow, now known as the Euler equations.

In 1781, Joseph-Louis Lagrange proposed a concept "stream function" for the first time.

In 1822, Claude-Louis Navier, a French engineer and physicist, and in 1845, George Gabriel Stokes, an English mathematician and physicist proposed Navier-Stokes equations.

In 1876, Osborne Reynolds, an English physicist, found two states of fluid flow: laminar flow and turbulent flow.

In 1886, Hermann von Helmholtz, a German physician and physicist, proposed Helmholtz's theorems for vortex dynamics in inviscid fluids.

In the late 19th century, the theory of similarity was proposed.

In 1904, Ludwig Prandtl, a German engineer, described the boundary layer in his groundbreaking paper "Fluid Flow in Very Little Separation."

Since 1960s, computational fluid mechanics was developed rapidly, and the connotation of fluid mechanics has also been continuously enriched and improved.

China has a long history of water conservancy. People gradually deepen the understanding of the law of fluid motion through long-term practice. For example:

"Dayu flood control" already knew water control should follow the nature of water before 4000 years.

Three water conservancy projects (Dujiang Weir, Zhengguo Canal and Ling Canal) were built from 256 B.C. to 210 B.C.

The Grand Canal was built from A.D. 587 to 610, indicating people knew the water flow and weir flow of open canal very well.

Depending on the cognition and utilization of ocean current and air currents, Zheng He's seven voyages to western countries was a miracle in the history of human voyage.

During Yong Zheng years of the Qing Dynasty, He Mengyao proposed the flow of water was equal to the area of the cross section multiplied by the average velocity of water in his book *Suandi*.

### 1.1.3 The methodology of fluid mechanics

#### 1. Theoretical methodology

Theoretical methodology is: reasonable theoretical model is proposed through scientific abstraction of physical properties and flow characteristic of fluid, and closed equations of controlling fluid motion are built according to universal law of mechanical motion. As a result, specific flow issues can be transformed to mathematical problems which can be worked out under corresponding boundary conditions and initial conditions. The key of theoretical methodology lies in proposing ideal model and working out theoretical results through mathematical methods in order to reveal the law of fluid motion. However, it is hard to solve many practical flow issues accurately due to mathematical difficulties.

Among the theoretical methods, major laws cited in fluid mechanics.

(1) Conservation of mass: $\frac{dm}{dt} = 0$

(2) Momentum law: $\sum F = \frac{d(mu)}{dt}$

(3) Newton's second law of motion: $F = ma$

(4) Mechanical energy transformation and conservation law: kinetic energy + pressure energy + potential energy + energy loss = constant

Because pure theoretical methodology suffers mathematical difficulties, a set of analytic methods which combines theory and experiment are gradually perfected during the era when computer has not yet developed. This set of analytic methods is simple and practical. Today, with the computer more and more sophisticated, it is still applicable.

2. Experimental methodology

Applied fluid mechanics is a basic subject which combines theory and practice. Many practical equation and factors are derived from experiments. Many engineering problems which can be worked out through modern theoretical analysis and numerical calculation should also be examined and corrected with the help of experiments.

Major ways of experimental study: a. Prototype measurement (directly observe practical engineering or natural flow distribution); b. Systematic experiment (study artificial flow phenomenon systematically in the laboratory); c. Model study (simulation of the conditions for practical engineering and study through previewing or repeating flow phenomenon).

Basic theories for experimental study: similarity theory and dimensional analysis.

3. Numerical methodology

Numerical methodology is to build various numerical models by means of various discretization methods, and then to obtain a collection of many data in time and space through numerical calculation and numerical experiment, and eventually obtain numerical solution which describes flow field quantitatively. In the past two or three decades, this methodology has been greatly developed to form a new subject—computational fluid mechanics.

4. Simplification of flow analysis

Due to the complexity of fluid motion state, it is very difficult and it is not necessary to take account of all the factors involved in fluid motion. Thus, it is essential to categorize and simplify various movement states. For example, flow can be categorized into one dimensional and multi-dimensional flow, constant flow and unsteady flow, uniform flow and non-uniform flow, rapidly varied flow and gradually varied flow, laminar flow and turbulent flow.

### 1.1.4  Fluid mechanics in civil engineering

**1. The nature and purpose of the course**

(1) Nature

This course is a compulsory professional basic course for the majors of civil engineering and water conservancy. The objects of study are water, air and compressible fluid. Research contents include fluid equilibrium, fluid mechanical motion law and engineering application.

(2) Purposes

This course provides students with basic concepts, theories, computational methods and experimental skills. This course helps the cultivation of students' ability to analyze problems and the creative ability required in their future study and work.

(3) Status

This course provides many courses such as hydrology, hydraulics, soil mechanics, engineering geology, civil engineering, hydraulic structure and construction equipment with fluid mechanics principles, and helps students understand the relationship between civil engineering, water conservancy and atmosphere, water environment deeply.

(4) Others

This course is a compulsory course for professionals in the field of construction engineering.

**2. Application of fluid mechanics in civil engineering**

Fluid is a physical form which comes across to human life and production. Many scientific and technical departments concern fluid mechanics. The solution of flow problems in the fields of irrigation works, civil engineering, traffic transportation, machinery manufacturing, oil production, chemical industry, and bioengineering depends on fluid mechanics. Therefore, the content of fluid mechanics course has been widely used in all areas of our life.

(1) Application in construction engineering such as water lowering of groundwork, roadbed drainage, groundwater seepage, stress analysis of underwater and underground structures, cofferdam construction, buoyancy and resistance to external disturbance of offshore platform.

(2) Application in municipal engineering such as aperture design of bridge and culvert, water supply and drainage, calculation of pipe network, design of pump station and water tower, ventilation of tunnel. The theories of fluid mechanics are theoretical basis for design and operation control of water supply and drainage system.

(3) Application in flood control works such as discharge capacity of river course, the force and seepage of dike and dam, discharge capacity of gate dam controlling flood.

(4) Application in building environment and equipment engineering such as heat supply,

ventilation and design of pump station.

(5) Application in water conservancy. Water conservancy is more dependent on fluid mechanics than other courses. For the professionals in the field of water conservancy, engineering hydraulics is also a compulsory course.

### 3. Basic requirements of this course

(1) Completely understand basic theories.

① Understand basic concepts in the field of fluid mechanics.

② Learn the analytic methods of gross flow, methods of combining dimension analysis and experiment and the ways to work out simple planar potential flow.

③ Understand the rules of energy transformation and head loss of fluid motion.

(2) Have analytical and computational skills for general flow problems, including:

① Calculation of hydraulic load.

② Calculation of discharge capacity of tunnel, canal, and weir as well as seepage of wells.

③ Analysis and calculation of head loss.

(3) Master the conventional methods for the measurements of water level, pressure, velocity, and flow rate well, and know how to observe flow phenomenon, analyze experimental data and write experimental report.

(4) Grasp basic concepts, basic equations and basic application, i.e. basic knowledge of fluid mechanics.

## 1.2 Fluid and modeling

### 1.2.1 Three states of matter

Matter generally exists in three states: gas, liquid and solid. Liquid and gas are collectively referred to as fluid. The main difference between fluid and solid is that they have different resistance to external forces in terms of mechanical analysis. Solid is characterized by constant shape and volume because the mean distance between the molecules is very small and cohesion between the molecules is great. Solids can bear the pressure, but also can bear the tension and resist tensile deformation. Fluids can bear the pressure, but cannot bear the tension and resist tensile deformation due to large distance and small cohesive force between the molecules. Under any tiny shear stress, fluids are prone to deformation or flow.

Although both liquid and gas are fluid, the two have a certain distinction. Liquid has smaller distance and greater cohesive force between the molecules than gas. As a result, liquids can maintain a relatively constant volume and develop a free surface. Another essential difference between them lies in the greater compressibility of gases.

### 1.2.2 Fluid particle and continuum model

From the point of view of molecular physics, fluid and other matters comprise of a large number of molecules in random motion, and space exists between molecules. Due to the space between molecules, technically speaking, fluid is not continuous and physical quantities describing fluid (mass density $\rho$, velocity $v$, pressure $p$) are not continuous in space, either. Randomness of fluid molecular motion also leads to discontinuity in time of any physical quantity in space.

Under standard conditions, 1 cm³ fluid roughly contains $3.3 \times 10^{22}$ molecules and mean distance between molecules is $3 \times 10^{-8}$ cm. The task of fluid mechanics is not to study microscopic motion of molecules, but to study macroscopic properties and macro mechanical motion law of the whole fluid. All characteristic scale and characteristic time employed in flow space and time are larger than molecular distance and collision time. Thus, two concepts of fluid particle and continuum model were introduced in order to study fluid motion laws.

### 1. Fluid particle

The volume which the cluster of molecules occupies must be small compared to the volume occupied by the whole part of the fluid under consideration. On the other hand, the number of molecules in the cluster must be large and the cluster have a certain mass. This fluid cluster is called a fluid particle.

### 2. Continuum model

Fluid continuum model is that fluid was assumed to be a continuum comprised of fluid particles filled with the whole space occupied and physical properties and physical quantities of fluid are continuous. Thus, physical quantities can be considered as continuous functions of place and time. Continuous function theory in mathematical analysis can be used to analyze fluid motion. This hypothesis is applicable for most fluids. However, when the air density is very low, the fluid should be considered as a discrete body and continuum model is not applicable.

## 1.3 Major physical properties of fluid

Important physical properties of fluid include inertia, compressibility, expansibility, viscosity and surface tension.

### 1.3.1 Inertia

Inertia is the resistance of any physical object to any change in its state of motion (this includes changes to its speed, direction or state of rest). It is the tendency of objects to keep moving in a straight line at constant velocity. Mass is the quantitative measure of inertia of a body. The greater the mass of a body, the greater is its inertia. Mass density of a substance is its mass per unit volume. The symbol most often used for density is $\rho$ with a unit of kg/m³. For homogeneous fluid, mass density is defined as mass divided by volume as described in following equation:

$$\rho = \frac{m}{V} \tag{1-1a}$$

where $m$ is mass and $V$ is volume.

If the fluid is not homogeneous, then its density varies between different regions of the fluid. In that case the density around any given location is determined by calculating the density of a small volume ($\Delta V$) around that location. In the limit of an infinitesimal volume, the density of an inhomogeneous fluid at a point becomes

$$\rho = \lim_{\Delta V \to 0} \frac{\Delta m}{\Delta V} = \frac{dm}{dV} \tag{1-1b}$$

where $dV$ is an elementary volume at position $r$.

The density of a material varies with temperature and pressure. This variation is typically small for liquids. The density of liquid can be considered as a constant. For example, water density is 1 000 kg/m³ and mercury density is 13 600 kg/m³.

The density of gas varies with temperature and pressure greatly. Air density is 1.29 kg/m³ at 0 ℃ and under a standard atmospheric pressure.

## 1.3.2 Compressibility

### 1. Compressibility of liquid

In fluid mechanics, compressibility is a measure of the relative volume change of a liquid as a response to a pressure (or mean stress) change. The compressibility of liquid can be quantified by bulk compressibility, which can be denoted by $\kappa$ with a unit of $Pa^{-1}$. The compressibility of liquid can be determined by the following equation:

$$\kappa = -\frac{1}{V}\frac{dV}{dp} \tag{1-2}$$

where $V$ is volume and $p$ is pressure.

With the enhancement of pressure, the volume of liquid become smaller and the mass remains unchanged ($dm = 0$), leading to the increase of density. The relationship between relative increments can be determined by $dm = d(\rho V) = \rho dV + V d\rho = 0$, and then we can write:

$$-\frac{dV}{V} = \frac{d\rho}{\rho}$$

which was substituted into equation (1-2), then equation (1-2) becomes

$$\kappa = \frac{1}{\rho}\frac{d\rho}{dp} \tag{1-3}$$

The smaller the value of $\kappa$, the more difficult is the liquid to be compressed.

The bulk modulus $K$ is used to represent the compressibility of the liquid in engineering. Bulk modulus is the reciprocal of coefficient of compressibility ($\kappa$).

$$K = \frac{1}{\kappa} = -V\frac{dp}{dV} = \rho\frac{dp}{d\rho} \tag{1-4}$$

The unit of bulk modulus is Pa (N/m²), which is same as that of pressure. The greater the value of $K$, the more difficult is the liquid to be compressed. When $K$ tends to infinity, the liquid is impossible to be compressed.

$K$ and $k$ values vary with the type of liquid. For example, the compressibility of mercury is around 8% of that of water and the compressibility of nitric acid is around 6 times greater than that of water. For the same liquid, the values of $k$ and $K$ vary with temperature and pressure. However, this variation is small.

Bulk modulus variation is not very large at fixed temperature and moderate pressure. Bulk modulus can be approximatively described as:

$$\frac{V_2-V_1}{V_1}=-\frac{p_2-p_1}{K} \quad \text{or} \quad \frac{\Delta V}{\Delta V_1}=-\frac{\Delta p}{K} \tag{1-5}$$

In the field of engineering design, bulk modulus of water is approximated to be $2.1 \times 10^9$ Pa, indicating the relative change of the volume $\left(\frac{\Delta V}{V_1}\right)$ is about $\frac{1}{20\,000}$ when the variation of the pressure is an atmospheric pressure. Thus, when $\Delta p$ is not large, the compressibility of water can be ignored and water density can be considered as a constant. However, the compressibility of water is beyond neglect when water-hammer problems of water flow in the pipeline are discussed.

Expansion coefficient $\alpha_V$ can be used to represent expansion of liquid, indicating the relative variation rate of the volume when pressure remains unchanged and variation of liquid temperature is 1 K, which can be described as follows:

$$\alpha_V=\frac{1}{V}\frac{dV}{dT}=-\frac{1}{\rho}\frac{d\rho}{dT} \tag{1-6}$$

where $dT$ is temperature variation and the unit of $\alpha_V$ is $K^{-1}$.

Expansion coefficient of liquid varies with pressure and temperature. At an atmospheric pressure and 20 ℃, $\alpha_V$ is $2.1 \times 10^{-4}$ $K^{-1}$, which can be ignored. However, expansion of water must be taken into account when the variation of working temperature is large.

### 2. Compressibility of gas

Gas density is determined by the equation of state.

An ideal gas is a theoretical gas composed of many randomly moving point particles that do not interact except when they collide elastically. Its equation of state can be described as follows:

$$\frac{p}{\rho}=RT \tag{1-7}$$

where $p$ is absolution pressure of gas, $R$ is ideal gas constant with the value of 287 N·m/(kg·K), and $T$ is absolute temperature.

Ideal gas does not exist. However, at normal conditions such as standard temperature and pressure, most real gases behave qualitatively like an ideal gas. Many gases such as nitrogen, oxygen, hydrogen, noble gases, and some heavier gases like carbon dioxide can be treated like ideal gases within reasonable tolerances.

**Example 1-1** The volume of water is 2.5 m³ at 20 ℃. How much is its volume increase when the temperature rises to 80 ℃?

**Solution:** Water density $\rho_1$ is 998.23 kg/m³ at 20 ℃ while water density $\rho_2$ is 971.83 kg/m³ at 80 ℃. Due to conservation of mass, the volume of water increases with the decrease of density when the temperature rises. Then, we can write

$$-\frac{dV}{V} = \frac{d\rho}{\rho}$$

Data were substituted into the equation above, then

$$\Delta V = -\frac{\Delta \rho}{\rho} V = -\frac{\rho_2 - \rho_1}{\rho_1} V_1$$
$$= -\frac{971.83 - 998.23}{998.23} \times 2.5 = 0.066\ 1\ \text{m}^3$$

Then,

$$\frac{\Delta V}{V_1} = \frac{0.066\ 1}{2.5} \times 100\% = 2.64\%$$

The volume of water increases by 2.64 percent.

**Example 1-2** How much is pressure increase in order that the relative reduction of water volume can reach 0.1% and 1%? Bulk modulus $K$ of water is 2 000 MPa.

**Solution:** Equation (1-5) was rearranged to become

$$\Delta p = -K \frac{\Delta V}{V_1}$$

When $\frac{\Delta V}{V_1} = -0.1\%$, $\Delta p = -2\ 000 \times (-0.1\%) = 2.0$ MPa

When $\frac{\Delta V}{V_1} = -1\%$, $\Delta p = -2\ 000 \times (-0.1\%) = 20$ MPa

Pressure increase is 2.0 MPa and 20 MPa respectively in order that the relative reduction of water volume can reach 0.1% and 1%.

**Example 1-3** The length and diameter of delivery pipeline are 200 m and 0.4 m, respectively. During hydraulic testing, stop pressurizing after pressure in pipe reaches $5.39 \times 10^6$ Pa. The pressure in pipe drops to $4.90 \times 10^6$ Pa one hour after stopping pressurizing. Without considering pipe deformation, how much is the volume of water leaking out of pipe crack per second? Water compressibility $\kappa$ is $4.83 \times 10^{-10}$ Pa$^{-1}$.

**Solution:** After water leak out of pipe crack, the pressure in the pipe drops, leading to water expansion. The volume of expanding water is

$$dV = -\kappa V dp$$
$$= (-4.83 \times 10^{-10}) \times \left(\frac{\pi}{4} \times 0.4^2 \times 200\right) \times (4.90 - 5.39) \times 10^6$$
$$= 5.95 \times 10^{-3} \text{ m}^3$$

The expansion volume of water is $5.95 \times 10^{-3} \text{ m}^3$, which is the volume of water leaking out of pipe crack in one hour. Then the volume of water leaking out of pipe crack per second is

$$q_v = \frac{5.95 \times 10^{-3}}{3\,600} = 1.65 \times 10^{-6} \text{ m}^3/\text{s}$$

### 1.3.3 Viscosity

#### 1. Viscosity and internal friction of fluid

Fluid is flowable. Even a very small shear stress will deform a fluid body at rest while fluid can resist deformation by shear stress in motion state. The viscosity of a fluid is a measure of its resistance to gradual deformation by shear stress or tensile stress.

To illustrate this viscosity of fluid, consider fluid between two parallel plates and adhering to them acted on by a constant shear stress $F$. The distance between two parallel plates is $Y$. The plates are assumed to be very large, so that one need not consider what happens near their edges. When the lower plate keeps fixed, the upper plate moves parallel to the lower one at constant speed $U$ under the action of external force (see Fig.1-1a). If the speed of the upper plate is small enough, the fluid particles will move parallel to it, and their speed will vary linearly from zero at the bottom to $U$ at the top. If the distance $Y$ between the plates is not big or the velocity $U$ is not high, the velocity distribution of the fluid in the direction normal to the plate is linear. Then, the following equation can be obtained as:

$$u(y) = \frac{U}{Y} y \tag{1-8}$$

or

$$\frac{du}{dy} = \frac{U}{Y} \tag{1-9}$$

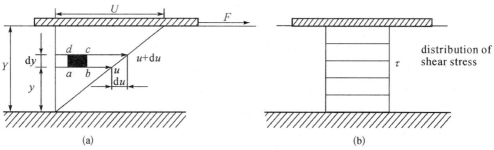

Fig.1-1  Internal friction of flowing fluid

The flowing fluid can be considered to be composed of a series of thin layers. Each layer of fluid will move faster than the one just below it, and friction between them will give rise to a force resisting their relative motion, i.e. internal friction, which is characterization of viscosity. The fast moving fluid layer exerts on the slow moving layer an internal friction in the direction same as its motion. The internal friction accelerates the movement. On the other hand, the slower fluid moving layer exerts on the fast moving layer an internal friction in the direction opposite to its motion. The internal friction slows down the movement. Thus, the flow pattern of the fluid can be changed. Due to the presence of viscosity, work was done and mechanical energy was consumed in order to overcome the frictional force in the process of fluid movement. Therefore, fluid viscosity is the reason why fluid mechanical energy was lost.

## 2. Law of Newton internal friction

In 1686, the law proposed by Isaac Newton and proved by the descendants through experiments is that for most fluids, the shear stress is in direct proportion to the contact area $A$ and the velocity gradient $\dfrac{du}{dy}$. Thus, we can write

$$F \propto A \frac{du}{dy}$$

The proportional factor $\mu$ is introduced, then the following equation can be obtained:

$$\tau = \frac{F}{A} = \mu \frac{du}{dy} \quad (1-10)$$

According to equation (1-10), viscous shear stress $\tau$ is a constant when flow velocity distribution is linear as shown in Fig.1-1a.

The velocity gradient $\dfrac{du}{dy}$ is the change rate of flow velocity in the direction perpendicular to the flow layer. In order to explain the physical implication of the velocity gradient, a rectangular fluid micelle ($abcd$) is considered between the two plates and the thickness of rectangle is $dy$. After $dt$, this fluid micelle moves from $abcd$ to $a'b'c'd'$, as shown in Fig.1-2. Due to the presence of velocity difference $du$ between the two moving layers, fluid micelle has angular deformation $d\theta$ besides translation movement. Because $dt$ is very small, $d\theta$ is also very small. Thus, we can write

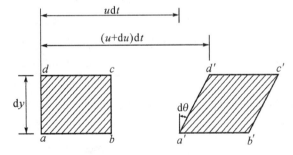

Fig.1-2  Shear deformation rate of fluid particles

$$d\theta \approx \tan(d\theta) = \frac{du\, dt}{dy}$$

$$\frac{du}{dy} = \frac{d\theta}{dt} \quad (1-11)$$

According to equation (1-11), $\dfrac{du}{dy}$ is local shear velocity of fluid micelle. Because shear stress leads to the generation of angular deformation, the rate of angular deformation is also called the rate of shear deformation.

Therefore, law of Newton internal friction can be rewritten as:

$$\tau = \mu \dfrac{du}{dy} = \mu \dfrac{d\theta}{dt} \tag{1-12}$$

indicating the shear stress of the fluid is proportional to the rate of shear deformation, which is a very important characteristic of a fluid that is different from the solid. The shear stress of solid is proportional to the angular deformation, and is independent of the deformation rate.

The proportional factor $\mu$ is known as dynamic viscosity (abbreviated as viscosity), which reflects the measurement of fluid viscosity. Its unit is Pa · s. $\tau$ is viscous shear stress, indicating internal friction per area. The unit of $\tau$ is Pa. The viscosity of fluid varies with the type of fluid. Under same conditions, the viscosity of the liquid is greater than that of gas and varies with temperature and pressure. And basically, the viscosity of the fluid is caused by the cohesive force of the flowing fluid and the molecular momentum transfer. The viscosity of liquids decreases with increasing temperature while the viscosity of gases increases with temperature. In liquids, where the molecules are packed as closely together as the repulsive forces will allow, each molecule is in the range of attraction of several others. The exchange of sites among molecules, responsible for the deformability, is impeded by the force of attraction from neighboring molecules, indicating intermolecular attraction (cohesive force) is large. The contribution from these intermolecular forces to the force on surface elements of fluid particles having different macroscopic velocities is greater than the contribution from the molecular momentum transfer. For gases, the mean distance between the molecules is great (about ten effective molecular diameters) and cohesive force is small. The viscosity of gases, where the momentum transfer is basically its only source, increases with temperature, since increasing the temperature increases the thermal velocity of the molecules, and thus the momentum exchange is favored. Although the viscosity of both gases and liquids are affected by temperature, the influence is different. The effect of temperature on the viscosity of liquid is greater than that on the gas viscosity. For example, the viscosity of water is reduced by about half and the viscosity of the air increase by about 9% when the temperature rises from 15 to 50 ℃. The pressure has little effect on the viscosity of the fluid. For common fluids such as water and air, the effect of pressure can be ignored.

The viscosity of fluid can also be represented by kinematic viscosity. The kinematic viscosity (also called "momentum diffusivity") is the ratio of the dynamic viscosity $\mu$ to the density of the fluid $\rho$. It is usually denoted by the Greek letter $\nu$ $\left(\nu = \dfrac{\mu}{\rho}\right)$ with a unit of $m^2/s$. Kinematic viscosity also varies with the temperature. The kinematic viscosity $\nu$ of water can be calculated by the following empirical equation:

$$\nu = \frac{0.01775 \times 10^{-4}}{1 + 0.0337t + 0.000221t^2} \quad (1-13)$$

where $t$ is centigrade temperature.

**Example 1-4** Try to plot distribution curve of liquid flow velocity and shear stresses. The flow velocity distribution of the upper and lower layers of the liquid between two horizontal plates is linear. The viscosity of the upper layer of liquid is $\mu_1$ and the height of this liquid layer is $h_1$. The viscosity of the lower layer of liquid is $\mu_2$ and the height is $h_2$. The movement speed of plate is $U$, as shown in Fig.1-3a.

**Solution:** The flow rate on the interface of the liquid layers is set as $u$.

The flow velocity distribution of liquid is linear. Thus, according to the law of Newton's internal friction, shear stress on the upper and lower layers can be described as:

Upper: $\tau_1 = \mu_1 \dfrac{U-u}{h_1}$

Lower: $\tau_2 = \mu_2 \dfrac{u-0}{h_2}$

On the interface of liquid layers, it is known: $\tau_1 = \tau_2 = \tau$.

Then, we can write

$$\mu_1 \frac{U-u}{h_1} = \mu_2 \frac{u-0}{h_2}$$

which was rearranged as:

$$u = \frac{\mu_1 h_2 U}{\mu_2 h_1 + \mu_1 h_2}$$

Thus, the flow velocity distribution is

Upper: $u_1 = u + \dfrac{U-u}{h_1}(y - h_2)$

Lower: $u_2 = \dfrac{u}{h_2} y$

Because the flow velocity distribution of each layer of fluid is linear, distribution curve of liquid flow velocity can be obtained and the shear stress is uniformly distributed as long as the flow velocity on the interface was plotted qualitatively, as shown in Fig.1-3b.

**Example 1-5** A wood board with the base area of 0.4 m × 0.45 m, the height of 0.01 m and the mass of 5 kg moves downward along a inclined plane coated with lubricant at a

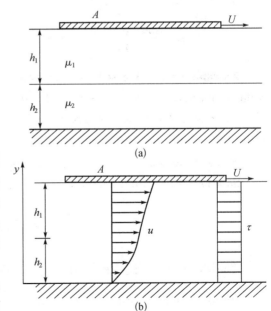

**Fig.1-3** Flow velocity and shear stress distribution of liquid between the parallel plates

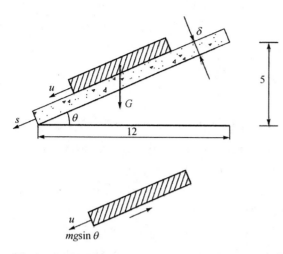

Fig.1-4  Analysis of forces acting on the wood board

constant velocity, as shown in Fig.1-4. The movement speed $u$ of this board is 1 m/s. The thickness $\delta$ of lubricant is 0.1 mm. The velocity distribution of oil movement driven by the wood board is linear. Calculate the viscosity of the lubricant.

**Solution:** Analyze the forces exerted on the wood board in the direction of motion.

The component force of gravity in the direction of motion is $mg\sin\theta$, and the shear stress is

$$F = -\tau A = -\mu\left(\frac{du}{dy}\right)A = -\mu\left(\frac{u}{\delta}\right)A$$

According to Newton's laws of motion, we can write

$$\sum F_s = ma_s = 0$$

Then the equation below can be derived:

$$mg\sin\theta - \mu\frac{u}{\delta}A = 0$$

where $\theta = \arctan\frac{5}{12} = 22.62°$.

Then $\mu$ can be calculated as follows:

$$\mu = \frac{mg\sin\theta \cdot \delta}{Au} = \frac{5\times 9.8\times \sin 22.62°\times 0.000\,1}{0.4\times 0.45\times 1} = 0.010\,5 \text{ Pa}\cdot\text{s}$$

**Example 1-6**  A disc (the diameter is 0.1 m), which is driven by a rotating shaft, rotates on a fixed plate. A layer of oil film with the thickness $\delta$ of 1.5 mm is trapped between disc and plate. When the disc rotated with the rate of 50 r/min, the torque $M$ was measured to be $2.94\times 10^{-4}$ N·m. The velocity distribution in the oil film is linear in the direction perpendicular to the disc. Determine the viscosity of oil and shear stress $\tau_0$ at the edge of disc.

**Solution:** For a microelement, the flow rate is

$$u = \omega r = \frac{n}{60}\cdot 2\pi r = \frac{\pi n}{30}r$$

The shear stress is: $dF = \tau dA = \mu dA\dfrac{u}{\delta} = \mu\cdot 2\pi r dr\,\dfrac{1}{\delta}\cdot\dfrac{\pi n}{30}r$

Viscous friction torque exerted on the microelement of this disc is:

$$dM = rdF = \frac{\mu\pi^2 r^3 n\,dr}{15\delta}$$

Total friction torque is

$$M = \int_0^R dM = \mu\pi^2 \frac{n}{15\delta} \int_0^R r^3 dr = \frac{\mu\pi^2 n R^4}{60\delta}$$

$$\mu = \frac{6\delta M}{\pi^2 n R^4} = \frac{60 \times 0.0015 \times 2.94 \times 10^{-4}}{3.14^2 \times 50 \times 0.05^4} = 85.8 \times 10^{-4} \text{Pa} \cdot \text{s}$$

And $\tau = \mu \frac{1}{\delta} \cdot \frac{\pi n}{30} r$, $\tau$ distribution is linear, as shown in Fig.1 – 5.

Thus, when $r = 0.05$ m, $\tau_0 = 1.50$ Pa.

### 1.3.4 Surface tension

Surface tension is the elastic tendency of liquids that makes them acquire the least surface area possible. At liquid-air interfaces, surface tension results from the greater attraction of liquid molecules to each other (due to cohesion) than to the molecules in the air (due to adhesion). The

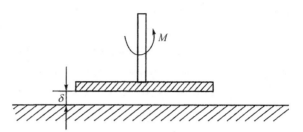

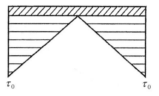

Fig.1 – 5  Rotating disc viscometer

net effect is an inward force at its surface that causes the liquid to behave as if its surface were covered with a stretched elastic membrane. Thus, the surface becomes under tension from the imbalanced forces, which is probably where the term "surface tension" came from. Surface tension can make the hemispherical water drop suspending at the outlet of the water tap without dropping. When tube is inserted into the liquid, surface tension can draw liquid up the tube in a phenomenon known as capillary action if the tube is sufficiently narrow and the liquid adhesion to its walls is sufficiently strong.

Surface tension, usually represented by the symbol $\sigma$, is measured in force per unit length. Its SI unit is Newton per meter (N/m). Surface tension $\sigma$ varies with the type of fluid and temperature. At 20 ℃, surface tension of water and mercury is 0.074 N/m and 0.51 N/m, respectively when exposure to air.

Surface tension is not great. Thus its influence can be ignored in practical engineering. For experiments regarding capillary rise, water droplet and bubble formation, the dispersion of liquid jet and small hydraulic model, surface tension is important.

## 1.4  Classification of fluid

### 1.4.1  Newtonian fluid and non-Newtonian fluid

Newton's law of internal friction is only applicable to the general fluid and not applicable to some special fluids. Generally, the fluids in accordance with the law of Newton internal friction are called "Newtonian fluid", otherwise the fluids are called "non-Newtonian fluid". In continuum mechanics, a Newtonian fluid is a fluid in which the viscous stresses arising from its flow, at every point, are linearly proportional to the local strain rate—the rate of change of its deformation over time. While no real fluid fits the definition perfectly, many common liquids and gases, such as water and air, can be assumed to be Newtonian fluid for practical calculations under ordinary conditions.

Fig.1 - 6 shows the curves describing the relationship between shear stress and shear strain rate of fluid. All these curves are called rheological curves. Slope of the curve is the viscosity $\mu$ of fluid. The viscosity $\mu$ of Newtonian fluid remains unchanged under constant temperature and pressure. The relationship curve between shear stress and shear strain rate of fluid has a fixed slope.

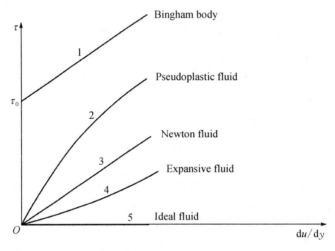

Fig.1 - 6  Rheological curves

Among non-Newtonian fluids, a typical example of these substances is Bingham plastic fluid which does not exhibit any shear rate (no flow and thus no velocity) until a certain stress is achieved. Slurry, plasma and tooth paste are all Bingham plastic fluid. Another type of non-Newtonian fluids is pseudo-plastic fluid such as nylon, rubber, paper pulp, blood, milk and cement paste, whose viscosity decreases with the increase of shear strain rate. In addition, expansive fluid is also a type of non-Newtonian fluid. The viscosity of expansive fluids including dough and thick starch paste, increases with the enhancement of shear strain rate. This book only discusses Newtonian fluids.

### 1.4.2  Ideal fluids

Viscosity, i.e. resistance to gradual deformation by shear stress, is a property of all real fluids. Taking into account the viscosity of fluid complicates the fluid problems. In order to simply the fluid problems, the concept "ideal fluid" was introduced. Ideal fluid is a hypothetical fluid with zero viscosity states. Ideal fluid is an idealized fluid, which does not exist. However, fluids in the flow field or flow zone, where effects of viscosity are small, can be considered as ideal fluids. The conclusions drawn on the basis of ideal fluids can be corrected in order that they are applicable in real fluids.

### 1.4.3 Compressible fluids and incompressible fluids

Fluids can be categorized into compressible and incompressible fluids according to the difference in the volume reduction of liquid under pressure. The variation of fluid density with pressure is very small. The fluid with a constant density can be considered as an incompressible fluid; otherwise it is a compressible fluid. Strictly speaking, there is no completely incompressible fluid. Under normal conditions, the density of liquid can be considered a constant and liquid can be regarded as incompressible fluid. Gas can be considered incompressible fluid when the variation of pressure is relatively small. When high speed flow of air or water vapor pass through long pipes, the pressure drop is very large and the density varies greatly. In this instance, gas must be considered compressible fluid.

## 1.5 Forces acting on the fluid

When analyzing the forces applied on the fluid, an isolated body taken from the fluid is a part of fluid surrounded by any closed surface. The forces acting on the isolated body of the fluid can be categorized into surface force and mass force.

### 1.5.1 Surface force

Surface force is contact force which adjacent fluid or other objects directly exert on the surface of isolated body of fluid. In other words, surface force denoted $F_s$ is the force that acts across an internal or external surface element in a material body. The magnitude of surface force is proportional to cross sectional area of the moving fluid. Surface force can be decomposed into two perpendicular components: pressure and stress forces. Pressure force acts normally over an area and stress force acts tangentially over an area. The isolated body is taken as the research object in the moving fluid as shown in Fig.1 - 7 and an area element $\Delta A$ was taken at the point $A$ on the surface of the isolated body. Then pressure force is

$$p = \lim_{\Delta A \to A} \frac{\Delta F_p}{\Delta A}$$

where $\Delta F_p$ is the pressure acting on the area element $\Delta A$.

And stress force is

$$\tau = \lim_{\Delta A \to A} \frac{\Delta F_\tau}{\Delta A}$$

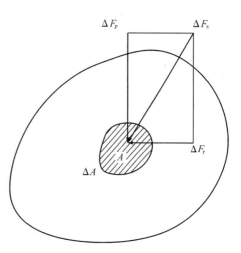

Fig.1 - 7 **Detached body and surface force**

where $\Delta F_r$ is the stress force acting on the area element $\Delta A$.

The SI unit of both pressure and stress forces is Pa ($N/m^2$).

### 1.5.2 Mass force

Mass force is a force which acts on each fluid particle in the isolated body. Mass force is proportional to mass.

Unit mass force is the mass force acting on unit mass fluid:

$$f = \frac{F}{m} = f_x i + f_y j + f_z k = \frac{F_x}{m} i + \frac{F_y}{m} j + \frac{F_z}{m} k$$

where $f_x$, $f_y$ and $f_z$ are mass forces which act on unit mass fluid in the direction of $x$, $y$ and $z$ coordinate, respectively. The unit of unit mass force is the same as that of acceleration. The most common mass force is gravity and inertia force.

For a static flow under gravity, if $z$ axis is perpendicular to the upper, the component of unit mass force of fluid in each direction is:

$$f_x = f_y = 0, \ f_z = -g$$

# Exercises

1. Does the viscous fluid have shear stress at rest? Does the ideal fluid have shear stress in motion? Is a stationary fluid viscous?

2. If the density of kerosene is 808 $kg/m^3$, how much is the mass and weight of the kerosene with the volume of $2 \times 10^{-3}$ $m^3$?

3. If pistol was pressurized, the volume of the liquid in the cylinder block is $10^{-3}$ $m^3$ when the pressure is 0.1 MPa and the volume is $9.94 \times 10^{-4}$ $m^3$ when the pressure is 10 MPa. Calculate the bulk modulus.

4. As shown in Fig.1-8, the distance between the two horizontal plates is 0.5 mm and liquid was trapped between the two horizontal plates. The upper plate moves at the velocity of 0.25 m/s in the horizontal direction and the pressure is 2 Pa. Calculate the viscosity of liquid.

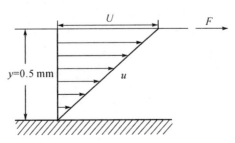

Fig.1-8

4. Calculate the fluid viscosity between the shaft and the shaft sleeve (see Fig.1 - 9).

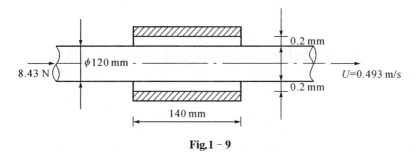

**Fig.1 - 9**

# Chapter 2  Hydrostatics

Hydrostatics is the branch of fluid mechanics that studies incompressible fluids at rest. It encompasses the study of the conditions under which fluids are at rest in stable equilibrium as opposed to fluid dynamics, the study of fluids in motion.

Stagnation is relative, indicating there is no relative motion between fluid particles which is in a relatively static or relatively balanced state. In a fluid at rest, all frictional stresses vanish. Hydrostatics focuses on the studies of the mechanical laws related to the fluids at rest. Hydrostatics elaborates the characteristics and distribution rules of hydrostatic pressure, Euler equations, calculation methods of total pressure of stationary water acting on the plane or curved surface, stability of submerged body and floating body, and so on.

Requirements: Grasp the basic concepts such as absolute pressure, relative pressure, vacuum degree, equipressure surface, piezometric head, pressure prism, center of pressure, metacenter radius and eccentricity; Know the characteristics of stress forces in a stationary fluid and the principles of hydrostatic pressure distribution; Understand the stability of submerged body and floating body; Understand the analytic methods for relative balance of liquid; Grasp the determination methods for equipressure surface, pressure distribution graph and how to plot pressure prism; Grasp the basic equations involved in hydrostatics and know how to use these equations and understand their physical meaning; Learn Euler's differential balance equation; Know how to calculate the total hydrostatic pressure acting on the plane and curved surface (analytic method and graphic method); and know how to use the fundamentals of fluid hydrostatics to solve engineering problems.

## 2.1  Stress characteristics in a stationary fluid

### 2.1.1  Two characteristics of stress forces in a stationary fluid

Compared with solids and flowing fluid, stress force at arbitrary point of a stationary fluid has two important characteristics:

(1) Surface stress in a stationary fluid is compression stress, i.e. pressure, and the direction of hydrostatic pressure is the same as that of the normal to the action surface.

In a stationary fluid, there is no relative movement between adjacent fluid layers. Thus, there is no shear stress in a fluid at rest. A stationary fluid was split into two parts by $N\text{-}N'$

surface. The shaded part was thought of as a detached body. For any point $A$ on the surface of the detached body, the shear stress $\tau$ is zero and the degree of angle $\alpha$ is 90 as shown in Fig.2 – 1. The pressure $p$ is perpendicular to the acting force.

Fluids cannot bear tensile stress due to their mobility. Thus, the direction of hydrostatic pressure can only point to its compressed face. It indicates that the surface stress of a stationary fluid can only be compressive stress and the direction is the same as that of the inward normal of the acting plane.

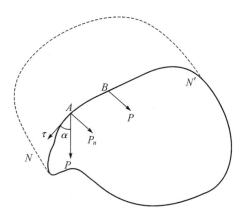

**Fig.2 – 1  Characteristic 1 of stress at any point in a stationary fluid**

(2) If a point in a stationary fluid is thought of as an infinitesimally small cube, then it follows the principles of equilibrium that the pressure on every side of this unit of fluid must be equal. Thus, the pressure on a fluid at rest is isotropic; i.e., it acts with equal magnitude in all directions.

This characteristic can be proven as follows:

An infinitesimally small tetrahedron including point $O$ was taken from a fluid at equilibrium as an object of study, and point $O$ was designated as the origin as shown in Fig.2 – 2. $n$ is the direction of exterior normal of angular surface $ABC$. Then its force balance was analyzed.

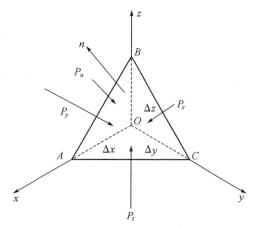

**Fig.2 – 2  Characteristic 2 of stress at any point in a stationary fluid**

Firstly, surface force was determined. $p_x$, $p_y$, $p_z$ and $p_n$ represent hydrostatic pressure on the three different coordinate surfaces and angular surface, respectively. Because the area of each lateral surface of this tetrahedron is infinitesimal, pressure at all points on the surface is same. Thus, surface force in the direction of $x$, $y$, $z$ and $n$ can be represented as $p_x \cdot \dfrac{dydz}{2}$, $p_y \cdot \dfrac{dzdx}{2}$, $p_z \cdot \dfrac{dxdy}{2}$ and $p_n \cdot dA_n$ ($dA_n$ is the area of angular surface).

In addition to the surface force exerted on the tetrahedron $OABC$, the tetrahedron suffers mass force. The density of fluid is $\rho$, $f_x$, $f_y$, $f_z$ represents the component of unit mass force in the direction of the corresponding coordinate axis. The component of mass force on the tetrahedron in the three coordinate axis is $f_x \cdot \dfrac{\rho dx dy dz}{6}$, $f_y \cdot \dfrac{\rho dx dy dz}{6}$ and $f_z \cdot \dfrac{\rho dx dy dz}{6}$, respectively.

Under the equilibrium conditions, $\sum F = 0$, indicating the sum of the projection of its anisotropic component in the $x$ axis is also zero. Thus, in the direction of $x$ axis, we can write

$$\sum F_x = p_x \cdot \frac{dydz}{2} - p_n dA_n \cos(n, x) + f_x \cdot \rho \frac{dxdydz}{6} = 0$$

where $\cos(n, x)$ represents the cosine of included angle between the exterior normal of angular plane and the positive $x$ axis, and $dA_n \cos(n, x)$ is the projection of $dA_n$ in $Oyz$ plane perpendicular to $x$ axis. Therefore,

$$dA_n \cos(n, x) = \frac{dydz}{2}$$

which was substituted into the equation above, the following equation can be obtained:

$$p_x - p_n + \frac{1}{3} dx \cdot \rho f_x = 0$$

The edge-length of tetrahedron tends to zero, and the third term in the equation above also tends to zero. Thus the equation above can be simplified as:

$$p_x = p_n$$

In a same way, we can write

$$p_y = p_n, \quad p_z = p_n$$

Thus, it can be concluded:

$$p_x = p_y = p_z = p_n$$

The inclined plane was selected at random, suggesting direction $n$ is also arbitrary. Therefore, it can be concluded the pressure on a fluid at rest is isotropic; i.e., it acts with equal magnitude in all directions. The hydrostatic pressure at different points is different and it is a continuous function of position, i.e. $p = p(x, y, z)$.

### 2.1.2 Inference—pressure in moving fluid and ideal fluid

For real fluid in motion, viscosity results in the generation of shear stresses due to the relative movement between fluid layers. The normal stress is not equal in different direction at the same point. The dynamic pressure of fluid can be determined by averaging three compressive stresses perpendicular to each other, i.e.

$$p = \frac{1}{3}(p_x + p_y + p_z)$$

However, if the moving fluids are ideal fluids, there is no shear stress. Thus, only compressive stress exists in ideal fluids, and

$$p_x = p_y = p_z = p$$

## 2.2 Differential balance equation

### 2.2.1 Differential balance equation of fluids—Euler equations

In order to analyze the law of fluid balance, an infinitesimally small cube was considered in a fluid at equilibrium as an object of study, and each side of the cube is parallel to the corresponding Cartesian coordinate axis, respectively (see Fig. 2 - 3). The pressure at the central point $M$ of cube was set as $p = p(x, y, z)$. According to the characteristics of continuous function, when the independent variables (coordinate position) even change slightly, the magnitude of function (pressure) would change accordingly. The magnitude of function can be calculated by a Taylor series with the ignorance of high-degree Taylor polynomials. Then, the pressure at the central point on the left and right surfaces can be determined as $\left(p - \dfrac{\partial p}{\partial x} \dfrac{\mathrm{d}x}{2}\right)$ and $\left(p + \dfrac{\partial p}{\partial x} \dfrac{\mathrm{d}x}{2}\right)$, respectively.

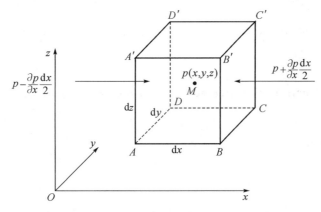

Fig.2 - 3  Derivation of Euler's equilibrium differential equation

Because each face of the cube is infinitesimally small, it can be approximated that the pressure applied at each point on the surface is the same as that at the central point on the surface. The surface force on the surface $AA'D'D$ and $BB'C'C$ can be inferred as follows:

$$\left(p - \frac{\partial p}{\partial x} \frac{\mathrm{d}x}{2}\right) \mathrm{d}y \mathrm{d}z$$

$$\left(p + \frac{\partial p}{\partial x} \frac{\mathrm{d}x}{2}\right) \mathrm{d}y \mathrm{d}z$$

The mass of the cube is $\rho \mathrm{d}x \mathrm{d}y \mathrm{d}z$, and $f_x$, $f_y$, $f_z$ represents the component of unit mass force in the direction of the corresponding coordinate axis. The component of mass force in the direction of $x$-axis is $f_x \rho \mathrm{d}x \mathrm{d}y \mathrm{d}z$.

Under the equilibrium conditions, $\sum F_x = 0$ in the direction of $x$-axis, i.e.

$$\left(p - \frac{\partial p}{\partial x} \frac{\mathrm{d}x}{2}\right) \mathrm{d}y \mathrm{d}z - \left(p + \frac{\partial p}{\partial x} \frac{\mathrm{d}x}{2}\right) \mathrm{d}y \mathrm{d}z + f_x \rho \mathrm{d}x \mathrm{d}y \mathrm{d}z = 0$$

which can be rearranged as:

$$f_x - \frac{1}{\rho}\frac{\partial p}{\partial x} = 0 \tag{2-1a}$$

In the same way, the following equations can be obtained in the direction of $y$ and $z$ axes:

$$f_y - \frac{1}{\rho}\frac{\partial p}{\partial y} = 0 \tag{2-1b}$$

$$f_z - \frac{1}{\rho}\frac{\partial p}{\partial z} = 0 \tag{2-1c}$$

Equations (2 - 1) are the differential balance equations of fluids, which is also called Euler equations because these equations were firstly deduced by Euler, a Swiss scholar in 1755. These equations indicate for the fluid at equilibrium, the surface forces $\left(\frac{1}{\rho}\frac{\partial p}{\partial x}, \frac{1}{\rho}\frac{\partial p}{\partial y}, \frac{1}{\rho}\frac{\partial p}{\partial z}\right)$ acting on the unit mass fluid and the components of unit mass force are equal to each other. The variation rate of pressure along the axial direction $\left(\frac{\partial p}{\partial x}, \frac{\partial p}{\partial y}, \frac{\partial p}{\partial z}\right)$ is equal to the component of mass force acting on unit volume in the axial direction ($\rho f_x$, $\rho f_y$, $\rho f_z$).

### 2.2.2 Integrated form of the differential balance equations of fluids

If equation (2 - 1) was multiplied by $dx$, $dy$ and $dz$, respectively and then added together, we can write

$$f_x dx + f_y dy + f_z dz = \frac{1}{\rho}\left(\frac{\partial p}{\partial x}dx + \frac{\partial p}{\partial y}dy + \frac{\partial p}{\partial z}dz\right)$$

Hydrostatic pressure $p$ is the function of coordinates, i.e. $p = p(x, y, z)$, the total differentiation of which is:

$$dp = \frac{\partial p}{\partial x}dx + \frac{\partial p}{\partial y}dy + \frac{\partial p}{\partial z}dz$$

When the two equations above were merged together, equation can be obtained as follows:

$$dp = \rho(f_x dx + f_y dy + f_z dz) \tag{2-2}$$

Equation (2 - 2) is called integrated form of the differential balance equations of fluids. When the mass forces exerted on the fluid are known, the pressure distribution in the fluid can be calculated.

The physical meaning of equation (2 - 2) can be described as follows.

For the fluid at equilibrium, $\rho$ is a constant. Make $\frac{p}{\rho} = \omega$. Because $p = p(x, y, z)$, $\omega = \omega(x, y, z)$. According to equation (2 - 2), we can write

$$d\left(\frac{p}{\rho}\right) = f_x dx + f_y dy + f_z dz = \frac{\partial \omega}{\partial x} dx + \frac{\partial \omega}{\partial y} dy + \frac{\partial \omega}{\partial z} dz$$

Thus, it can be concluded that

$$f_x = \frac{\partial \omega}{\partial x}, \ f_y = \frac{\partial \omega}{\partial y}, \ f_z = \frac{\partial \omega}{\partial z}$$

If partial derivatives of a coordinate function $\omega(x, y, z)$ and the component of force in the force field in the corresponding coordinate axis are equal to each other, the function is called force function or potential function. The force is called potential force. Thus, the mass force in equation (2-2) is potential force. In other words, the fluid with a constant density can maintain the balance only under the action of potential mass.

**Inference:** Fluid can be at equilibrium under the action of gravity or inertia force, indicating gravity and inertia force are potential force.

## 2.2.3 Equipressure surface

### 1. The concept of equipressure surface

Equipressure surface is a surface comprised of the points, at which the pressure equals. For example, the interface between liquid and gas and the interface between two liquids at equilibrium are equipressure surfaces.

On equipressure surface, $p = \text{const}$, $dp = 0$. Thus,

$$f_x dx + f_y dy + f_z dz = 0 \tag{2-3}$$

### 2. Characteristics of equipressure surface

Mass force at any point on equipressure surface is eternally orthogonal to equipressure surface, which can be proven as follows.

There is a fluid particle $M$ on equipressure surface, as shown in Fig.2-4. Unit mass force of this fluid particle is: $\boldsymbol{f} = f_x \boldsymbol{i} + f_y \boldsymbol{j} + f_z \boldsymbol{k}$, and the linear vector of this fluid particle is: $d\boldsymbol{s} = dx\boldsymbol{i} + dy\boldsymbol{j} + dz\boldsymbol{k}$. According to equation (2-3), we can write

$$\boldsymbol{f} \cdot d\boldsymbol{s} = f_x dx + f_y dy + f_z dz = 0$$

indicating $\boldsymbol{f}$ is orthogonal to $d\boldsymbol{s}$, $\theta = 90°$, where the direction of $d\boldsymbol{s}$ is arbitrary on equpressure surface. Thus, it can be concluded that the mass force exerted on any point in stationary fluid must be perpendicular

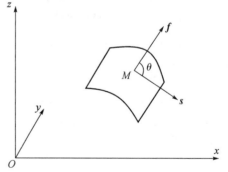

Fig.2-4 Characteristics of equipressure surface

to the equipressure surface passing through this point. This is an important characteristic of equipressure surface. According to this characteristic, the shape of equipressure surface can be determined if the direction of mass force is already known, or the direction of mass force can be determined if the shape of equipressure surface is already known.

## 2.3 Fluid hydrostatic pressure distribution under the action of gravity

Mass force includes gravity and inertia force. The distribution law of hydrostatic pressure is different under the action of different mass forces. In practical engineering or daily life, fluid equilibrium means the fluids are motionless relative to the earth, or the mass force exerted on the fluids is only gravity. So this section focuses on hydrostatic pressure distribution under the action of gravity.

### 2.3.1 Basic equations of hydrostatics

#### 1. Three expressions for the basic equations of hydrostatics

A stationary fluid, for which the mass force only includes gravity, was considered as the object of study. In a rectangular coordinate system, the $z$ axis is vertically upward as shown in Fig.2-5. The pressure on the liquid surface is $p_0$ and the pressure at an arbitrary point $A$ was is $p$. And unit mass force is: $f_x = f_y = 0$, $f_z = -g$. Thus, integrated form of the differential balance equations of fluids [equation (2-2)] can be rewritten as:

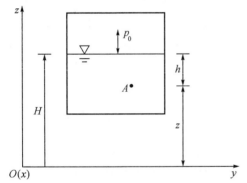

Fig.2-5 Pressure at a point in a stationary liquid

$$dp = -\rho g \, dz$$

After this equation is integrated, equation can be obtained as follows:

$$p = -\rho g z + C' \qquad (2-4)$$

On the free surface, $z = H$, $p = p_0$. Then, we can write

$$C' = p_0 + \rho g H$$

And $h = H - z$. By substituting these two equations above into equation (2-4), equation (2-4) can thus be written as:

$$p = p_0 + \rho g h \qquad (2-5)$$

After each term being divided by $\rho g$, equation (2-4) becomes

$$z + \frac{p}{\rho g} = C \qquad (2-6)$$

where $p$ is the pressure at a point in the stationary fluid, and $p_0$ is liquid surface pressure. For the liquid in an open-atmospheric container, $p_0$ is atmospheric pressure, which is denoted as $p_a$, $h$ is submerged depth and $z$ is the height of this point above the coordinate plane.

Equation (2-5) and equation (2-6) are two expressions for basic equations of hydrostatics. equation (2-5) indicates: (1) For a stationary fluid, the pressure increases linearly with depth and the hydrostatic pressure of each point on the same horizontal plane is equal to each other, which is independent of the shape of container; (2) The pressure $p$ at an arbitrary point in the fluid is equal to the addition of liquid surface pressure $p_0$ and the weight $\rho g h$ of vertical liquid column per an unit area from this point to the liquid surface.

$C$ is a constant in equation (2-6), the magnitude of which can be determined on the basis of boundary conditions. As shown in Fig. 2-6, a reference plane 0-0 was taken at random. The height of the two points (1) and (2) away from the reference plane 0-0 is $z_1$ and $z_2$, respectively. The hydrostatic pressure of these two points is $p_1$ and $p_2$, respectively. According to equation (2-6), equation can be deduced as follows:

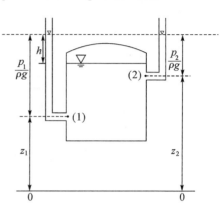

$$z_1 + \frac{p_1}{\rho g} = z_2 + \frac{p_2}{\rho g} \qquad (2-7)$$

Fig.2-6  Relationship between the height and hydrostatic pressure

Equation (2-7) is the third expression for basic equations of hydrostatics. This equation is applicable to stationary, continuous and homogeneous fluids, in which mass force is only gravity.

### 2. Physical meaning of each term in the basic equations of hydrostatics and basic concepts

A few important basic concepts were derived from equation (2-6), which are described as follows:

**Piezometric tube**  When the absolute pressure at the measuring point is greater than the local atmospheric pressure, a transparent tube with an upward opening was set up at this measuring point. This transparent tube is called piezometric tube.

**Elevation head**  $z$ is the position height at an arbitrary point away from reference plane 0-0. $z$ also indicates the position potential energy of a unit weight fluid from a datum plane.

**Height of piezometric tube**  $\frac{p}{\rho g}$ is rising height of liquid in the piezometric tube, which is also called pressure head, indicating the pressure potential energy (pressure energy) of the unit weight fluid from atmosphere pressure.

**Piezometric head**  $\left(z + \frac{p}{\rho g}\right)$ is the total potential energy of a unit weight fluid from a reference plane.

The physical meaning of equation (2-6) can be summarized as follows: Under the action of gravity, total potential energy of unit weight fluid at an arbitrary point in a continuous and homogeneous fluid at rest is same. In other words, the piezometric head is equal everywhere. This is the law of energy distribution in a stationary fluid.

### 3. Inference

(1) Discriminance of equipressure surface

According to equation (2 - 7), the horizontal surface of a stationary and continuous fluid, in which mass force is only gravity, is equipressure surface. In other words, it is necessary to analyze whether the fluid in the horizontal plane satisfies the following four conditions: stationary, continuous, homogeneous and mass force is only gravity, in order to distinguish whether a horizontal surface is an equipressure surface. If the fluids are discontinuous, or two or more fluids coexist, or the fluids are in a magnetic field, or the fluids are in relative balance (refer to section 2.4), the horizontal surface is not necessarily equipressure surface. For example, the horizontal surface $B$-$B'$ is an equipressure surface while surface $C$-$C'$ is not an equipressure surface as illustrated in Fig.2 - 7.

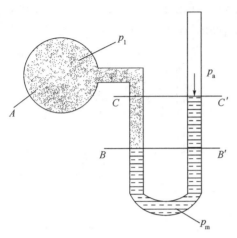

Fig.2 - 7  Determination of equipressure surface

(2) For the pressure distribution of gas in limited space, the pressure at an arbitrary point in the gas is equal ($p_{gas}=C_0$) because the density of gas is small, the height $z$ is limited and the influence gravity exerts on gas pressure can be neglected.

## 2.3.2 Measurement of pressure

### 1. The representation of pressure

The pressure can be calculated from different reference plane. Thus, there are different expression ways such as: absolute pressure, relative pressure and vacuum degree.

Absolute pressure is zero-referenced against a perfect vacuum, using an absolute scale, so it is equal to gauge pressure plus atmospheric pressure.

In the hydraulic engineering, the pressure on the surface of water flow and building is ambient air pressure. As a result, ambient air pressure is considered as a reference. The pressure which is calculated with the ambient air pressure as a reference is called relative pressure denoted as $p$. Relative pressure is also called gauge pressure. The pressure measured by a pressure gauge is the relative pressure since ambient air pressure is assumed to be zero in the pressure gauge. Gauge pressure is zero-referenced against ambient air pressure, so it is equal to absolute pressure minus atmospheric pressure. Obviously, absolute pressure is greater than or equal to zero while relative pressure can be positive, negative or zero. The relationship between absolute pressure, relative pressure and ambient air pressure can be determined as:

$$p = p_{abs} - p_a \tag{2-8}$$

As shown in Fig.2-8, the pressure at the measuring point 1 is negative. The value of pressure may be appended with the word "vacuum", indicating absolute pressure is smaller than ambient air pressure. Vacuum degree is usually represented by which can be calculated according to the following equation:

$$p_v = p_a - p_{abs} \quad (2-9)$$

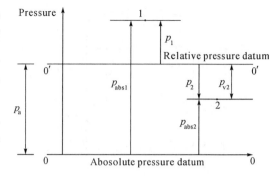

Fig.2-8 Measurement of pressure

As illustrated in Fig.2-8, at the measuring point, the vacuum degree can be determined as: $p_{v2} = p_a - p_{abs2}$.

Vacuum degree can be expressed by the water column height as follows:

$$p_v = -p = \rho g h_v \quad (2-10)$$

where $h_v$ is called as vacuum height.

The relationship between absolute pressure, relative pressure and vacuum degree is described in Fig. 2-8. The reference for absolute pressure and the reference for relative pressure differ by a local atmospheric pressure $p_a$. It should be noted the pressure mentioned in this book means relative pressure unless otherwise stated.

### 2. Measurement unit of pressure

(1) The SI unit of pressure is Pa (N/m²).

(2) Expression methods of atmospheric pressure include standard atmosphere and engineering atmosphere. Engineering atmosphere is slightly different from standard atmosphere. A standard atmosphere (atm) equals 101.325 kPa while an engineering atmosphere (at) is 98 kPa. Engineering sciences usually use engineering atmosphere in place of standard atmosphere. Atmospheric pressure slightly varies in different locations.

(3) The height of water column or mercury column is usually used to represent the height of liquid column. The common unit of the height of liquid column is mH₂O or mmHg.

**Example 2-1** Try to mark the elevation head, the height of piezometric tube and piezometric head at points $A$, $B$ and $C$ in the container filled with liquid as shown in Fig.2-9. The plane 0-0 was set as reference plane.

**Solution:** The elevation head, the height of piezometric tube and the piezometric head at point $A$ is 3 m, 2 m and 5 m, respectively (see Fig.2-10).

Because $z + \dfrac{p}{\rho g} = C$, according to the piezometric head at point $A$, the elevation head and the height of piezometric tube at point $B$ can be determined to be 2 m and 3 m, respectively.

At point $C$, the piezometric head is $z_C + \dfrac{p_C}{\rho g} = z_A + \dfrac{p_A}{\rho g} = 5$ m. The elevation head is $z_C = 6$ m. The height of piezometric tube is $\dfrac{p_C}{\rho g} = -1$ m. $p_C < 0$, indicating point $C$ is in a vacuum.

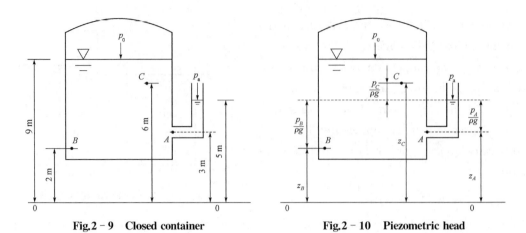

Fig.2-9  Closed container   Fig.2-10  Piezometric head

**Example 2-2**  As shown in Fig.2-11, $h_v = 2$ m, the liquid in the container $B$ is water. Calculate the vacuum degree in the sealed container $A$. If the vacuum degree is fixed and oil with a density $\rho'$ of 820 kg/m³ replaces water, calculate the height $h'_v$ of oil column in the piezometric tube.

**Solution:** (1) Calculate the vacuum degree $p_v$ in the sealed container $A$.

According to equation (2-10), $p_v = \rho g h_v = 9\ 800 \times 2 = 19.6$ kN/m²

The vacuum degree in container $A$ is 19.6 kN/m².

(2) Calculate the height of oil column.

$$h_v \rho g = h'_v \rho' g$$

$$h'_v = \frac{h_v \rho}{\rho'} = \frac{2 \times 1\ 000}{820} = 2.44 \text{ m}$$

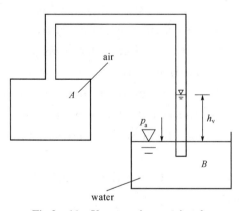

Fig.2-11  Vacuum piezometric tube

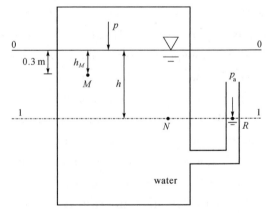

Fig.2-12  Calculation of pressure at measurement point

**Example 2-3**  There is a sealed water tank as shown in Fig.2-12. If the relative pressure on the water surface is $-44.5$ kPa, calculate (1) $h$; (2) relative pressure, absolute pressure and vacuum degree with the unit of at and water column height at point $M$ underwater 0.3 m; (3) the piezometric head at point $M$ with plane 0-0 as a reference.

**Solution:** (1) Calculate $h$

On the equipressure surface 1 - 1, $p_N = p_R = p_a$. Relative pressure was used in the following calculation.

$$p_0 + \rho g h = 0$$
$$-44.5 \times 10^3 + 9\,800 \times h = 0$$
$$h = 4.54 \text{ m}$$

(2) Calculate $p_M$

Relative pressure:

$$p_M = p_0 + \rho g h_M = -44.5 \times 10^3 + 9\,800 \times 0.3 = -41.56 \text{ kPa}$$

or
$$p_M = \frac{-41.56}{98} \times 1 = -0.424 \text{ at}(1 \text{ at} = 98 \text{ kPa})$$

or
$$p_M = \frac{-41.56}{9.8} \times 1 = -4.24 \text{ mH}_2\text{O}$$

Absolute pressure:

$$p_{M\text{abs}} = p_M + p_a = -41.56 + 98 = 56.44 \text{ kPa} = 0.576 \text{ at} = 5.76 \text{ mH}_2\text{O}$$

Vacuum degree:

$$p_v = -p = 41.56 \text{ kPa} = 0.424 \text{ at} = 4.24 \text{ mH}_2\text{O}$$

Vacuum degree expressed by the height of water column:

$$h_v = \frac{p_v}{\rho g} = \frac{41.56 \times 10^3}{9\,800} = 4.24 \text{ m}$$

(3) The piezometric head at point $M$ with plane 0 - 0 as a reference is:

$$z_M + \frac{p_M}{\rho g} = -0.3 + \frac{-41.56 \times 10^3}{9\,800} = -0.3 - 4.24 = -4.54 \text{ m}$$

**Example 2 - 4**  As shown in Fig.2 - 13, the difference of piezometric head between point $A$ and $B$ was measured by a mercury pressure gauge. Try to write its expression.

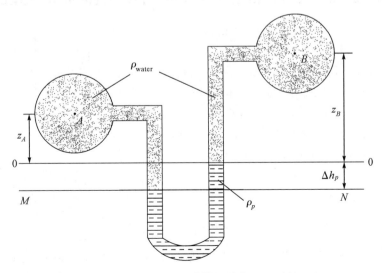

**Fig.2 - 13  Mercury differential manometer**

**Solution:** The horizontal plane 0 - 0 in which upper liquid surface of mercury column locates was taken as reference. The plane where the lower liquid surface locates was taken as equipressure plane. Then,

$$p_M = p_A + \rho_{water} g z_A + \rho_{water} g \Delta h_p$$
$$p_N = p_B + \rho_{water} g z_B + \rho_p g \Delta h_p$$

Because $p_M = p_N$, we can write

$$p_A - p_B = (\rho_p g - \rho_{water} g) \Delta h_p + \rho_{water} g (z_B - z_A)$$

$$\left(z_A + \frac{p_A}{\rho_{water} g}\right) - \left(z_B + \frac{p_B}{\rho_{water} g}\right) = \left(\frac{\rho_p - \rho_{water}}{\rho_{water}}\right) \Delta h_p$$

Because the ratio of mercury density to water density is 13.6, the following expression can be obtained:

$$\left(z_A + \frac{p_A}{\rho_{water} g}\right) - \left(z_B + \frac{p_B}{\rho_{water} g}\right) = 13.6 \Delta h_p$$

## 2.4 Relative balance of liquid

The liquid in a container makes relative movement with the earth as a reference, and there is no relative movement between different parts of the liquid or between the liquid and the container. Thus, liquid is in a state of equilibrium relative to the coordinate system if the coordinate system was built in the container. For the fluids in a relative balance, mass forces include gravity and inertia force. According to D'Alembert's principle, the relative motion of a liquid can be transformed into an energy balance problem in form and equation (2 - 2) can be applied to analyze the relative motion of the liquid if inertia force is included in mass force.

### 2.4.1 Relative balance of the liquid in a uniformly accelerated rectilinear motion

As shown in Fig. 2 - 14, the liquid makes a uniformly accelerated rectilinear motion along with the container. The central point of the original liquid surface is designated as the origin. $z$ axis is vertically upward and the included angle between the acceleration and the $x$ axis is $\alpha$. Unit mass force includes gravity and inertia force of acceleration. The unit mass force at an arbitrary point $M(x, y, z)$ is:

$$f_x = -a \cos \alpha, \quad f_y = 0, \quad f_z = -g - a \sin \alpha$$

The following equation can be obtained by substituting the equations above into equation

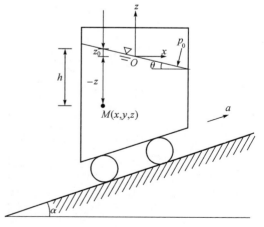

Fig. 2 - 14  Balance of liquid in a uniformly accelerated motion

(2-2):
$$dp = \rho[-a\cos\alpha\, dx - (g + a\sin\alpha)dz]$$

After being integrated, the following equation can be arrived at:
$$p = -\rho[a\cos\alpha \cdot x + (g + a\sin\alpha)z] + C$$

On the liquid surface, $x = z = 0$, and $p = p_0$, which was substituted into the equation above, $C = p_0$ can be obtained.

Then, we can write
$$p = -\rho(g + a\sin\alpha)\left(\frac{a\cos\alpha}{g + a\sin\alpha}x + z\right) + p_0 \qquad (2-11)$$

This is general expression for the law of pressure distribution when the liquid in uniformly accelerated rectilinear motion along with the container is in a state of relative equilibrium.

Based on equation (2-11), isobaric equation can be obtained as:
$$\frac{a\cos\alpha}{g + a\sin\alpha}x + z = C_1 = \text{constant} \qquad (2-12)$$

Thus, equipressure plane is an inclined plane at an angle $\theta$ to the horizontal plane. $\theta$ can be calculated as:
$$\theta = \arctan\left(-\frac{a\cos\alpha}{g + a\sin\alpha}\right)$$

The value of vertical coordinate of liquid surface can be calculated as:
$$z_0 = -\frac{a\cos\alpha}{g + a\sin\alpha}x \qquad (2-13)$$

The following equation can be obtained by substituting equation (2-13) into equation (2-11):
$$p = p_0 + \rho(g + a\sin\alpha)(z_0 - z)$$

where $z_0 - z$ is the submergence depth of the particle below the liquid level, denoted as $h$. So that
$$p = p_0 + \rho(g + a\sin\alpha)h \qquad (2-14)$$

which indicates the variation of the pressure at an arbitrary point with water depth is linear for the liquid in uniformly accelerated rectilinear motion.

Equation (2-14) also suggests the pressure is $p_0$ at an arbitrary point of the liquid and the pressure is zero at an arbitrary of the liquid in an open container if the liquid goes into free fall along with the container.

### 2.4.2 Relative balance of rotating liquid

In the case of a upright cylindrical container filled with liquid rotating at the

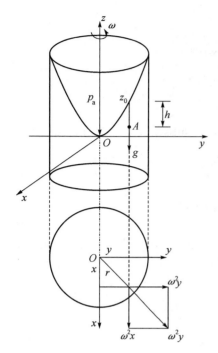

**Fig.2-15** Relative balance of rotating liquid

uniform angular speed $\omega$ as shown in Fig. 2-15, liquid close to the wall of the container rotates with the container at the beginning, and subsequently all liquid rotates with the cylindrical container at the uniform angular speed $\omega$. A state of relative equilibrium can be achieved. In this case, free surface of the liquid transforms from a horizontal plane to a paraboloid of revolution. A coordinate axis is built in the rotating cylinder, with origin coinciding with the vertex of rotating paraoloid and $z$ axis vertically upward. The centripetal acceleration of any particle $A$ in a rotating object is $-\omega^2 r$, and the component on the $x$ and $y$ axis is $-\omega^2 x$ and $-\omega^2 y$, respectively. Unit mass force of gravity is $f_x = f_y = 0$, and $f_z = -g$. Unit mass force of inertia force is $f_x = \omega^2 x$, $f_y = \omega^2 y$, and $f_z = 0$. Thus, the total unit mass force is $f_x = \omega^2 x$, $f_y = \omega^2 y$, and $f_z = -g$.

The following equation can be obtained by substituting the equations above into Euler equations:

$$dp = \rho(\omega^2 x \, dx + \omega^2 y \, dy - g \, dz)$$

After being integrated, the equation above can be transformed as follows:

$$p = \rho\left(\frac{1}{2}\omega^2 x^2 + \frac{1}{2}\omega^2 y^2 - gz\right) + C$$

At the origin point ($x = y = z = 0$), $p = p_0$, so $C = p_0$. Then, we can write

$$p = p_0 + \rho g \left(\frac{\omega^2 r^2}{2g} - z\right) \tag{2-15}$$

This is the general expression for the law of pressure distribution of the liquid at relative equilibrium in a vertical container rotating at the uniform angular speed.

The equation of isobaric cluster including free surface is:

$$\frac{\omega^2 r^2}{2g} - z = C_1 = \text{constant} \tag{2-16}$$

which indicates isobaric cluster is a cluster of rotating paraoloids with central axis.

On the free surface of liquid, $p = p_a = p_0$. The equation of free surface can be expressed by relative pressure:

$$z_0 = \frac{\omega^2 r^2}{2g} \tag{2-17}$$

where $z_0$ is vertical coordinate of free surface. The following equation can be obtained by substituting equation (2-17) into equation (2-15):

$$p = p_0 + \rho g(z_0 - z)$$

where $z_0 - z$ is the depth of the particle $A$ below the free surface of liquid, denoted as $h$. Then we can write

$$p = p_0 + \rho g h \qquad (2-18)$$

Equation (2-18) indicates the variation of pressure at an arbitrary point with the depth of the liquid is linear in the rotating liquid at a relative equilibrium. However, the piezometric head at an arbitrary is not a constant in a rotating container.

## 2.5 Total hydrostatic pressure of liquid acting on a horizontal surface

Determining the magnitude, direction and action point of the total hydrostatic pressure of liquid acting on a horizontal surface is a mechanical problem which many engineering techniques have to solve. The methods used to calculate the total hydrostatic pressure on a horizontal surface include analytic method and graphic method.

### 2.5.1 Analytic method

As shown in Fig.2-16, $MN$ is a projection of an inclined plane at an angle $\theta$ with the horizontal plane. The area of the compressed face in the horizontal plane is $A$. Atmospheric pressure is exerted on the surface of water. The intersection between the extended plane of $MN$ and free surface of liquid is designated as $Ox$ axis. $Oy$ axis is downward and perpendicular to $Ox$ axis. The coordinate plane, in which the horizontal surface locates, rotates 90° around $Oy$ axis in order to exhibit the geometric relationship of plane $xOy$. $h$ is the depth of water at an arbitrary point. $y$ is the distance from this point to $Ox$ axis. $C$ represents the centroid of the compressed face.

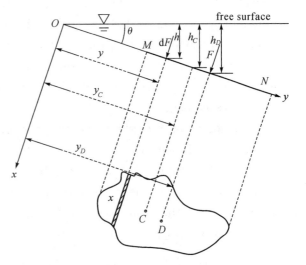

Fig.2-16 Derivation of total hydrostatic pressure on a plane by analytic method

#### 1. Magnitude and direction of force

An infinitesimal area $dA$ in a horizontal strip was considered as the object of study at the depth $h$ of water. The pressure distribution is uniform in this infinitesimal surface. Thus, the pressure exerted on the infinitesimal surface is:

$$dF = p\,dA = \rho g h\,dA = \rho g y \sin\theta\,dA$$

The direction of the pressure is orthogonal to dA and same as that of interior normal.

Because MN is a surface, the direction of the pressure acting on any infinitesimal area is parallel to each other. Their algebraic sum can be calculated by integrating the equation above. The total pressure acting on the surface is

$$F = \int dF = \int_A \rho g y \sin\theta \, dA = \rho g \sin\theta \int_A y \, dA$$

where $\int_A y \, dA$ is the static moment of the compressed face with respect to $Ox$ axis. Its magnitude equals the area $A$ of the compressed face multiplied by the coordinate $y_C$ of centroid.

With $\int_A y \, dA = y_C \cdot A$ being substituted into the equation above, and $h_C = y_C \sin\theta$, we can write

$$F = \rho g \sin\theta \cdot y_C \cdot A = \rho g h_C A = p_C A \qquad (2-19)$$

where $F$ is the total hydrostatic pressure acting on the surface, $h_C$ is the submerged depth of the centroid of the compressed face, and $p_C$ is the pressure at the centroid of the compressed face.

Equation (2-19) indicates the magnitude of the total hydrostatic pressure acting on the surface of arbitrary shape in arbitrary direction is equal to the hydrostatic pressure at the centroid of the compressed face multiplied by the area. In other words, the average pressure acting on any compressed face is equal to the pressure exerted on its centroid. Note that the magnitude of the total hydrostatic pressure is independent of the inclined angle $\theta$ and depends only on the specific weight of the fluid, the area and the depth of the centroid below the surface. The direction of total hydrostatic pressure is the same as that of interior normal of the compressed face.

### 2. Acting point of total hydrostatic pressure—center of pressure

The point through which the total hydrostatic pressure acts is called center of pressure denoted as $D$, as shown in Fig.2-16. The acting point can be located according to the law of resultant moment, i.e. the moment of the resultant force on an arbitrary axis is equal to the algebraic sum of the moment of each component force on the axis. On the $x$ axis, the equation can be established as follows:

$$F \cdot y_D = \int y \cdot dF = \rho g \sin\theta \int_A y^2 \, dA$$

where $\int_A y^2 \, dA$ is the moment of inertia of the compressed face $A$ with respect to $Ox$ axis, i.e. $\int_A y^2 \, dA = I_x$. After this equation being substituted into the equation above, we can write

$$F \cdot y_D = \rho g \sin\theta \cdot I_x$$

After $F = \rho g \sin\theta \cdot y_C \cdot A$ being substituted into the equation, we can write

$$y_D = \frac{\rho g \sin\theta \cdot I_x}{F} = \frac{I_x}{y_C A}$$

According to the law of parallel motion, we can write

$$I_x = I_{Cx} + A y_C^2$$

$I_{Cx}$ is the moment of inertia of the compressed face with respect to the centroidal axis which passes through its centroid and parallel to $Ox$ axis. Thus, the distance from the acting point of the total hydrostatic pressure on the surface to $Ox$ axis can be calculated as:

$$y_D = \frac{I_x}{y_C A} = y_C + \frac{I_{Cx}}{y_C A} \tag{2-20}$$

This equation indicates the location of the center of pressure is independent of the inclined angle $\theta$ of the compressed face. Because $\dfrac{I_{Cx}}{y_C A} \geqslant 0$, the center of pressure is always below the centroid ($y_D \geqslant y_C$) and move closer to the centroid with the increase of the immersion depth. Only when the compressed face is horizontal, the center of pressure and centroid overlap.

In practical engineering, compressed face is generally axis-symmetric plane, where the axis is parallel to $Oy$ axis. The acting point of the total pressure $F$ must locate on the symmetrical axis, i.e. $x_D = x_C$. Thus, the center of pressure can be determined as long as the position of the center of pressure in the direction of $y$ is calculated.

**Example 2 - 5**  A vertical rectangular door was placed in the water (see Fig. 2 - 17). The distance from the gate top to the water surface is $h_1 = 1$ m. The height of the gate is $h_2 = 2$ m. The width of the gate is $b = 1.5$ m. Calculate the total hydrostatic pressure and acting point on the gate.

**Solution:** According to equation (2 - 19), the total hydrostatic pressure is

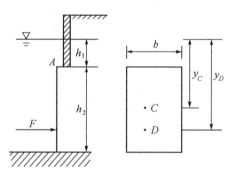

Fig.2 - 17  Rectangular gate

$$F = p_C A = \rho g h_C A = \rho g \left(h_1 + \frac{h_2}{2}\right) \cdot b h_2$$

After the data being substituted, $F$ can be calculated as follows:

$$F = 9\,800 \times \left(1 + \frac{2}{2}\right) \times 1.5 \times 2 = 58\,800 \text{ N} = 58.8 \text{ kN}$$

According to equation (2 - 20), the location of the center of pressure can be determined by

$$y_D = y_C + \frac{I_{Cx}}{y_C A}$$

The moment of inertia with respect to the rectangular centroidal axis is $I_{Cx} = \dfrac{1}{12} b h_2^3$.

After the data being substituted into the equation above, $y_D$ can be calculated as follows:

$$y_D = \left(h_1 + \frac{h_2}{2}\right) + \frac{\frac{1}{12}bh_2^3}{\left(h_1 + \frac{h_2}{2}\right)bh_2}$$

$$= \left(1 + \frac{2}{2}\right) + \frac{\frac{1}{12} \times 1.5 \times 2^3}{2 \times 1.5 \times 2} = 2.17 = h_D$$

The location of the center of pressure is 2.17 m below the water surface, and the direction is horizontal to the right.

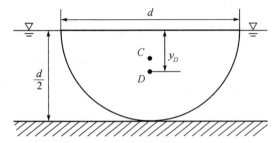

Fig.2 - 18  Water retaining semi-circle plate

**Example 2 - 6**  There is a vertical semi-circle plane. As shown in Fig.2 - 18, the front is facing water. The diameter of water attaining surface exactly lies in the surface of liquid. Calculate the magnitude and acting point of the total hydrostatic pressure $\left(h_C = \frac{4r}{3\pi},\ I_{Cx} = \frac{9\pi^2 - 64}{72\pi}r^4\right)$.

**Solution:** $F = p_C A = \rho g h_C A = \rho g \times \frac{4d}{6\pi} \times \frac{1}{8}\pi d^2 = \rho g \times \frac{1}{12}d^3 = \frac{1}{12}\rho g d^3$

Substituting $y_C = h_C = \frac{4d}{6\pi}$, $I_{Cx} = \frac{9\pi^2 - 64}{1\,152\pi}$ and $A = \frac{1}{8}\pi d^2$ into equation (2 - 20), then, we can write

$$y_D = y_C + \frac{I_{Cx}}{y_C A} = \frac{4d}{6\pi} + \frac{(9\pi^2 - 64)d^4}{1\,152\pi \times \frac{4d}{6\pi} \times \frac{\pi d^2}{8}} = \frac{64d}{96\pi} + \frac{9\pi^2 d - 64d}{96\pi} = \frac{3}{32}\pi d$$

The total hydrostatic pressure is $\frac{1}{12}\rho g d^3$ and the center of pressure is located $\frac{3}{32}\pi d$ below water surface.

### 2.5.2  Graphic method

Graphic method is more suitable for the calculation of total hydrostatic pressure and acting point on the specification plane (for example, rectangle). Pressure distribution graph need be plotted in order to calculate the total hydrostatic pressure.

#### 1. Hydrostatic pressure distribution map

The law of hydrostatic pressure distribution can be described by geometrical figure. The length of the line indicates the magnitude of the pressure at a point. The arrow at the line ends represents the direction of the pressure at a point, i.e. the direction of the interior normal of the compressed face. The figure is a combination of lines perpendicular to the action surface, which is called hydrostatic pressure distribution map. It is necessary to note the hydrostatic pressure in the hydrostatic pressure distribution map is relative pressure. Because all around the building is in the atmosphere, the atmospheric pressure in each direction counteracts each other. It is known the hydrostatic pressure is directly proportional to the depth, i. e. the relationship between the hydrostatic pressure and depth is linear correlation. If the compressed face is a plane, the pressure distribution map must be described by straight lines and two points can determine a straight line. If the compressed face is a curved surface, the distance between envelope and the curved surface represents the magnitude of pressure, which is directly proportional to the depth of the action point on the curved surface. If the curved surface is circular arc, each action line of pressure must pass through the center of the circle. Fig.2 - 19 shows different pressure distribution maps for different scenarios.

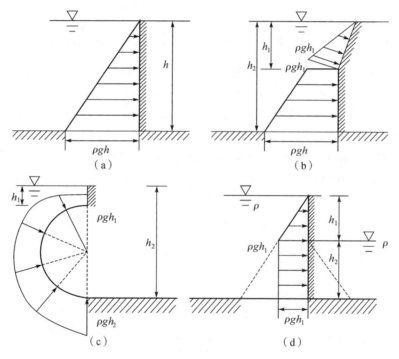

Fig.2 - 19  Hydrostatic pressure graph

## 2. Graphic method

There is a vertical rectangular gate with the height of $h$ and the width of $b$. The top edge is on an even height with the water surface. The plane below the water is $ABCD$, as shown in Fig.2 - 20.

According to equation (2 - 19),

$$F = p_c A = \rho g h_c \cdot A = \rho g \times \frac{1}{2} h \cdot bh$$

$$= \frac{1}{2} \rho g h^2 \cdot b$$

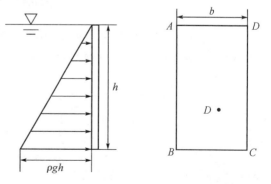

Fig.2 - 20  Derivation of total hydrostatic pressure on a plane by graphic method

where the magnitude of $\frac{1}{2}\rho g h^2$ is equal to the area of the hydrostatic pressure distribution graph $\Omega$, and the unit of $\Omega$ is N/m.

Thus, the equation above can be rewritten as follows:

$$F = \Omega \cdot b \qquad (2-21)$$

Equation (2 - 21) indicates the magnitude of the total hydrostatic pressure is equal to the volume of the distribution graph of pressure acting on the plane.

$$h_D = h_C + \frac{I_{Cx}}{h_C A} = \frac{1}{2}h + \frac{\frac{1}{12}bh^3}{\frac{1}{2}h \cdot bh} = \frac{2}{3}h$$

where $\frac{2}{3}h$ is the depth of the centroid of the pressure distribution graph below the water surface. The action line of the total hydrostatic pressure passes through the centroid of the pressure distribution graph and points to the action surface.

**Example 2 - 7** Use graphic method to calculate the total hydrostatic pressure and the location of the center of pressure in Example 2 - 5.

**Solutions:** Plot the pressure distribution graph of rectangular gate: the bottom is the area of the compressed face and the height is the pressure at each point as shown in Fig.2 - 21b.

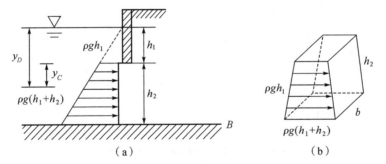

Fig.2 - 21  Calculation of total hydrostatic pressure on a rectangular gate by graphic method

According to the principle of graphic method, the magnitude of the total pressure is the volume of pressure distribution graph. Then,

$$F = \frac{1}{2}[\rho g h_1 + \rho g (h_1 + h_2)] \cdot b \cdot h_2$$

$$= \frac{1}{2} \times 9\,800 \times (1+1+2) \times 1.5 \times 2$$

$$= 58\,800 \text{ N} = 58.8 \text{ kN}$$

Action line passes through the center of the pressure distribution graph, i.e. the centroid of the trapezoid. The equation for the centroid of the trapezoid is

$$y_C = \frac{h}{3} \cdot \frac{a+2b}{a+b}$$

where $a$, $b$ and $h$ is the width of the upper and the bottom and the height of the trapezoid, respectively ($a = \rho g h_1$, $b = \rho g (h_1 + h_2)$, $h = h_2$). Then,

$$y_D = h_1 + y_C = h_1 + \frac{h_2}{3} \cdot \frac{\rho g h_1 + 2\rho g (h_1 + h_2)}{\rho g h_1 + \rho g (h_1 + h_2)}$$

$$= 1 + \frac{2}{3} \times \frac{1 + 2 \times (1+2)}{1 + 1 + 2} = 2.17 \text{ m} = h_D$$

**Example 2-8** A rectangular plane is tilted in the water (Fig. 2-22a). The top of the rectangular plane is 1 m below the water surface ($h = 1$ m) and the bottom is 3 m below the water surface ($H = 3$ m). The width of the rectangle is $b = 5$ m. Use analytic method and graphic method to calculate the total hydrostatic pressure acting on the plane and locate the action point of the total hydrostatic pressure.

**Solution:** (1) Analytic method

The length of the rectangular plane along the $y$ axis is

$$l = \frac{H-h}{\sin 30°} = \frac{3-1}{0.5} = 4 \text{ m}$$

and

$$h_C = h + \frac{H-h}{2} = 1 + \frac{3-1}{2} = 2 \text{ m}$$

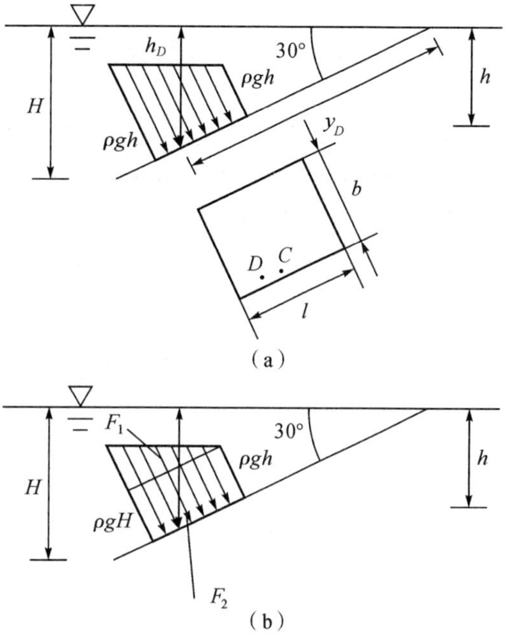

**Fig.2-22 Calculation of total hydrostatic pressure on a rectangular inclined plane**

According to equation (2-19),

$$F = \rho g h_C \cdot A = 9\,800 \times 2 \times 20^2 = 392\,000 \text{ N} = 392 \text{ kN}$$

and

$$y_D = y_C + \frac{I_{Cx}}{y_C A}$$

where

$$y_C = \frac{h_C}{\sin 30°} = \frac{2}{0.5} = 4 \text{ m, and } I_{Cx} = \frac{1}{12} b l^3 = \frac{1}{12} \times 5 \times 4^3 = \frac{80}{3} \text{ m}^4$$

Then,

$$y_D = 4 + \frac{\frac{80}{3}}{4 \times 20} = 4.33 \text{ m}$$

$$h_D = y_D \sin 30° = 2.17 \text{ m}$$

The total hydrostatic pressure is 392 kN. The center of the pressure locates 2.17 m below water surface and the direction of the pressure is the direction of interior normal of the compressed surface.

(2) Graphic method

Because this is a regular rectangular plane, graphic method can be used to calculate the total hydrostatic pressure. The pressure distribution graph was plotted and divided into two parts including a rectangle and triangle, as shown in Fig.2-22b.

It is known $l = \dfrac{H-h}{\sin 30°} = 4$ m. The area of the trapezoid can be calculated as follows:

$$\Omega = \dfrac{\rho g}{2}(H+h)l = \dfrac{\rho g}{2} \times (3+1) \times 4 = 8 \text{ m}^2 \times \rho g$$

Then, $F = \Omega b = 8 \times \rho g \times 5 = 40 \times 9\,800 = 392\,000$ N $= 392$ kN

Because the equation for the centroid point coordinate of the trapezoid is complicated, the trapezoid can be divided in to a simple triangle and rectangle in order to calculate its area and the center of pressure.

For the triangle,

$$F_1 = \dfrac{\rho g}{2}(H-h)lb = 20 \text{ m}^3 \times \rho g$$

$$y_{D1} = \dfrac{2}{3}l + \dfrac{h}{\sin 30°} = \dfrac{2}{3} \times 4 + 2 = \dfrac{14}{3} \text{ m}$$

For the rectangle,

$$F_2 = \rho g h b l = 20 \text{ m}^3 \times \rho g$$

$$y_{D2} = \dfrac{1}{2}l + \dfrac{h}{\sin 30°} = \dfrac{1}{2} \times 4 + 2 = 4 \text{ m}$$

According to the law of resultant moment, $F \cdot y_D = F_1 y_{D1} + F_2 y_{D2}$

After the data being substituted into this equation, $y_D$ and $h_D$ can be calculated as follows:

$$y_D = \dfrac{1}{2}(y_{D1} + y_{D2}) = \dfrac{1}{2} \times \left(\dfrac{14}{3} + 4\right) = 4.33 \text{ m}$$

$$h_D = y_D \sin 30° = 2.17 \text{ m}$$

The results are the same as that calculated by means of the analytic method.

## 2.6 Total hydrostatic force on curved surface

In engineering practice, some action surfaces bearing water pressure are curved planes such as the surface of arch dams, arcuate gate and U-shaped aqueduct. The common curved surfaces are two-directional curved surfaces, i.e. cylinders with parallel generatrix. Thus, this book focuses on the total hydrostatic pressure exerted on the two-directional curved surface.

There is a two-directional curved surface MN (cylinder), with one side pressure-bearing. The generatrix is perpendicular to the map. The area of the curved plane is $A$. A coordinate system is built, in which $Oxy$ plane and liquid surface overlap, and $Oz$ axis downward, as illustrated in Fig.2-23a. In order to calculate the total hydrostatic pressure $F$ on the two-directional curved plane, the components $F_x$ and $F_y$ of $F$ need be calculated, respectively.

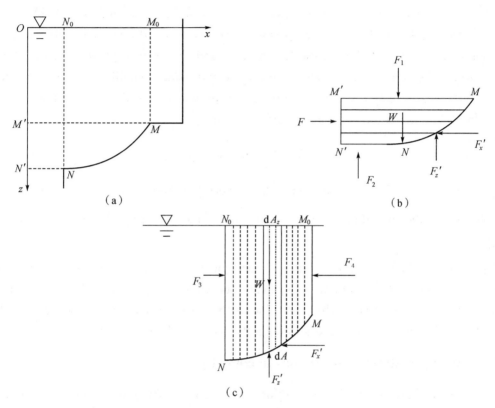

Fig.2-23  Total hydrostatic pressure on curved surface

## 2.6.1  Horizontal component

The projection plane of the curved surface $MN$ on the vertical plane is $M'N'$. The area of the projection plane is $A_x$. The liquid $M'N'NM$ was taken as a detached body. The equilibrium of forces acting on the detached body was analyzed as shown in Fig.2-23b. There are three forces exerted on the detached body, including $F'$, $F_1$ and $F_2$. The counterforce exerted by the liquid on the curved surface includes horizontal force $F'_x$, vertical force $F'_z$ and the gravity $W$. Thus, the horizontal forces exerted on the detached body only include $F'$ and $F'_x$, which are equal in magnitude and opposite in direction.

According to the relationship between action and reaction, we can write

$$F_x = F' = p_C A_x \qquad (2-22)$$

where $F_x$ is the horizontal component of the force which the liquid exerts on the curved surface and $p_C$ is the pressure at the centroid of the projection plane $A_x$.

Equation (2 - 22) indicates the horizontal component $F_x$ of the force exerted on the curved surface is equal to the force $F'$ exerted on the vertical projection plane of the curved surface. The action line of $F_x$ must be the action line of $F'$.

### 2.6.2 Vertical component

The vertical component on the curved surface can be calculated on the basis of the volume of the liquid (called pressure prism) surrounded by the curved surface MN, the projection plane $M_0 N_0$ of MN on the free surface and the vertical plane at the edge of the curved surface (see Fig.2 - 23c). This liquid is at hydrostatic equilibrium. The pressure on the free surface is zero. Vertical forces only include gravity $W$ and the reaction force $F'_z$ of the vertical force $F_z$ exerted on the curved surface. Thus,

$$F_z = W = \int_{A_z} \rho g h \, \mathrm{d}A_z = \rho g V_p \qquad (2-23)$$

where $F_z$ is the vertical force, which the liquid exerts on the curved plane; $A_z$ is the area of the projection of the curved plane on the horizontal plane; $\int_{A_z} h \, \mathrm{d}A_z = V_p$ is the volume of the pressure prism.

Equation (2 - 23) indicates the vertical component of the hydrostatic pressure exerted on the curved plane is equal to the weight of the liquid surrounded by MN, $M_0 N_0$ and the vertical plane at the edge of MN. The location of the action line of the vertical component is the same as that of gravity $W$ and the action line must pass through the center of the pressure prism.

### 2.6.3 Total hydrostatic pressure on the curved plane

In general, there is no single resultant force on the irregular plane. The horizontal component and the vertical component may not be on the same plane. For a two-directional plane, the horizontal and vertical components are on the same plane, which can be combined into a force $F$:

$$F = \sqrt{F_x^2 + F_z^2} \qquad (2-24)$$

The included angle between $F$ and horizontal line is

$$\theta = \arctan \frac{F_z}{F_x} \qquad (2-25)$$

The action line of the total hydrostatic pressure must pass through the intersection point of the action lines of $F_x$ and $F_z$. The acting point of $F$ locates at the intersection point of the curved plane and the action line of $F$. It is not necessary to calculate $F$ in many practical engineering. Calculating the components of $F$, including the magnitude, direction and acting point, can meet the needs of many practical engineering.

## 2.6.4 Pressure prism

In equation (2-23), the integral $\int_{A_z} h\, dA_z = V_p$ represents geometric volume which is called pressure prism. In fact, the calculation of vertical component of fluid hydrostatic pressure on the curved plane is the calculation of pressure prism. So the concept of pressure prism is very important.

Pressure prism is a cylinder with the action surface as lower surface and the free surface as upper surface, which is cut out by a vertical line moving a circle along the edge of the action face. According to the definition, pressure prism is surrounded by three curved surfaces: (1) compressed surface; (2) vertical surface along the edge of the curved surface; (3) free surface or the extended surface of free surface.

### 1. Real pressure prism

When the liquid and pressure prism are on the same side of the curved surface as shown in Fig.2-24a, the direction of $F_z$ is downward and $F_z$ is equal to the water weight of the pressure prism in magnitude. This pressure prism is called real pressure prism.

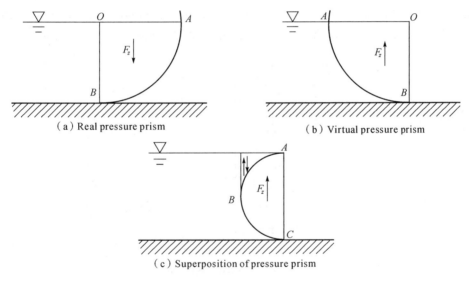

(a) Real pressure prism  (b) Virtual pressure prism

(c) Superposition of pressure prism

Fig.2-24  Pressure prism

### 2. Virtual pressure prism

When the liquid and pressure prism are on the different sides of the curved surface as shown in Fig.2-24b, the direction of $F_z$ is upward and $F_z$ is equal to the water weight of the pressure prism in magnitude. This imaginary pressure prism is called virtual pressure prism.

### 3. Superposition of pressure prism

For the curved surfaces whose horizontal projections overlap, separate the defined pressure prisms and superpose them. The pressure prism of half cylinder can be determined according the curved surfaces $\overset{\frown}{AB}$ and $\overset{\frown}{BC}$ (Fig. 2-24c). For the virtual pressure prism derived from superposition $ABC$, the direction of $F_z$ is upward.

## 2.6.5 Inference

(1) In a limited space, the variation of gas pressure with the depth is slight and the weight of gas is negligible. After analyzing the forces acting on the detached body, the pressure exerted by gas on the curved surface is:

Horizontal component $\qquad F_x = p \cdot A_x \qquad$ (2-26)

Vertical component $\qquad F_z = p \cdot A_z \qquad$ (2-27)

where $A_x$ and $A_z$ are the projection areas of the curved surface on the horizontal and vertical plane, respectively.

(2) If the liquid surface is not free surface, but in a closed container and the surface pressure is $p_0$, the total pressure is:

Horizontal component $\qquad F_x = (p_C + p_0)A_x \qquad$ (2-28)

Vertical component $\qquad F_z = \rho g V_p + p_0 A_z \qquad$ (2-29)

**Example 2-9** Two hemispherical surfaces were riveted to form a spherical container. There are $n$ rivets. The spherical container is filled with the liquid with density of $\rho$ as shown in Fig.2-25a. Calculate the pulling force exerted on each rivet.

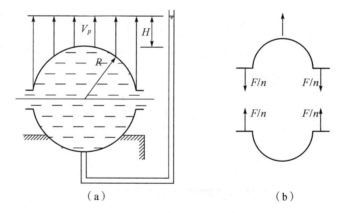

Fig.2-25 Spherical container

**Solution:** The upper hemisphere of the spherical container was considered as compressed curve. The pressure prism acting on the curved surface is described in Fig.2-25b. Then,

$$nF = \rho g V_p = \rho g \left[ \pi R^2 (R+H) - \frac{2}{3}\pi R^3 \right]$$

$$= \rho g \left( \frac{1}{3}\pi R^3 + \pi R^2 H \right)$$

Thus,

$$F = \frac{\rho g}{n}\left( \frac{1}{3}\pi R^3 + \pi R^2 H \right)$$

**Example 2-10** A steel plate with allowable stress of $[\sigma] = 150$ MPa, was used to make a water pipe with inner diameter of $D(D=1\text{ m})$ as shown in Fig. 2 - 26a. The pressure inside the water pipe is $500\text{ mH}_2\text{O}$. Calculate the wall thickness of the water pipe. (Note: The pressure difference can be ignored due to the height difference at different points inside the pipe.)

**Solution:** A pipe segment with length of 1 m is considered as the object of study. The pressure at arbitrary point on the wall of the pipe is equal.

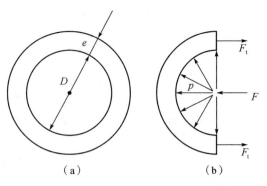

Fig.2 - 26  Pressure on the wall of pressure pipe

The pipe wall was cut open along the pipe diameter. The half pipe segment was taken as a detached body. The forces exerted on the detached body were analyzed (see Fig.2 - 26b). The horizontal pressure exerted on the inner surface of the semi-cycle is equal to the pressure on the vertical projection plane of semi-cycle, i.e. $F=p \cdot A_z=p \cdot D \times 1$. The pressure keeps balance with the tensile stress on the wall of semi-cycle, i.e. $2F_t=F=pD$. $F_t$ is assumed to be uniformly distributed along the wall thickness of the pipe. Then,

$$F_t \leqslant [\sigma] \cdot e \times 1$$

Thus,

$$e \geqslant \frac{F_t}{[\sigma]} = \frac{\dfrac{pD}{2}}{[\sigma]} = \frac{9\ 800 \times 500 \times 1}{2 \times 150 \times 10^6} = 0.016\ 3\text{ m}$$

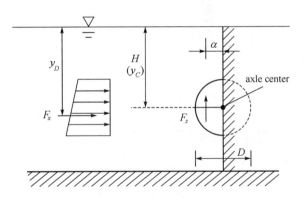

Fig.2 - 27  Water pressure on the half cylinder

**Example 2 - 11** There is a half cylinder with a unit width $(b=1\text{ m})$. Whether the joint effects of buoyant force $F_z$ and horizontal water pressure $F_x$, can create rotational moment with respect to the axle center or not(see Fig.2 - 27)?

**Solution:** (1) Concept analysis

Because the action line of water pressure acting on the half cylinder is always perpendicular to the action surface and passes through the center of circle, the half cylinder below the water surface does not generate rotational moment.

(2) Prove it by means of calculation

$$F_z = \rho g V_p = \rho g \times \frac{\pi}{4}D^2 \times \frac{1}{2} \times 1 = \frac{\pi}{8}D^2 \rho g$$

$$F_x = \rho g h_c A = \rho g H D \times 1 = \rho g H D$$

$$y_D = y_C + \frac{I_{Cx}}{y_C A}$$

$$y_D - y_C = \frac{I_{Cx}}{y_C A} = \frac{1 \times \frac{D^3}{12}}{H \times D \times 1} = \frac{D^2}{12H}$$

$$M_x = -F_x \cdot \frac{D^2}{12H} = -\frac{\rho g D^3}{12}$$

The distance from the acting point of vertical force to the axle center is

$$\alpha = \frac{2D}{3\pi}$$

Then,

$$M_z = F_z \cdot \alpha = \frac{\pi}{8} D^2 \rho g \times \frac{2D}{3\pi} = \frac{\rho g D^3}{12}$$

$$\sum M = M_x + M_z = 0$$

So, the water pressure exerted on the half cylinder does not generate rotational moment with respect to the axle center.

**Example 2-12** The diameter of cylinder is 2 m, which is horizontally-positioned. The measurements of various parts are shown in Fig.2-28a. There is water on the left side and

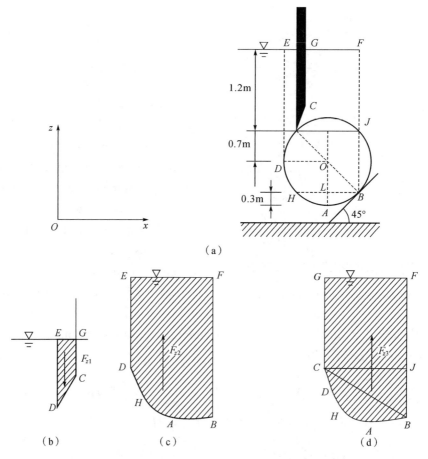

Fig.2-28 Total hydrostatic pressure on the cylinder

no water on the right side. Calculate the horizontal component $F_x$ and vertical component $F_z$ of the total hydrostatic pressure exerted on the cylinder per meter.

**Solution:** CDHAB is the compressed plane of the cylinder. The horizontal component on the two sides of HAB plane offset each other. The horizontal component on the compressed surface CDH is

$$F_x = \rho g h_c A = 9\,800 \times (1.2 + 0.7) \times 1.4 \times 1 = 26\,068 \text{ N} = 26.068 \text{ kN}$$

The vertical component $F_z$ can be calculated through plotting the pressure prism of curved surface CDHAB. The curved surface CDHAB was divided into two sections (CD and DHAB). The pressure prisms of the two sections were plotted as shown in Fig.2-28b and Fig.2-28c. For the pressure prism of CD, the direction of $F_{z1}$ is downward. For the pressure prism of DHAB, the direction of $F_{z2}$ is upward. The two can counteract to each other partly. The pressure prism finally obtained is the hatching part as shown in Fig.2-28d. The vertical component $F_z$ of the total hydrostatic pressure acting on the cylinder per meter is equal to water weight of DHABJFGCD in magnitude. In order to facilitate the calculation, DHABJFGCD was divided into three simple geometric figures such as rectangle, triangle and semi-circle. Then,

$$F_z = \text{Water weight of (Rectangle } JFGC + \text{Triangle } CJB\\ + \text{Semi-circle } CDHABC)$$

$$F_z = 9\,800 \times \left(1.2 \times 1.4 + \frac{1}{2} \times 1.4 \times 1.4 + \frac{1}{2} \pi \times 1^2\right) \times 1$$
$$= 9\,800 \times (1.68 + 0.98 + 1.57) \times 1 = 41\,454 \text{ N} = 41.454 \text{ kN}$$

**Example 2-13** There is a rectangular orifice on a clapboard. The height of the orifice is $a = 1.0$ m, and the width is $b = 3$ m. A cylinder with a diameter of $d = 2$ m was used to block the orifice. Water is filled in both sides of the clapboard. $h = 2$ m and $z = 0.6$ m. Calculate the total hydrostatic pressure exerted on the cylinder.

**Solution:** Calculation by plotting pressure distribution curve and pressure prism.

With hydrostatic pressure acting on the both sides of clapboard, the horizontal pressure distribution of clapboard is described in Fig.2-19d. The pressure distribution of cylinder is rectangular distribution after the left and right pressure counteract to each other, as shown in Fig.2-29.

The left part of pressure prism is equal to area $BAA'B'$ multiplied by the width $b$ of the cylinder. Thus, the sum of the pressure prisms of the compressed curve on the both sides of the clapboard is equal to the volume of the cylinder, which is virtual pressure prism. The direction of $F_z$ is upward.

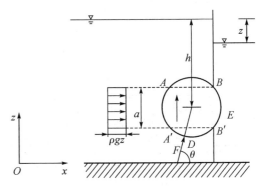

Fig.2-29 Water-resisting cylinder

After pressure distribution figure and the pressure prism were plotted, the horizontal component of the total hydrostatic pressure is:

$$F_x = ab\rho g z = 1.0 \times 3 \times 1 \times 9\,800 \times 0.6 = 17\,640 \text{ N} = 17.64 \text{ kN}$$

The direction of $F_x$ is rightward.

The vertical component of the total hydrostatic pressure is

$$F_z = \rho g V_p = \rho g \left(\frac{\pi}{4} d^2 \times b\right) = 9\,800 \left(\frac{\pi}{4} \times 2^2 \times 3\right)$$
$$= 92\,316 \text{ N} \approx 92.32 \text{ kN}$$

The direction of $F_z$ is upward.

Thus, the total hydrostatic pressure acting on the cylinder is

$$F = \sqrt{F_x^2 + F_z^2} = \sqrt{17.64^2 + 92.32^2} = 93.99 \text{ kN}$$

The pressure at arbitrary point on the surface of the cylinder passes through the center axis of the cylinder. Thus, the resultant force must pass through the center axis of the cylinder. The included angle between action line and the horizontal plane is

$$\theta = \arctan \frac{F_z}{F_x} = \arctan \frac{92.32}{17.64} = 79.18°$$

The depth of the acting point $D$ below the water surface is

$$h_D = h + \frac{d}{2} \sin \theta = 2 + \frac{2}{2} \sin 79.18° = 2.98 \text{ m}$$

## 2.7 Buoyancy force and stability of snorkeling

### 2.7.1 Buoyancy force and three states of submerged body

#### 1. Buoyancy force and buoyancy center of a submerged body

The object floating on the surface of the water or submerged below the water surface also bears hydrostatic pressure, which is the sum of hydrostatic pressure at each point on the surface of the object.

As illustrated in Fig. 2-30, vertical tangents ($AA'$, $BB'$, ⋯) of the submerged body were drawn. These tangents are the generatrix of vertical column tangent to the surface of the submerged body. The intersection of the column surface and the surface of the submerged body decompose the surface

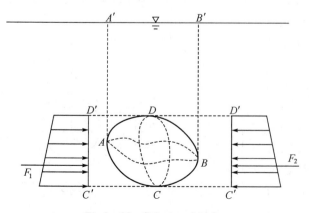

**Fig. 2-30  Submerged body**

of the submerged body into two parts: $ADB$ and $ACB$. The vertical component $F_{z1}$ of the total hydrostatic pressure acting on the surface of the submerged body above the intersection is equal to the weight of the pressure prism above the curved surface $ADB$ in magnitude. The direction is downward. The vertical component $F_{z2}$ of the total hydrostatic pressure acting on the surface of the submerged body below the intersection is equal to the weight of the pressure prism above the curved surface $ACB$, and the direction is upward.

The vertical component $F_z$ acting on the whole surface of the submerged body is

$$F_z = F_{z2} - F_{z1} = \rho g V_{AA'CBB'} - \rho g V_{A'ADBB'} = \rho g V_{ACBDA} \qquad (2-30)$$

where $V_{ACBDA}$ is the volume of the liquid displaced by the submerged body.

Equation (2-30) indicates the vertical force acting on the submerged body is equal to the weight of the displaced fluid.

In the same way, the submerged body can be divided into two parts: $CAD$ (left) and $CBD$ (right). The horizontal component $F_x$ acting on the body is the sum of $F_1$ and $F_2$. $F_1$ and $F_2$ are equal to the hydrostatic pressure acting the vertical projection of the curved surface $CAD$ and $CBD$ in magnitude. The projection areas of these two parts on the vertical plane are equal. The positions of the projection are the same. Thus, $F_1$ and $F_2$ are equal in magnitude, but opposite in direction. These two forces counteract to each other. As a result, the horizontal component acting on the submerged body is zero.

In summary, the total hydrostatic pressure acting on the object in liquid only has vertical component, which is equal to the weight of the liquid displaced by the object in magnitude. This is Archimedes principle.

Because $F_z$ tends to push the object to the surface of liquid, $F_z$ is called buoyancy. The point through which the buoyant force acts is called the center of buoyancy. The center of buoyancy overlaps with the centroid of the displaced volume.

### 2. Three states of submerged body

When the object immersed in liquid is not supported by other objects, it is only subjected to gravity $G$ and buoyant force $F_z$. According to the relative magnitude of gravity and buoyancy, the immersed object has three states:

(1) Sinking body

When $G > F_z$, the object sinks to the bottom.

(2) Submerged body

When $G = F_z$, the object is in suspended state.

(3) Floating body

When $G < F_z$, the object surfaces and keeps balance till the weight of the liquid displaced by the part below the liquid surface is equal to the weight of the object. The object is in a floating state.

## 2.7.2 The balance and stability of submerged body

There is a submerged object with the weight of $G$. The barycenter and center of buoyancy of this object locate at point $C$ and $D$, respectively. According to the relative position of the center of gravity and buoyancy in the same vertical line, the equilibrium stability, i.e. the ability to restore to its original equilibrium state when the submerged body is tilted due to external disturbances, can be categorized as follows.

(1) Indifferent equilibrium

The center $D$ of buoyancy overlaps with the center $C$ of gravity (see Fig.2-31a). The orientation of the object in a liquid is arbitrary, i.e. the submerged body is balanced at any position.

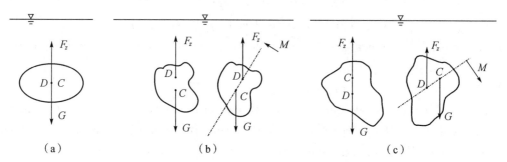

Fig.2-31　Balance and stability of submerged body

(2) Stable equilibrium

For the completely submerged body, which has a center $D$ of buoyancy is above the center $C$ of gravity (see Fig.2-31b), a rotation from its equilibrium position will create a restoring couple formed by the gravity $G$ and buoyant force, $F_z$, which causes the body rotate back to its original position. If the external disturbances are displaced, it automatically returns to its equilibrium position. It is to be noted that as long as the center of gravity falls below the center of buoyancy, this will always be true; that is, the body is in a stable equilibrium position.

(3) Unstable equilibrium

If the center $D$ of buoyancy is below the center $C$ of gravity (see Fig.2-31c), the resulting couple formed by the gravity $G$ and buoyancy $F_z$ will cause the submerged body to overturn. Even if the external disturbances are displaced, it moves to a new equilibrium position. Thus, a completely submerged body with its center of gravity above its center of buoyancy is in an unstable equilibrium position.

Therefore, the center $D$ of buoyancy must be above the center $C$ of gravity in order to keep the balance of the submerged body.

### 2.7.3　The balance and stability of floating bodies

The conditions for the equilibrium of the floating bodies and that of submerged bodies are the same. The equilibrium conditions are: (1) Gravity $G$ and buoyancy $F_z$ are equal in magnitude, but opposite in direction; (2) Gravity $G$ and buoyancy $F_z$ act on the same vertical line. However, the conditions for stability of them are different. For floating objects, it can be stable even though the center $C$ of gravity lies above the center $D$ of buoyancy. This is true since as the body rotates the buoyant force, $F_z$, shifts to pass through the centroid of the newly formed displaced volume and combines with the gravity $G$ to form a couple which will cause the body to return to its original equilibrium position. However, for the relatively tall, slender body, a small rotational displacement can cause the buoyant force and the gravity to form an overturning couple.

### 1. Metacenter radius and eccentricity

As shown in Fig.2-32a, there is a symmetrical floating body. After it tilts, its center of gravity keeps unchanged. The center of buoyancy moves from $D$ to $D'$ due to the shape change of the part immersed in water as described in Fig. 2-32b. There are a few new concepts. The line $H$-$H$ passes through the center $D$ of buoyancy and the center $C$ of gravity is called floating axle. The crossover point $M$ of the action line of buoyancy $F'_z$ passing through $D'$ and the floating axle is called metacenter. The distance from metacenter $M$ to the original center $D$ of buoyancy is metacenter radium, denoted as $\rho$. The distance between the center $C$ of gravity and the center $D$ of buoyance is called eccentricity, denoted as $e$.

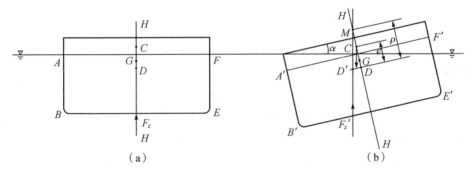

Fig.2-32  Balance and stability of floating body

### 2. Balance of floating body

Whether the floating bodies can restore to their original equilibrium position after they tilt depends on the relative position of the center $C$ of gravity and metacenter $M$.

It can be seen from Fig.2-32: if the metacenter $M$ of the floating body is above the center $C$ of gravity, i.e. $\rho > e$, the joint effect of gravity $G$ and buoyancy $F'_z$ acting on the floating body after inclination generates a rotational moment which helps the floating body restore to the original equilibrium position. The floating body is in stable equilibrium. On the contrary, if the metacenter $M$ of the floating body is below the center $C$ of gravity, i.e. $\rho < e$, the joint effect of gravity $G$ and buoyancy $F'_z$ generates a rotational moment which makes the floating body tend to be more inclined. If point $M$ and point $C$ overlap, i.e. $\rho = e$, the joint effect of gravity $M$ and buoyance $F_z$ does not generate moment of force. The floating body is in indifferent equilibrium.

### 3. Stability of floating body

Therefore, the conditions for floating body to keep balance are that the center $C$ of gravity must be below the metacenter $M$, i.e. the metacenter radius $\rho$ is larger than the eccentricity $e$.

For a symmetrical floating body with a fixed center of gravity, the eccentricity $e$ can be determined if the shape and weight of the floating body are definite. Thus, whether the floating body is stable depends on the magnitude of metacenter radius.

For the floating body with a small inclination angle ($\alpha < 10°$), we can write

$$\rho = \frac{I_0}{V} \tag{2-31}$$

where $I_0$ is the moment of inertia of the floatation plane with respect to the central vertical axle, and $V$ is the volume of the liquid displaced by the floating body.

The intersection plane between the floating body and water surface is called floatation plane.

Equation (2-31) indicates the magnitude of metacenter radium $\rho$ is related to the moment of inertia of the floatation plane on the central vertical axle and the volume of the liquid displaced by the floating body. If $\rho$ is larger than the eccentricity $e$, the floating body is stable, otherwise it is unstable. The larger the metacenter radius, the more stable the floating body.

# Exercises

1. As shown in the Fig. 2-33, there are two types of liquids ($\rho_1 < \rho_2$) in the same container. Two pressure gauges have been installed on the side walls of the container. Is the water level inside the pressure gauge indicated in the following figure correct?

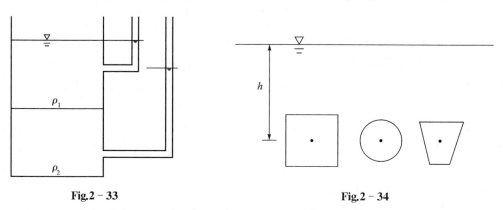

Fig.2-33　　　　　　　　　　Fig.2-34

2. As shown in the Fig. 2-34, there are three flat objects in the water, with different shapes and one side blocking the water. Are the total static pressures acting on these three objects equal when their areas are the same and the water depth at their centroids is the same? Which of these three objects has the deepest pressure center position?

3. As shown in the Fig. 2-35, a pipeline is filled with oil in a stationary state ($\rho_g = 8.5 \text{ kN/m}^3$). Please calculate the pressure at points $A$ and $B$ in $\text{mH}_2\text{O}$.

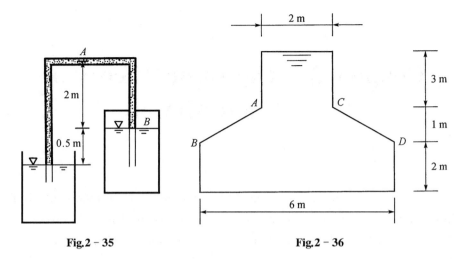

Fig.2 – 35  Fig.2 – 36

4. As shown in the Fig.2 – 36, the variable section cylinder container is filled with water. Please calculate the vertical separation force acting on the conical tube segment $ABCD$.

5. As shown in the Fig.2 – 37, please calculate (1) the magnitude and action line of horizontal force acting on the radial gate; (2) the magnitude and action line of vertical force components; (3) the magnitude and direction of the resultant force.

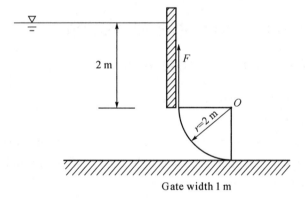

Fig.2 – 37

# Chapter 3 The Basic Theory of Fluid Motion

This chapter mainly elaborates on the basic viewpoints and methods of studying fluid motion. This chapter focuses on some of the basic concepts used in fluid mechanics analysis and how they are discribed. The key concepts should be understood, including: steady and unsteady flow, pathline, streamline, streamtube, flow rate, one-dimensional flow, two-dimensional flow, three-dimensional flow, flow net, potential flow, etc. The caculation of rotational flow and irrotational flow, the differential continuity equations, and the Bernoulli equation have been derived. It is required to master the physical meaning and solution method of velocity potenial function and flow function, and to understand the physical meaning of flow net and the superposition principle of potential flow.

## 3.1 Description method of fluid motion

### 3.1.1 Two description methods of fluid motion

There are two ways to describe fluid motion.

#### 1. Lagrangian approach

The Lagrangian description is based on the motion of fluid particles. It is the direct extension of single particle kinematics to a whole field of fluid particles that are labeled by their location at a reference time.

Since the trajectory of fluid particles is very complex, the Lagrange approach rarely used in fluid mechanics. This book uses Eulerian approach.

#### 2. Eulerian approach

In the Eulerian description, a flow field's characteristics are monitored at fixed locations or in stationary regions of space.

The Eulerian description focuses on flow field properties at locations or in regions of interest, and involves four independent variables: the three spatial coordinates represented by the position vector $x$, $y$, $z$, and time $t$. Such as

$$\boldsymbol{u}=\boldsymbol{u}(x, y, z, t), \text{ or } \begin{cases} u_x = u_x(x, y, z, t) \\ u_y = u_y(x, y, z, t) \\ u_z = u_z(x, y, z, t) \end{cases} \quad (3-1)$$

$$p = p(x, y, z, t) \quad (3-2)$$

$$\rho = \rho(x, y, z, t) \qquad (3-3)$$

### 3.1.2 Euler acceleration

#### 1. Composition of Euler acceleration

In fluid motion, the acceleration can be divided into two parts:
(1) local acceleration;
(2) convection acceleration.

#### 2. Mathematical description

The flow rate is a complex function of time. Euler acceleration can be obtained from velocity with respect to time.

$$a = \frac{du}{dt} = \frac{\partial u}{\partial t} + \frac{\partial u}{\partial x}\frac{dx}{dt} + \frac{\partial u}{\partial y}\frac{dy}{dt} + \frac{\partial u}{\partial z}\frac{dz}{dt} = \frac{\partial u}{\partial t} + u_x\frac{\partial u}{\partial x} + u_y\frac{\partial u}{\partial y} + u_z\frac{\partial u}{\partial t} \qquad (3-4)$$

so

$$a = \frac{du}{dt} = \frac{\partial u}{\partial t} + (u \cdot \nabla)u \qquad (3-5)$$

where $\nabla$ is differential operator of Hamilton.

$$\nabla = i\frac{\partial}{\partial x} + j\frac{\partial}{\partial y} + k\frac{\partial}{\partial z}$$

The component form:

$$\begin{cases} a_x = \dfrac{\partial u_x}{\partial t} + u_x\dfrac{\partial u_x}{\partial x} + u_y\dfrac{\partial u_x}{\partial y} + u_z\dfrac{\partial u_x}{\partial z} \\ a_y = \dfrac{\partial u_y}{\partial t} + u_x\dfrac{\partial u_y}{\partial x} + u_y\dfrac{\partial u_y}{\partial y} + u_z\dfrac{\partial u_y}{\partial z} \\ a_z = \dfrac{\partial u_z}{\partial t} + u_x\dfrac{\partial u_z}{\partial x} + u_y\dfrac{\partial u_z}{\partial y} + u_z\dfrac{\partial u_z}{\partial z} \end{cases} \qquad (3-6)$$

## 3.2 Some basic concepts of fluid motion

### 3.2.1 Euler method on the classification of fluid

#### 1. Steady flow and unsteady flow

(1) Steady flow is the flow whose motion factors (such as flow rate, density, pressure and viscosity) don't change with time. That is

$$\frac{\partial u}{\partial t} = 0 \qquad (3-7)$$

So Euler acceleration at steady flow is $a = (u \cdot \nabla)u$.

(2) Unsteady flow is the flow that at least one of its motion factors changes with time. That is

$$\frac{\partial u}{\partial t} \neq 0 \qquad (3-8)$$

As can be seen in Fig. 3 - 1a, when the level of tank, the trajectory of jet, the velocity and direction remain constant, it is steady flow. As can be seen in Fig.3 - 1b, when the level of tank, the trajectory of jet, the velocity and direction changes with time, it is unsteady flow.

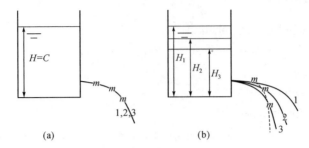

Fig.3 - 1  Steady flow and unsteady flow

## 2. Uniform flow and non-uniform flow

(1) Uniform flow

The flow has the same velocity including the direction and magnitude at various points following the flow direction. That is

$$(u \cdot \nabla)u = 0 \tag{3-9}$$

(2) non-uniform flow

The flow changes the same velocity including the direction or magnitude at various points following the flow direction. That is

$$(u \cdot \nabla)u \neq 0 \tag{3-10}$$

## 3. Gradually varied flow and rapidly varied flow

(1) Gradually varied flow

Gradually varied flow: each flow line close to a straight line parallel to the flow.

(2) Rapidly varied flow

Rapidly varied flow: flow line drastically changed along the non-uniform flow.

## 4. 1-D, 2-D and 3-D flow

(1) One-dimensional flow: fluid motion factors are function of a space coordinate.

So $u = u(s, t)$, Euler acceleration is $a = \dfrac{du}{dt} = \dfrac{\partial \bar{u}}{\partial t} + \bar{u}\dfrac{\partial \bar{u}}{\partial s}$ (see Fig.3 - 2).

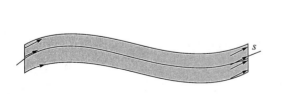

Fig.3 - 2  One-dimensional flow

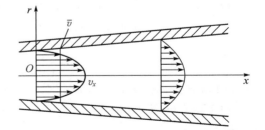

Fig.3 - 3  Two-dimensional flow

(2) Two-dimensional flow: fluid motion factors are function of two space coordinates. (Not only limited to rectangular coordinates, see Fig.3 - 3)

(3) Three-dimensional flow: fluid flow's motion factors are functions of three space coordinates. For example: water flow in a natural river whose cross section shape and magnitude change along the direction of flow; water flows around the ship.

### 3.2.2 The description of flow

Three types of curves are commonly used to describe fluid motion—streamlines, pathlines, and streaklines. These are defined and described here assuming that the fluid velocity vector, $v$, is known at every point of space and instant of time throughout the region of interest. Streamlines, pathlines, and streaklines all coincide when the flow is steady. These curves are often valuable for understanding fluid motion and form the basis for experimental techniques that track seed particles or dye filaments.

#### 1. Streamline

(1) A streamline is a curve that is instantaneously tangent to the fluid velocity throughout the flow field. In unsteady flows the streamline pattern changes with time. The curve indicates the velocity vectors of any points occupying on the streamline (see Fig.3-4).

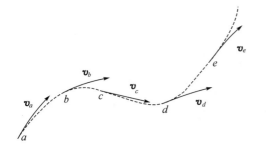

Fig.3-4  The streamline in a moment

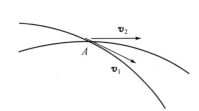

Fig.3-5  Streamline can not intersect

(2) Character of streamline

A. Streamline: many fluid particles, one instant of time (pathline: a fluid particle, a period of time), as can be seen in Fig.3-5.

B. At the same instant of time, streamlines can not intersect.

C. Streamline can't be a folding line, but a smooth curve.

D. In steady flow, streamlines and pathlines coincide.

(3) Equation of streamline

Select point $A$ in streamline, $ds$ is a differential arc length, $v$ is the velocity at point $A$. In Cartesian coordinates, if $d\boldsymbol{s} = (dx, dy, dz)$ is an element of arc length along a streamline and $\boldsymbol{v} = (u, v, w)$ is the local fluid velocity vector, then the tangency requirement on $d\boldsymbol{s}$ and $\boldsymbol{u}$ leads to

$$d\boldsymbol{s} = dx\boldsymbol{i} + dy\boldsymbol{j} + dz\boldsymbol{k}, \quad \boldsymbol{v} = u\boldsymbol{i} + v\boldsymbol{j} + w\boldsymbol{k}$$

so
$$\frac{u}{dx} = \frac{v}{dy} = \frac{w}{dz} = \frac{v}{ds} \qquad (3-11)$$

#### 2. Pathline

A pathline is the trajectory of a fluid particle of fixed identity. And a pathline is the trace

after a single particle travels in a field of flow over a period of time(see Fig.3 - 6).

The equation of pathline is: $\dfrac{dx}{u}=\dfrac{dy}{v}=\dfrac{dz}{w}=dt$,

$u$, $v$, $w$ are functions of both time $t$ and space ($x$, $y$, $z$). Here $t$ is an independent variable.

### 3. Streakline

A streakline is the curve obtained by connecting all the fluid particles that will pass or have passed through a fixed point in space. Streaklines may be visualized in experiments by injecting a passive marker, like dye or smoke, from a small port and observing where it goes as it is carried through the flow field by the moving fluid.

Fig.3 - 6　Pathline

**Example 3 - 1**　The velocity of a 2 - D flow field is $u=x+1$, $v=-y$, find the streamline equation of point (1, 2).

**Solution:**
$$\frac{dx}{u}=\frac{dy}{v}$$

So
$$\frac{dx}{x+1}=\frac{dy}{-y}$$

Integrate it: $\ln(x+1)=-\ln y+\ln C$ or $(x+1)y=C$

The streamline equation of point (1, 2) is

$$(x+1)y=4$$

### 3.2.3　Streamtube, tube flow, cross section, and discharge

#### 1. Streamtube

Take any closed curve (not streamline) in the flow field, then draw streamlines through every point on it, so as to form a tube-shaping space whose walls are streamlines. This tube is called the streamtube(see Fig.3 - 7).

#### 2. Tube flow

Fluid fulling the stream tube is called the tube flow and the limit of a tube flow is a streamline.

#### 3. Cross section

The section is perpendicular to the direction of fluid flow (such as pipe flow and channel flow) (see Fig.3 - 8).

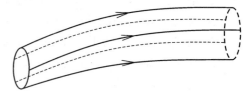

Fig.3 - 7　Stream tube and tube flow

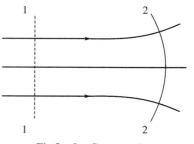

Fig.3 - 8　Cross section

## 4. flow rate

Amount of fluid pass through a cross section (such as the section in the channel or pipe) per unit time.

Volume flow rate (m³/s) $$Q_v = \int_A V dA \qquad (3-12)$$

Mass flow rate (kg/s) $$Q_m = \int_A \rho V dA \qquad (3-13)$$

Weight flow rate (N/s) $$Q_g = \int_A \rho g V dA \qquad (3-14)$$

Volume flow rate is often called discharge. Because these two terms are synonyms, this text uses both terms interchangeably.

### 3.2.4 The character of gradually varied flow in cross section

Two important properties of gradually varied flow in cross section:

(1) Flow cross-section of gradually varied flow is approximately flat. Direction of velocity approximately parallel to each point in cross section.

(2) The hydrodynamic pressure of gradually varied flow in cross section approximately distribute according to static pressure law (see Fig.3-9). That is $z + \dfrac{p}{\rho g} =$ constant.

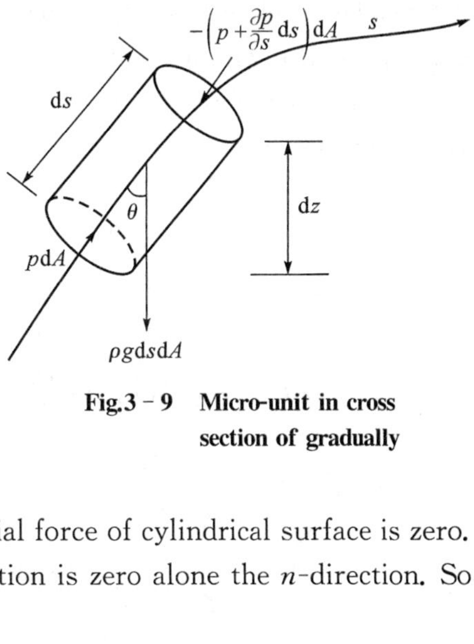

① Gravity $G = \rho g \, dA \, dn$

$$-G\cos\alpha = -\rho g \, dA \, dn \, \frac{z_2 - z_1}{dn} = \rho g \, dA (z_1 - z_2)$$

② Pressure on the bottom and up the bottom are $-p_2 dA$ and $p_1 dA$, respectively.

③ Because the flow velocity perpendicular to the plane, velocity along the $n$ direction is zero, so tangential force of cylindrical surface is zero.

④ Under the gradient flow condition, the acceleration is zero alone the $n$-direction. So there is no inertia force along the direction $n$.

According to Newton's second law, $\sum F_n = 0$

$$\sum F_n = p_1 dA - p_2 dA + \rho g \, dA (z_1 - z_2) = 0$$

Simplify $$z_1 + \frac{p_1}{\rho g} = z_2 + \frac{p_2}{\rho g}$$

$$z + \frac{p}{\rho g} = \text{constant}$$

Fig.3-9 Micro-unit in cross section of gradually

## 3.3 Motion analysis of fluid parcel

### 3.3.1 Composition of fluid parcel motion

The particle is micelles, micellar motions were decomposited the Helmholtz. Motion of fluid particle may be subjected to translation, rotation and deformation (including linear deformation and angular deformation). Actually, many movement of fluid was combined with motion translation, rotation and deformation (see Fig.3-10).

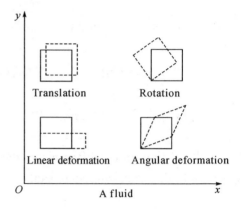

Fig.3-10　Motion types of an ideal fluid element

### 3.3.2 Rotation motion of fluid parcel

It is antisymmetric so its diagonal elements are zero and its off-diagonal elements are equal and opposite. Furthermore, its three independent elements can be put in correspondence with a vector.

Defining $O: (u, v)$, $A: (u_A, v_A)$ (as shown in Fig.3-11)

$$v_A = v + \frac{\partial v}{\partial x} dx$$

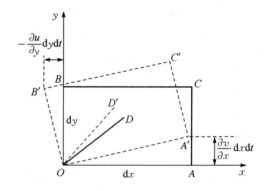

Fig.3-11　Rotation

$dt:$

$$OA \rightarrow OA', \angle AOA' \approx \frac{\partial v}{\partial x} dt$$

Similarly

$$\angle BOB' \approx -\frac{\partial u}{\partial y} dt$$

So

$$\angle DOD' = \frac{\angle AOA' + \angle BOB'}{2} = \frac{1}{2}\left(\frac{\partial v}{\partial x} dt - \frac{\partial u}{\partial y} dt\right)$$

Defining

$$\omega_z = \frac{1}{2}\left(\frac{\partial v}{\partial x} - \frac{\partial u}{\partial y}\right) \quad (3-15a)$$

Similarly

$$\omega_x = \frac{1}{2}\left(\frac{\partial w}{\partial y} - \frac{\partial v}{\partial z}\right) \quad (3-15b)$$

$$\omega_y = \frac{1}{2}\left(\frac{\partial u}{\partial z} - \frac{\partial w}{\partial x}\right) \quad (3-15c)$$

### 3.3.3 Rotational flow and irrotational flow

Rotational flow is also called vortex flow. The fluid particles of rotational flow rotate about its axis, and is independent of motion path. One of the three components ($\omega_x$, $\omega_y$, $\omega_z$) of the rotation is not zero at least.

Rotational flow or irrotational flow depends on whether fluid particles rotate about its axis, and is independent of motion path(see Fig.3 - 12a).

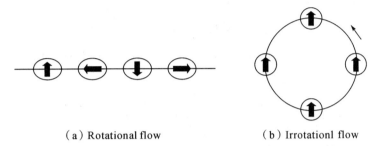

(a) Rotational flow  (b) Irrotationl flow

**Fig.3 - 12  Rotational flow and irrotational flow**

Irrotational flow is also called potential flow, in which elements will translate or deform but not rotate, that is, fluid particles will not rotate about its any axes(see Fig.3 - 12b).

If
$$\omega_x=0,\ \omega_y=0,\ \omega_z=0 \tag{3-16}$$

or

$$\begin{cases} \dfrac{\partial w}{\partial y}-\dfrac{\partial v}{\partial z}=0 \\ \dfrac{\partial u}{\partial z}-\dfrac{\partial w}{\partial x}=0 \\ \dfrac{\partial v}{\partial x}-\dfrac{\partial u}{\partial y}=0 \end{cases} \tag{3-17}$$

then the flow is irrotational flow, otherwise, it is rotational flow.

## 3.4  The basic equation for steady flow

The basic equation for steady flow are included continuity equation, motion differential equation for idea flow and motion differential equation for viscosity flow.

### 3.4.1  Continuity equation for steady flow

#### 1. The general form of basic equation for steady flow

Take a six-sided infinitesimal as control body in the flow field, which side length are $dx$, $dy$, $dz$. Control body is a designated space in the flow field, its shape, position fixed so that fluid may not be affected. For example, analyzing the quality control in the body in $x$ direction.

By taking the first-order trace with Taylor series expansion, then

left surface velocity is: $$u_M = u_x - \frac{1}{2}\frac{\partial u_x}{\partial x}dx$$

right surface velocity is: $$u_N = u_x + \frac{1}{2}\frac{\partial u_x}{\partial x}dx$$

Quality flow from micro area in $dt$:

$$\Delta M = \rho u \Delta A\, dt$$

Poor quality of inflow and outflow from control body in the $x$ direction:

$$\Delta M_x = M_{\text{right}} - M_{\text{left}}$$
$$= \left[\rho u_x + \frac{1}{2}\frac{\partial(\rho u_x)}{\partial x}dx\right]dy\,dz\,dt - \left[\rho u_x - \frac{1}{2}\frac{\partial(\rho u_x)}{\partial x}dx\right]dy\,dz\,dt$$
$$= \frac{\partial(\rho u_x)}{\partial x}dx\,dy\,dz\,dt$$

Similarity

$$\Delta M_y = \frac{\partial(\rho u_y)}{\partial y}dx\,dy\,dz\,dt$$

$$\Delta M_z = \frac{\partial(\rho u_z)}{\partial z}dx\,dy\,dz\,dt$$

According to the law of conservation of mass, we have

$$\left[\frac{\partial(\rho u_x)}{\partial x} + \frac{\partial(\rho u_y)}{\partial y} + \frac{\partial(\rho u_z)}{\partial z}\right]dx\,dy\,dz\,dt = -\frac{\partial \rho}{\partial t}dx\,dy\,dz\,dt$$

The general form of basic equation for steady flow is

$$\frac{\partial \rho}{\partial t} + \frac{\partial(\rho u_x)}{\partial x} + \frac{\partial(\rho u_y)}{\partial y} + \frac{\partial(\rho u_z)}{\partial z} = 0 \tag{3-18}$$

or
$$\frac{\partial \rho}{\partial t} + \operatorname{div}(\rho \boldsymbol{u}) = 0 \tag{3-19}$$

The differential equations of continuity is derived without introducing any constraints. Therefore, there is no limit for the scope of equation.

### 2. Different scope of application of the form

(1) Differential constant flow continuity

$$\frac{\partial(\rho u_x)}{\partial x} + \frac{\partial(\rho u_y)}{\partial y} + \frac{\partial(\rho u_z)}{\partial z} = 0 \tag{3-20}$$

or
$$\operatorname{div}(\rho \boldsymbol{u}) = 0 \tag{3-21}$$

The scope: ideal, practical, compressible and incompressible steady flow.

(2) Incompressible fluid continuity differential equation

Because $\rho = $ constant, then

$$\frac{\partial u_x}{\partial x} + \frac{\partial u_y}{\partial y} + \frac{\partial u_z}{\partial z} = 0 \qquad (3-22)$$

or
$$\text{div}(\boldsymbol{u}) = 0 \qquad (3-23)$$

The physical meaning of incompressible fluid continuity differential equation is that the volume of fluid flowing into the unit space equal to the volume of outflow in unit time.

### 3.4.2 Differential equation of ideal fluid motion

#### 1. Stress analysis

As can be seen in Fig.3-13, take a differential element hexahedral with a center $(x, y, z)$ in ideal fluid. Its side length are $dx$, $dy$, $dz$, and its center of pressure is $p(x, y, z)$. Pressure of $M$, $N$ points are through the center of pressure by Taylor series expansion to take the amount of the first order.

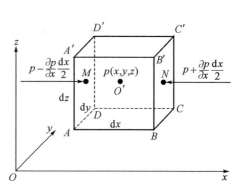

Fig.3-13 Differential equation of ideal fluid

(1) Surface force

Since ideal fluid has no viscosity, shear stress of each side are zero in infinitesimal hexahedron. And because $p_x = p_y = p_z = p$, then

Left force in $x$ direction is:
$$p_M A = \left(p - \frac{\partial p}{\partial x}\frac{dx}{2}\right) dy dz$$

Right force in $x$ direction is:
$$p_N A = \left(p + \frac{\partial p}{\partial x}\frac{dx}{2}\right) dy dz$$

(2) Mass force

Unit mass forces on each component of the coordinates are $f_x$, $f_y$, $f_z$. So mass force in $x$ direction is $f_x \rho dx dy dz$.

#### 2. Motion analysis

Newton's second law $\sum F_x = ma_x$ is applied in $x$ direction, then

$$\left(p - \frac{\partial p}{\partial x}\frac{dx}{2}\right) dy dz - \left(p + \frac{\partial p}{\partial x}\frac{dx}{2}\right) dy dz + f_x \rho dx dy dz = \rho dx dy dz \frac{du_x}{dt}$$

Both sides of the body weight are divided by the infinitesimal $\rho dx dy dz$.

$$f_x - \frac{1}{\rho}\frac{\partial p}{\partial x} = \frac{\partial u_x}{\partial t} + u_x \frac{\partial u_x}{\partial x} + u_y \frac{\partial u_x}{\partial y} + u_z \frac{\partial u_x}{\partial z} \qquad (3-24a)$$

Similarity

$$f_y - \frac{1}{\rho}\frac{\partial p}{\partial y} = \frac{\partial u_y}{\partial t} + u_x \frac{\partial u_y}{\partial x} + u_y \frac{\partial u_y}{\partial y} + u_z \frac{\partial u_y}{\partial z} \qquad (3-24b)$$

$$f_z - \frac{1}{\rho}\frac{\partial p}{\partial y} = \frac{\partial u_y}{\partial t} + u_x \frac{\partial u_z}{\partial x} + u_y \frac{\partial u_z}{\partial y} + u_z \frac{\partial u_z}{\partial z} \qquad (3-24c)$$

This equation is differential equation of ideal fluid.

If the acceleration $\dfrac{du_x}{dt}$, $\dfrac{du_y}{dt}$, $\dfrac{du_z}{dt}$ are zero, this equation can be transformed into Euler equilibrium equations.

### 3.4.3 Differential equations of viscous fluid motion

Viscous fluid, also known as the actual fluid. Derivation of differential equations of motion of a viscous fluid the same as above, but due to the presence of viscous stress state than the more complex an ideal fluid. Here will not be detailed derivation, only the dynamics concept as outlined.

1. **The area force of the viscous fluid**

The area force of the viscous fluid includes compressive stress and shear stress caused by viscosity.

(1) shear stress caused by friction within the generalized Newton's law (Newton friction law in the promotion of ternary stream).

$$\begin{cases} \tau_{xy} = \mu \left( \dfrac{\partial u_x}{\partial y} + \dfrac{\partial u_y}{\partial x} \right) = \tau_{yx} \\ \tau_{yz} = \mu \left( \dfrac{\partial u_y}{\partial z} + \dfrac{\partial u_z}{\partial y} \right) = \tau_{zy} \\ \tau_{zx} = \mu \left( \dfrac{\partial u_z}{\partial x} + \dfrac{\partial u_x}{\partial z} \right) = \tau_{xz} \end{cases} \quad (3-25)$$

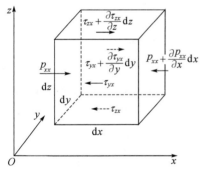

Fig.3 - 14  Viscous fluid motion differential equations

where, $\tau$ represents shear stress, the first subscript indicates the role of the surface normal direction, the second index indicates the effect of shear stress direction, as shown in Fig.3 - 14.

(2) the actual flow of the fluid dynamic pressure at any point, due to the presence of viscous shear stress, each of the dynamic pressure of varying sizes, namely $p_{xx} \neq p_{yy} \neq p_{zz}$. The dynamic pressure of any point

$$p = \dfrac{1}{3}(p_{xx} + p_{yy} + p_{zz}) \quad (3-26)$$

where the subscripts of $p$ and those of shear stress $\tau$ have the same meaning.

Hydrodynamic theory and analysis shows that each of incompressible fluid pressure to move the following relationship:

$$\begin{cases} p_{xx} = p - 2\mu \dfrac{\partial u_x}{\partial x} \\ p_{yy} = p - 2\mu \dfrac{\partial u_y}{\partial y} \\ p_{zz} = p - 2\mu \dfrac{\partial u_z}{\partial z} \end{cases} \quad (3-27)$$

## 2. The differential equations of motion of incompressible actual fluid

Also take a six-sided as infinitesimal stress analysis, surface forces in the $x$ direction as shown in Fig.3 – 14.

Over the same force and fluid quality. Thus, in the $x$ direction, according to Newton's second law $\sum F_x = ma_x$, we can write

$$f_x \rho dx dy dz + \left[ p_{xx} dy dz - \left( p_{xx} + \frac{\partial p_{xx}}{\partial x} dx \right) dy dz \right] -$$
$$\left[ \tau_{yx} dx dz - \left( \tau_{yx} + \frac{\partial \tau_{yx}}{\partial y} dy \right) dx dz \right] - \left[ \tau_{zx} dy dx - \left( \tau_{zx} + \frac{\partial \tau_{zx}}{\partial z} dz \right) dy dx \right]$$
$$= \rho dx dy dz \frac{du_x}{dt}$$

The relationship between the expression of shear stress and compressive stress, and incompressible fluid continuity differential

$$\frac{\partial u_x}{\partial x} + \frac{\partial u_y}{\partial y} + \frac{\partial u_z}{\partial z} = 0$$

Substituting simplification, available incompressible viscous fluid equations of motion, namely

$$f_x - \frac{1}{\rho} \frac{\partial p}{\partial x} + \nu \mathbf{\nabla}^2 u_x = \frac{du_x}{dt} = \frac{\partial u_x}{\partial t} + u_x \frac{\partial u_x}{\partial x} + u_y \frac{\partial u_x}{\partial y} + u_z \frac{\partial u_x}{\partial z} \quad (3-28a)$$

Similarly

$$f_y - \frac{1}{\rho} \frac{\partial p}{\partial y} + \nu \mathbf{\nabla}^2 u_y = \frac{du_y}{dt} = \frac{\partial u_y}{\partial t} + u_x \frac{\partial u_y}{\partial x} + u_y \frac{\partial u_y}{\partial y} + u_z \frac{\partial u_y}{\partial z} \quad (3-28b)$$

$$f_z - \frac{1}{\rho} \frac{\partial p}{\partial z} + \nu \mathbf{\nabla}^2 u_z = \frac{du_z}{dt} = \frac{\partial u_z}{\partial t} + u_x \frac{\partial u_z}{\partial x} + u_y \frac{\partial u_z}{\partial y} + u_z \frac{\partial u_z}{\partial z} \quad (3-28c)$$

Or represented by a vector

$$\boldsymbol{f} - \frac{1}{\rho} \mathbf{\nabla} p + \nu \mathbf{\nabla}^2 \boldsymbol{u} = \frac{\partial \boldsymbol{u}}{\partial t} + (\boldsymbol{u} \cdot \mathbf{\nabla}) \boldsymbol{u} \quad (3-29)$$

where $\mathbf{\nabla}$ is Laplace operator, $\mathbf{\nabla}^2 = \frac{\partial^2}{\partial x^2} + \frac{\partial^2}{\partial y^2} + \frac{\partial^2}{\partial z^2}$, e.g.

$$\mathbf{\nabla}^2 u_x = \frac{\partial^2 u_x}{\partial x^2} + \frac{\partial^2 u_x}{\partial y^2} + \frac{\partial^2 u_x}{\partial z^2} \quad (3-30)$$

Incompressible viscous fluid motion equations, also known as Navier-Stokes equations (briefly denoted as N – S equations). This is one of the fundamental equations of the theory of fluid mechanics. Because $\rho$ is constant, differential equations and continuity of the N – S equations may be composed of differential equations containing four unknowns. Combined with the boundary conditions, in theory, the solution of N – S equations exist. However, to describe directly solve complex and subtle viscous fluid motion, it is very complicated and difficult, research in this area is still the problem of frontier.

## 3.5 Integration of Euler differential equation

Euler equation in differential form given is used in the situations that the ideal fluid flows along a streamline in steady. The Euler equation are multiplied by $dx$, $dy$ and $dz$, and then added:

$$(f_x dx + f_y dy + f_z dz) - \frac{1}{\rho}\left(\frac{\partial p}{\partial x}dx + \frac{\partial p}{\partial y}dy + \frac{\partial p}{\partial z}dz\right) = \frac{du_x}{dt} + \frac{du_y}{dt}dy + \frac{du_z}{dt}dz \quad (3-31)$$
$$\langle 1 \rangle \qquad\qquad\qquad \langle 2 \rangle \qquad\qquad\qquad \langle 3 \rangle$$

### 3.5.1 Integration under steady potential flow, gravity, incompressible condition

#### 1. Bernoulli equation

Because the ideal fluid only have gravity, $f_x = f_y = 0$, $f_z = -g$, then

$$\langle 1 \rangle = -g\,dz \quad (3-31\text{a})$$

Because the ideal fluid is incompressible, then

$$\langle 2 \rangle = -\frac{1}{\rho}dp = -d\left(\frac{p}{\rho}\right) \quad (3-31\text{b})$$

Because the ideal fluid is potential flow, $\dfrac{\partial u_x}{\partial y} = \dfrac{\partial u_y}{\partial x}$, $\dfrac{\partial u_y}{\partial z} = \dfrac{\partial u_z}{\partial y}$, $\dfrac{\partial u_x}{\partial z} = \dfrac{\partial u_z}{\partial x}$, then

$$\langle 3 \rangle = \frac{du_x}{dt}dx + \frac{du_y}{dt}dy + \frac{du_z}{dt}dz$$

$$= \left(u_x\frac{\partial u_x}{\partial x} + u_y\frac{\partial u_x}{\partial y} + u_z\frac{\partial u_x}{\partial z}\right)dx + \left(u_x\frac{\partial u_y}{\partial x} + u_y\frac{\partial u_y}{\partial y} + u_z\frac{\partial u_y}{\partial z}\right)dy$$

$$+ \left(u_x\frac{\partial u_z}{\partial x} + u_y\frac{\partial u_z}{\partial y} + u_z\frac{\partial u_z}{\partial z}\right)dz$$

$$= \frac{\partial}{\partial x}\left[\frac{1}{2}(u_x^2 + u_y^2 + u_z^2)\right]dx + \frac{\partial}{\partial y}\left[\frac{1}{2}(u_x^2 + u_y^2 + u_z^2)\right]dy + \frac{\partial}{\partial z}\left[\frac{1}{2}(u_x^2 + u_y^2 + u_z^2)\right]dz$$

$$= \frac{\partial}{\partial x}\left(\frac{u^2}{2}\right)dx + \frac{\partial}{\partial y}\left(\frac{u^2}{2}\right)dy + \frac{\partial}{\partial z}\left(\frac{u^2}{2}\right)dz$$

$$= d\left(\frac{u^2}{2}\right) \quad (3-31\text{c})$$

So
$$d\left(gz + \frac{p}{\rho} + \frac{u^2}{2}\right) = 0 \quad (3-32)$$

Integration
$$gz + \frac{p}{\rho} + \frac{u^2}{2} = C' \quad (3-33)$$

or
$$z + \frac{p}{\rho g} + \frac{u^2}{2g} = C \quad (3-34)$$

$$z_1+\frac{p_1}{\rho g}+\frac{u_1^2}{2g}=z_2+\frac{p_2}{\rho g}+\frac{u_2^2}{2g} \qquad (3-35)$$

The Bernoulli equation can be obtained with an integral along a streamline as equation (3-34).

### 2. The physical meaning of Bernoulli equation

The physical meaning and geometric meaning of Bernoulli equation is shown in Tab.3-1.

Table 3-1  The physical meaning and geometric meaning of Bernoulli equation

| Term | Name | Physical meaning | Geometric meaning |
| --- | --- | --- | --- |
| $z$ | elevation head | location potential of unit weight fluid | position height |
| $\dfrac{p}{\rho g}$ | frictional head | pressure potential of unit weight fluid | piezometric height |
| $\dfrac{u^2}{2g}$ | velocity head | kinetic energy of unit weight fluid | |
| $z+\dfrac{p}{\rho g}$ | piezometric head | the total potential energy unit weight of fluid | |
| $z+\dfrac{p}{\rho g}+\dfrac{u^2}{2g}$ | energy grade | mechanical energy of unit weight fluid | |

## 3.5.2  Integration along the stream under steady, gravity, incompressible condition

Integration result in the potential flow only apply to potential flow. Along with a stream line is also available for integration to rotational flow.

$$\frac{\mathrm{d}x}{\mathrm{d}t}=u_x,\ \frac{\mathrm{d}y}{\mathrm{d}t}=u_y,\ \frac{\mathrm{d}z}{\mathrm{d}t}=u_z$$

$$\langle 3 \rangle = u_x\mathrm{d}u_x+u_y\mathrm{d}u_y+u_z\mathrm{d}u_z=\frac{1}{2}\mathrm{d}(u_x^2+u_y^2+u_z^2)=\mathrm{d}\!\left(\frac{u^2}{2}\right)$$

Energy equation is
$$z+\frac{p}{\rho g}+\frac{u^2}{2g}=C \qquad (3-36)$$

Equations (3-36) and (3-34) are exactly same in formation. However, integration constant are different.

## 3.6  Steady plane potential flows

### 3.6.1  Velocity potential function and Laplace equation

In plane potential flows, $u_x$, $u_y$ are equation of $x$ and $y$, and $\omega_z=0$ or

$$\frac{\partial u_x}{\partial y} = \frac{\partial u_y}{\partial x}$$

So
$$d\varphi = u_x dx + u_y dy = \frac{\partial \varphi}{\partial x} dx + \frac{\partial \varphi}{\partial y} dy \qquad (3-37)$$

which is
$$u_x = \frac{\partial \varphi}{\partial x}, \ u_y = \frac{\partial \varphi}{\partial y} \qquad (3-38)$$

where $\varphi$ is velocity potential function. Definition: irrotational flow.

The characteristic of velocity potential function are:

(1) Three components of velocity equal to the partial derivative of velocity potential $\varphi$ at respective coordinates. Or the gradient of velocity potential $\varphi$ equals to the velocity.

(2) Irrotational flow exist $\varphi$, and velocity potential $\varphi$ is irrotational flow.

Differential continuity equation for incompressible flow:

$$\frac{\partial^2 \varphi}{\partial x^2} + \frac{\partial^2 \varphi}{\partial y^2} = 0 \qquad (3-39)$$

This equation is the Laplace equation.

### 3.6.2 Stream function

#### 1. Stream function and laplace equation

Because the flow is constant plane flow, and they are only a function of $x$ and $y$. Differential continuity equation for incompressible flow:

$$\frac{\partial u_x}{\partial x} = -\frac{\partial u_y}{\partial y}$$

So
$$d\psi = -u_y dx + u_x dy = \frac{\partial \psi}{\partial x} dx + \frac{\partial \psi}{\partial y} dy \qquad (3-40)$$

which is
$$u_x = \frac{\partial \psi}{\partial y}, \ u_y = -\frac{\partial \psi}{\partial x} \qquad (3-41)$$

where $\psi$ is stream function. Definition: impressible flow.

The stream function is applied to the plane potential flow to obtain Laplace equation.

Plane potential flow: $\dfrac{\partial u_x}{\partial y} = \dfrac{\partial u_y}{\partial x}$

then
$$\frac{\partial^2 \psi}{\partial x^2} + \frac{\partial^2 \psi}{\partial y^2} = 0 \qquad (3-42)$$

or
$$\nabla^2 \psi = 0 \qquad (3-43)$$

The applicable conditions of equation (3-42) are incompressible fluid, steady flow, plane flow, and potential flow.

### 2. Physical meaning of stream function

(1) The lines of constant stream function $\psi(x, y) = C$ are streamlines.

Solution:
$$\psi(x, y) = C$$
$$d\psi = -u_y dx + u_x dy = 0$$

Obtain plane stream function:
$$\frac{dx}{u_x} = \frac{dy}{u_y}$$

(2) The discharge per unit width through the cross section $AB$ is
$$Q = \int_{y_1}^{y_2} u dy = \int_{y_1}^{y_2} \frac{\partial \psi}{\partial y} dy = \int_{\psi_1}^{\psi_2} d\psi = \psi_2 - \psi_1$$

In incompressible plane flow, the difference in the stream functions between two different streamlines is equal to the discharge per unit width between the two streamlines.

### 3. Inference—the relationship between stream function and potential function

$$\begin{cases} \dfrac{\partial \varphi}{\partial x} = \dfrac{\partial \psi}{\partial y} \\ \dfrac{\partial \varphi}{\partial y} = -\dfrac{\partial \psi}{\partial x} \end{cases} \quad (3-44)$$

The above equation is Cauchy-Riemann condition. $\varphi$, $\psi$ satisfy Laplace equation and Cauchy-Riemann condition.

**Example 3-2** There are two flow: (a) $u_x = 1$, $u_y = 2$; (b) $u_x = 4x$, $u_y = -4y$.

(1) Whether the flow of (a) is stream function?

(2) Whether the flow of (b) is potential function?

**Solution:** (1) $\dfrac{\partial u_x}{\partial x} + \dfrac{\partial u_y}{\partial y} = 0$

$$\psi = \int u_x dy - u_y dx = \int dy - 2dx = \int d(y - 2x) = y - 2x + C_1$$

(2) $\dfrac{\partial u_x}{\partial y} = \dfrac{\partial (4x)}{\partial y} = 0$, $\dfrac{\partial u_y}{\partial x} = \dfrac{\partial (-4x)}{\partial x} = 0$

So $\dfrac{\partial u_x}{\partial y} = \dfrac{\partial u_y}{\partial x}$, the flow is irrotational.

$$\varphi = \int u_x dx + u_y dy = \int 4x dx - 4y dy = 2\int d(x^2 - y^2) = 2x^2 - 2y^2 + C_2$$

### 3.6.3 Flow net

#### 1. The nature of flow nets

(1) Other streaming line of plane irrotational flows are perpendicularly intersecting with

equipotential line.

**Prove:** Equipotential line is stream line, $d\psi = u_x dy - u_y dx = 0$.

Slope of flow line at any point is

$$m_1 = \frac{dy}{dx} = \frac{u_y}{u_x}$$

$$d\varphi = u_x dx + u_y dy = 0$$

Slope of potential line at same point is $m_2 = \frac{dy}{dx} = -\frac{u_x}{u_y}$

Because $m_1 m_2 = -1$, so flow lines are perpendicularly intersecting with potential line.

(2) The ratio of each mesh side length in flow net is equal to $\frac{\Delta\varphi}{\Delta\psi}$. If $\Delta\varphi = \Delta\psi$, mesh of flow nets are curve square.

### 2. The draw of flow net

There are two ways to draw flow nets. One is graphic, another is electrical analogy method.

The principles of graphic:

(1) The equipotential lines are orthogonal with boundary.

(2) If flow rate at the free surface perpendicular to a surface equal is zero, the free surface is stream line.

(3) Draw stream lines and potential line based on the proportion of pre-selected grid lines.

### 3. The application of flow net

The flow net is the only solution in border at super position. So the flow net can be applied in all similar flow field.

$$u_{si} \approx \frac{\Delta C}{\Delta s_i} = \frac{\Delta C}{\Delta n_i}$$

The theory of flow net has been widely used over potential flow of ideal fluid.

### 3.6.4 The superposition principle of potential flow

If the potential function satisfy the Laplace equation, results of their superposition also satisfy the Laplace equation. For example, $\nabla^2 \varphi_1 = 0$ and $\nabla^2 \varphi_2 = 0$, results of their superposition are $\varphi = \varphi_1 + \varphi_2$.

$$\nabla^2 \varphi = \nabla^2 (\varphi_1 + \varphi_2) = \nabla^2 \varphi_1 + \nabla^2 \varphi_2 = 0 \qquad (3-45)$$

# Exercises

1. Given a two-dimensional flow, and its velocity distribution is $u_x = x^2 t$, $u_y = 2xyt$.

Try to find the streamline equation and pathline equation at $t=1$, point $(-2, 1)$.

2. Knowing the stream function $\psi = 2(x^2 - y^2)$, try to find the velocity potential function $\varphi$.

3. Assuming that $\psi_1$ and $\psi_2$ both satisfy the Laplace equation, try to prove that $\mathbf{V}^2\psi=0$, $\psi=\psi_1+\psi_2$.

4. The known flow field $u_x = -\dfrac{Cy}{r^2}$, $u_y = \dfrac{Cx}{r^2}$. where $C$ is a constant, $r^2 = x^2 + y^2$. Try to draw a schematic diagram of the flow net.

5. The flat rectangular section elbow is shown in Fig.3 – 15, its outer radius is $r_1$, the inner radius is $r_2$, and the center of the circle is $M$. Assuming that the flow velocity in the straight pipe is evenly distributed, its value is $u_0$, the dynamic pressure on the 0 – 0 section is $P_0$, the center of the elbow is symmetrical to the $A$-$A$ section, and the streamline is an arc with $M$ as the center, and conforms to the law of potential flow, $u_\theta r =$ constant. Try to find the distribution of flow velocity $u$ and dynamic pressure $p$ on the $A$-$A$ section.

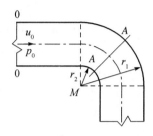

Fig.3 – 15

# Chapter 4   Fundamental Equations of Steady Total Flow

In practical engineering, many fluid problems (such as pipe flow, open channel flow, etc.) can be considered as a whole, treating it as a single entity without considering lateral variations in velocity, pressure, etc., along the main flow. Instead, only the average values for the corresponding cross-section are taken, thereby simplifying the issue and achieving better calculation results. Based on the exposition of the basic laws of one-dimensional flow, this chapter derives and analyzes the three basic equations of total flow motion, namely the continuity equation, energy equation, and momentum equation, and their applications, which are also the key learning points of this chapter.

## 4.1   Analysis method of total flow

In actual flow, even the simplest flow, the velocity distribution at various points of its cross section is not uniform. Thus kinetic energy, momentum and pressure are also not uniform. In total flow analysis, using cross-section average method to process the physical quantities based on the element flow.

### 4.1.1   Element flow, total flow and control cross section of total flow

#### 1. Element flow and total flow

Element flow is mini-stream tube in flow cross-section.

If the walls of the stream tube are extended to the flow field boundary, the fluid flow within the boundary is the total flow.

In the analyses of the variation of motion parameters such as flow velocity, flow rate, pressure and so on, the total flow can be divided into countless mini-stream tube. The motion parameters at every point on section d$A$ can be considered to be uniform due to the very small section of mini-stream tube, therefore the flow parameters of total flow can be obtained by the integral method.

#### 2. Control cross section

The control cross section is a section that every area element in the section is normal to mini-stream tube or streamline. The control cross section is denoted as d$A$ or $A$. Two control cross section $m$-$m$ and $c$-$c$ are given in Fig.4 – 1.

### 4.1.2 The analysis of total flow

#### 1. Based on the element flow

The total flow is constituted by element flow. Thus, the movement of element flow is analysis at first, and then extended to the total flow.

#### 2. Choose control cross section in gradually varied flow

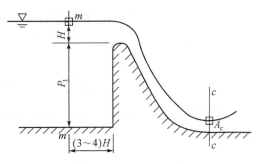

Fig.4-1 Control cross section

In the analysis of total flow, the control cross section is taken in the gradually varied flow. The reasons are two characteristics of gradually varied flow. One is that cross-section is a flat. Another is that the velocity directions of each point are almost parallel.

Why not choosing rapidly varied flow? Case in pipe flow, cross-section is not a flat in rapidly varied flow. Besides, the velocity distribution in section is very uneven.

#### 3. Physical quantities average of cross section

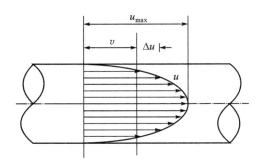

Fig.4-2 Mean velocity

(1) Mean velocity of cross-section

The velocity at every point in the cross section is different and its distribution is a parabola. The velocity $u$ takes the maximum $u_{max}$ on the pipe axle and the zero on the boundary as shown in Fig.4-2. According to the equivalency of flow rate, the fluid volume that pass through a cross section $A$ as an average $v$ must be equal to the fluid volume passing through the section as an actual velocity in unit time, namely, there with

$$v = \frac{\int_A u \, dA}{A} = \frac{q_v}{A} \tag{4-1}$$

(2) Momentum and correction coefficient of momentum

Momentum is a measure of the motion of an object and an important physical quantity that describes the mechanical motion of matter.

In words, for any direction the sum of variation of the momentum within a control volume plus the net outflow of momentum from the control volume is equal to the component of the resultant external force acting on a control volume at that direction in unit time.

For a steady flow of incompressible fluid, if the control surfaces, normal to the velocity wherever it cuts across the flow, have only the two surfaces, i.e. the outflow surface and the inflow surface, and if the velocities in the surfaces are uniform, the surface integral in equation could be removed. It reduces to

$$dK = dmu = (\rho u\,dA \cdot \Delta t)u = \rho uu\,dA \cdot \Delta t$$

For total flow,

$$\sum dK = \int_A \rho uu\,dA \cdot \Delta t = \left(\int_A \rho u\,dA_u\right) i \cdot \Delta t \qquad (4-2)$$

Leading into momentum correction factor $\beta$, and definition as follows:

$$\beta = \frac{\int_A u^2\,dA}{q_v v} = \frac{\int_A u^2\,dA}{Av^2} \qquad (4-3)$$

For incompressible constant flow, we have

$$\sum dK = \left(\rho \int_A u^2\,dA\right) i \cdot \Delta t = \beta \rho q_v v \cdot i \cdot \Delta t = \beta \rho q_v v \cdot \Delta t \qquad (4-4)$$

(3) Kinetic energy and correction coefficient of kinetic energy

The kinetic energy is object due to the motion of the mechanical energy. The kinetic energy of micro cell area $dA$ at unit of time is

$$dE_k = \frac{1}{2} dm \cdot u^2 = \frac{1}{2} \rho u^3\,dA$$

For incompressible steady flow,

$$\sum dE_k = \frac{1}{2}\rho \int_A u^3\,dA = \rho q_v \frac{\alpha v^2}{2} \qquad (4-5)$$

Leading into kinetic energy correction factor $\alpha$, and definition as follow:

$$\alpha = \frac{\int_A u^3\,dA}{q_v v^2} = \frac{\int_A u^3\,dA}{Av^3} \qquad (4-6)$$

## 4.2 Continuity equation of incompressible fluid

Select a control volume 1-1, 2-2 in a flow field (see Fig.4-3). For incompressible fluid:

$$\int_V \left(\frac{\partial u_x}{\partial x} + \frac{\partial u_y}{\partial y} + \frac{\partial u_z}{\partial z}\right) dV = \int_A u_n\,dA = 0 \qquad (4-7)$$

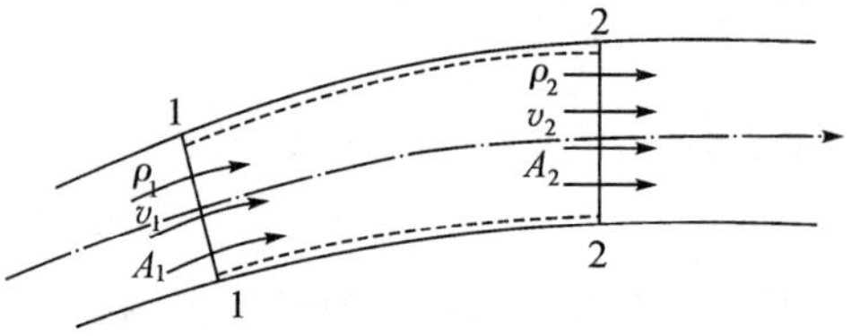

Because $\rho = $ const, then simplify:

$$\int_{A_1} u_1\,dA = \int_{A_2} u_2\,dA \qquad (4-8)$$

$$v_1 A_1 = v_2 A_2 \qquad (4-9)$$

**Fig.4-3  Continuity equation total flow**

Physical meaning is that for the incompressible

fluid, mean velocity is inverse proportional to the cross-section area.

Fit for: incompressible fluid, including steady and unsteady flow, idea and real fluid.

When there is inflow or outflow of fluid between the two sections, the flow of junction is

$$\sum_{i=1}^{n} q_{v_i} = 0 \qquad (4-10)$$

Such as Fig.4 - 4, $q_{v_1} - q_{v_2} - q_{v_3} = 0$
or $A_1 v_1 - A_2 v_2 - A_3 v_3 = 0$

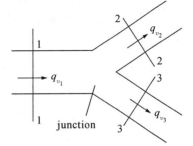

Fig.4 - 4  The flow of junction

## 4.3  The Bernoulli equation

### 4.3.1  The Bernoulli equation of element in real flow

We've got the Bernoulli equation of element flow from Section 3.5.

$$z + \frac{p}{\rho g} + \frac{u}{2g} = C$$

However, the actual fluid has a viscosity with resistance exercise. According to the energy equation, the Bernoulli equation of element in real flow can be obtained.

$$z_1 + \frac{p_1}{\rho g} + \frac{u_1^2}{2g} = z_2 + \frac{p_2}{\rho g} + \frac{u_2^2}{2g} + h'_w \qquad (4-11)$$

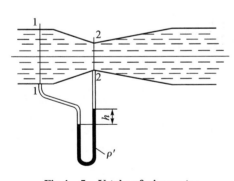

Fig.4 - 5  U-tube of piezometer

**Example 4 - 1**  As shown in Fig. 4 - 5, $d_1 = 250$ mm, $d_2 = 100$ mm, the mercury U-tube piezometer $h = 800$ mm, what is the discharge in the pipe?

**Solution:** Pressure difference between 1 - 1 and 2 - 2 sections:

$$p_1 - p_2 = (\rho_{Hg} - \rho_{H_2O})gh = (13\,600 - 1\,000) \times 9.8 \times 0.8$$
$$= 98\,784 \text{ Pa}$$

and $\quad \dfrac{A_1}{A_2} = \left(\dfrac{d_1}{d_2}\right)^2$

so

$$Q = A_1 \sqrt{2(p_1 - p_2)/\rho\left[\left(\frac{A_1}{A_2}\right)^2 - 1\right]}$$

$$= \pi \times \frac{0.25^2}{4} \times \sqrt{2 \times 98\,784 / [1\,000 \times (2.5^4 - 1)]} = 0.112 \text{ m}^3/\text{s}$$

### 4.3.2 The Bernoulli equation along a streamline

#### 1. The deduction of the Bernoulli equation along a streamline

The element flow is defined as $dq_v$. The equation is multiplied by $\rho g\, dq_v$, the energy relations of element flow can be obtained.

$$\left(z_1+\frac{p_1}{\rho g}+\frac{u_1^2}{2g}\right)\rho g\, dq_v = \left(z_2+\frac{p_2}{\rho g}+\frac{u_2^2}{2g}\right)\rho g\, dq_v + h'_w \rho g\, dq_v \qquad (4-12)$$

Attention: $dq_v = u_1 dA_1 = u_2 dA_2$, the energy relations of total flow is:

$$\int_{A_1}\left(z_1+\frac{p_1}{\rho g}+\frac{u_1^2}{2g}\right)\rho g u_1 dA_1 = \int_{A_2}\left(z_2+\frac{p_2}{\rho g}+\frac{u_2^2}{2g}\right)\rho g u_2 dA_2 + \int_{q_v} h'_w \rho g\, dq_v \qquad (4-13)$$

The items of equation are integral.

(1) The integral of potential energy

$$\rho g \int_A \left(z+\frac{p}{\rho g}\right) u\, dA = \rho g\left(z+\frac{p}{\rho g}\right)\int_A u\, dA = \left(z+\frac{p}{\rho g}\right)\cdot \rho g q_v \qquad (4-14)$$

(2) The integral of kinetic energy

$$\rho g \int_A \frac{u^3}{2g} dA = \frac{\alpha v^2}{2g}\rho g q_v \qquad (4-15)$$

(3) The integral of loss part

$$\int_{q_v} h'_w \rho g\, dq_v = h'_w \rho g q_v \qquad (4-16)$$

Attention: $q_{v_1} = q_{v_2} = q_v$, with divided $\rho g q_v$

$$z_1+\frac{p_1}{\rho g}+\frac{\alpha_1 v_1^2}{2g} = z_2+\frac{p_2}{\rho g}+\frac{\alpha_2 v_2^2}{2g}+h_w \qquad (4-17)$$

This is the Bernoulli equation of real total flow.

#### 2. Conditions and applications

(1) Conditions

The Bernoulli equation have several conditions of using:

① The flow is steady flow;
② The flow is incompressible fluid;
③ The flow is ideal flow;
④ Along a streamline;
⑤ The only mass force is gravity.

(2) Applications

① Select the plane;
② Select computing section;
③ Select calculation points;
④ List solution of Bernoulli equation.

## 3. The physical meaning and geometric meaning of the Bernoulli equation

(1) The geometric meaning

$z$—the elevation height above datum surface $0-0$, elevation head;

$\dfrac{p}{\rho g}$—rising height of fluid with unit weight under the action of pressure $p$, pressure head;

$\dfrac{\alpha v^2}{2g}$—rising height of fluid with unit weight under the action of velocity $v$, velocity head;

$h_w$—the loss of mechanical energy, the total head loss;

$z + \dfrac{p}{\rho g} + \dfrac{\alpha v^2}{2g}$—total head.

(2) The physical meaning

$z$—elevation potential energy per unit weight of fluid;

$\dfrac{p}{\rho g}$—pressure potential energy per unit weight of fluid;

$\dfrac{\alpha v^2}{2g}$—kinetic energy per unit weight of fluid;

$z + \dfrac{p}{\rho g} + \dfrac{\alpha v^2}{2g}$—mechanical energy per unit weight of fluid.

In a steady flow of incompressible ideal fluid under gravity, the mechanical energy per unit weight of fluid along a streamline is the same. Generally different for different streamlines.

**Example 4 - 2** As shown in Fig. 4 - 6, water is siphoned out of a tank by a bent pipe, the top section $h = 3.6$ m, if the evaporating pressure for this water is 0.32 atm, what is the distance $h$ when the discharge in the pipe is the most?

**Solution:** Select $B$ section as the base section, for the free surface and $B$ section

$$v_B = \sqrt{2gh}$$

from continuity equation:

$$v_A = v_B = \sqrt{2gh} = v$$

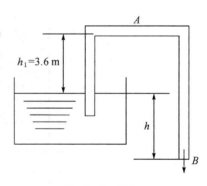

Fig.4 - 6  Siphon

Write Bernoulli equation from point $A$ to point $B$:

$$h_1 + h + \dfrac{p_{MA}}{\rho g} = 0$$

$$v = \sqrt{2gh}$$

where $h_1$ keep a constant,

$$p_{A\min} = 0.32 \text{ atm}, \quad p_{MA\min} = (0.32 - 1) \text{atm}$$

$$p_{MA} = (0.32 - 1)\text{atm} = -0.68 \times 101\,325 = -68\,901 \text{ Pa}$$

$$h = -h_1 - \frac{p_{MA}}{\rho g} = -3.6 - \frac{-68\,901}{1\,000 \times 9.81} = 3.42 \text{ m}$$

### 4.3.3 Head line

The sum of elevation head, pressure head and velocity head is called the total head in Bernoulli equation and is expressed as $H$. Because of each team of Bernoulli equation represents an elevation, the relationship of them could be indicated in a geometric graph. In Fig.4 - 7, the datum line for the elevation is 0 - 0; the line segment $amb$ is the line which connects each point $z$, it is called elevation head line; the segment $enf$ is the line which connected each vertex point of $z + \frac{p}{\rho g}$, it is called the piezometer head line; the segment $gkh$ is the line which connected each vertex point of $\frac{u^2}{2g}$, it is called the total head line. The meaning of Bernoulli equation for ideal fluid in geometry is as follows: the total head line is a horizontal line, each head could increase or decrease but the total head remains constant. In Fig.4 - 7, the total heads at the three points $a$, $m$ and $b$ are $H_1 = z_1 + \frac{p_1}{\rho g} + \frac{u_1^2}{2g}$, $H = z + \frac{p}{\rho g} + \frac{u^2}{2g}$ and $H_2 = z_2 + \frac{p_2}{\rho g} + \frac{u_2^2}{2g}$. Three total heads are all equal, namely.

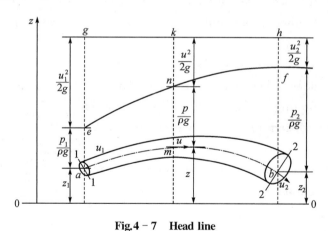

**Fig.4 - 7  Head line**

The slope of total head line is hydraulic slope:

$$J = \frac{\mathrm{d}h_w}{\mathrm{d}s} = -\frac{\mathrm{d}H}{\mathrm{d}s} = -\frac{\mathrm{d}\left(z + \frac{p}{\rho g} + \frac{\alpha v^2}{2g}\right)}{\mathrm{d}s} \tag{4-18}$$

The change of piezometer head line can use piezometer head line slope to express:

$$J_p = -\frac{\mathrm{d}\left(z + \frac{p}{\rho g}\right)}{\mathrm{d}s} \tag{4-19}$$

Because of no head loss, the total head line of ideal flow is a horizontal line.

## 4.3.4 Expand of Bernoulli equation

### 1. The Bernoulli equation along with flow separation or import

(1) Separation

The fluid is incompressible and the flow must be a steady flow. The flow rate remains a constant along the flow path. If there is output of shaft work between two sections, the calculation can be carried out with the following equations in Fig.4-8:

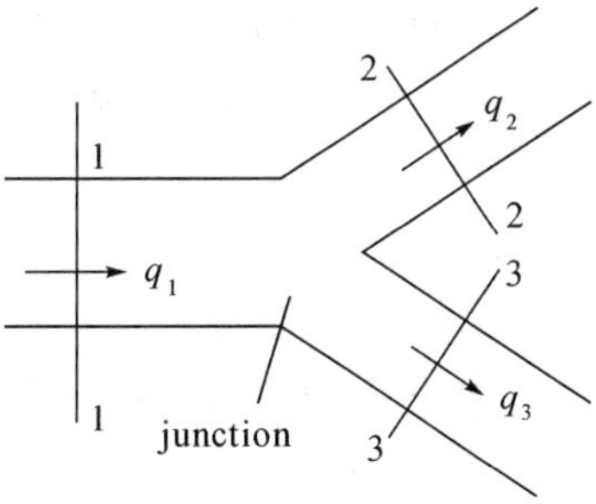

$$z_1+\frac{p_1}{\rho g}+\frac{\alpha_1 v_1^2}{2g}=z_2+\frac{p_2}{\rho g}+\frac{\alpha_2 v_2^2}{2g}+h_{wl,2} \quad (4-20)$$

$$z_1+\frac{p_1}{\rho g}+\frac{\alpha_1 v_1^2}{2g}=z_3+\frac{p_3}{\rho g}+\frac{\alpha_3 v_3^2}{2g}+h_{wl,3} \quad (4-21)$$

Fig.4-8  Separation

(2) Import

When the two fluids are confluence, the mechanical energy per unit weight of fluid are not equal except causing head loss. The calculation can be carried out with the following equation:

$$\rho g q_{v_1}\left(z_1+\frac{p_1}{\rho g}+\frac{\alpha_1 v_1^2}{2g}\right)+\rho g q_{v_2}\left(z_2+\frac{p_2}{\rho g}+\frac{\alpha_2 v_2^2}{2g}\right)$$
$$=\rho g q_{v_3}\left(z_3+\frac{p_3}{\rho g}+\frac{\alpha_3 v_3^2}{2g}\right)+\rho g q_{v_1} h_{wl,3}+\rho g q_{v_2} h_{w2,3} \quad (4-22)$$

### 2. The Bernoulli equation along with energy separation or import

The total Bernoulli equation is obtained with no energy separation or import except causing head loss in the two cross section. When there are pumps etc. between the two cross-section, the fluid can obtain additional energy or lose energy. So the Bernoulli equation is changed.

$$z_1+\frac{p_1}{\rho g}+\frac{\alpha_1 v_1^2}{2g}\pm H=z_2+\frac{p_2}{\rho g}+\frac{\alpha_2 v_2^2}{2g}+h_w \quad (4-23)$$

### 3. The Bernoulli equation of gas

The Bernoulli equation of total flow is deducted by incompressible flow. Gas is a compressible fluid, but the velocity is not great. So the Bernoulli equation are also suitable for gas.

From Fig.4-9, list Bernoulli equation by choosing the cross sections 1-1 and 2-2.

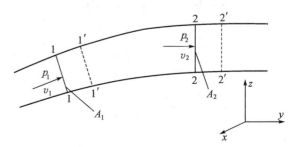

Fig.4-9  The Bernoulli equation of gas

$$z_1 + \frac{p_{1abs}}{\rho g} + \frac{v_1^2}{2g} = z_2 + \frac{p_{2abs}}{\rho g} + \frac{v_2^2}{2g} + h_w \quad (\alpha_1 = \alpha_2 = 1)$$

$$\rho g z_1 + p_{1abs} + \frac{\rho v_1^2}{2} = \rho g z_2 + p_{2abs} + \frac{\rho v_2^2}{2} + p_w \tag{4-24}$$

Because 
$$p_{1abs} = p_1 + p_a$$
$$p_{2abs} = p_2 + p_a - \rho_a g (z_2 - z_1)$$

then
$$p_1 + \frac{\rho v_1^2}{2} + (\rho_a - \rho) g (z_2 - z_1) = p_2 + \frac{\rho v_2^2}{2} + p_w \tag{4-25}$$

When $\rho_a = \rho$, $z_2 = z_1$, then

$$p_1 + \frac{\rho v_1^2}{2} = p_2 + \frac{\rho v_2^2}{2} + p_w \tag{4-26}$$

When $\rho \gg \rho_a$, then $\rho_a$ can be ignored.

$$p_1 + \frac{\rho v_1^2}{2} - \rho g (z_2 - z_1) = p_2 + \frac{\rho v_2^2}{2} + p_w$$

Divided by $\rho g$:

$$z_1 + \frac{p_1}{\rho g} + \frac{v_1^2}{2g} = z_2 + \frac{p_2}{\rho g} + \frac{v_2^2}{2g} + h_w$$

**Example 4-3** A centrifugal water pump with a suction pipe is shown in Fig. 4-10. Pump output is $Q = 0.03 \text{ m}^3/\text{s}$, the diameter of suction pipe $d = 150 \text{ mm}$, the vacuum that the pump can reach is $\dfrac{p_v}{\rho g} = 6.8 \text{ mH}_2\text{O}$. Determine the utmost elevation $h_e$ from the pump shaft to the water surface on the pond.

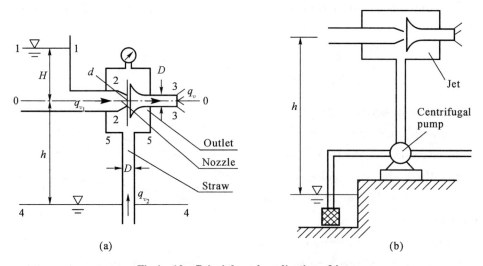

Fig.4-10 **Principle and application of jet**

**Solution:** Firstly, two cross sections here are: (1) the water surface 0 - 0 on the pond; (2) the section 1 - 1 on the inlet of pump. Meanwhile, the section 0 - 0 is taken as the datum plane, $z_0 = 0$, so

$$z_0 + \frac{p_0}{\rho g} + \frac{\alpha_0 v_0^2}{2g} = z_1 + \frac{p_1}{\rho g} + \frac{\alpha_1 v_1^2}{2g} + h_{w0\text{-}1}$$

Secondly, the parameters in the Bernoulli equation are determined.

The pressure $p_0$ and $p_1$ are expressed in relative pressure (guage pressure).

$$\frac{p_0}{\rho g} = 0, \text{ and } \frac{p_1}{\rho g} = -\frac{p_v}{\rho g} = -6.8 \text{ mH}_2\text{O}$$

Since the velocity $v_0$ at the free surface of pond is far less than the velocity $v_1$ at the section 1 - 1 in the pipe, so $v_0 = 0$. The velocity $v_1$ is

$$v_1 = \frac{Q}{A} = \frac{0.03}{\pi \times \frac{0.15^2}{4}} = 0.17 \text{ m/s}$$

Let the kinetic-energy correction factor $\alpha_1$ be 1.

The losses in the flow: the energy losses between two sections for fluid per unit weight passing through are $h_{w0\text{-}1} = 1 \text{ mH}_2\text{O}$.

Finally, calculation for the unknown parameter is carried out.

By substituting $v_0 = 0$, $p_0 = 0$, $z_0 = 0$, $p_1 = -p_v$, $z_1 = h_e$, $\alpha = 1$ and $v_1 = 0.17$ into the Bernoulli equation, i.e.

$$0 + 0 + 0 = \frac{v_1^2}{2g} + \frac{p_v}{\rho g} + h_e + h_{w0\text{-}1}$$

namely

$$h_e = \frac{p_v}{\rho g} - \frac{v_1^2}{2g} - h_{w0\text{-}1}$$

The values of $\frac{p_v}{\rho g}$ and $v_1$ are substituted into above equation and it gives

$$h_e = 6.8 - 0.15 - 1.0 = 5.65 \text{ m}$$

The utmost elevation $h_e$ from the pump shaft to the water surface is 5.65 meters. If exceeded, the boiling of water would occur and the pump can not run well.

## 4.4　Momentum equation

The momentum equation ia another basic equation of 1 - D flow after Bernoulli equation and continuity equation. We always need to calculate forces between fluids and solid boundaries in engineering practice. In addition, Bernoulli equation and continuity equation can not reflect the relationship of forces between fluids and solid boundaries. And Bernoulli equation includes the head loss. But it is difficult to confirm the head loss in some fluids. The momentum equation can make up for these shortcomings.

### 4.4.1 The deduction of momentum equation

For any direction the sum of variation of the momentum within a control volume plus the net outflow of momentum from the control volume is equal to the component of the resultant external force acting on a control volume at that direction in unit time.

The rate of change of momentum of system about time is equal to the vector sum of all forces on the system, that is, it refers to

$$\sum \boldsymbol{F} = \frac{\mathrm{d}\boldsymbol{K}}{\mathrm{d}t} = \frac{\mathrm{d}(\sum m\boldsymbol{u})}{\mathrm{d}t}$$

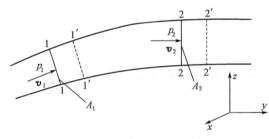

Fig.4-11  The deduction of momentum equation

Select a tube flow in a steady total flow as the control volume, as is shown in Fig.4-11. After time $\mathrm{d}t$, fluid flows from $11-22$ to $1'1'-2'2'$.

Left surface flow in $(\mathrm{d}t)$: $\left(\int_{A_1} \boldsymbol{v}_1 \rho_1 v_1 \mathrm{d}A_1\right) \mathrm{d}t$

Right surface flow in $(\mathrm{d}t)$: $\left(\int_{A_2} \boldsymbol{v}_2 \rho_2 v_2 \mathrm{d}A_2\right) \mathrm{d}t$

The change of momentum of tube flow in $\mathrm{d}t$ steady flow is

$$\mathrm{d}(M\boldsymbol{v}) = \left(\int_{A_2} \boldsymbol{v}_2 \rho_2 v_2 \mathrm{d}A_2\right) \mathrm{d}t - \left(\int_{A_1} \boldsymbol{v}_1 \rho_1 v_1 \mathrm{d}A_1\right) \mathrm{d}t$$

The rate of change of momentum:

$$\frac{\mathrm{d}(M\boldsymbol{v})}{\mathrm{d}t} = \int_{A_2} \boldsymbol{v}_2 \rho_2 v_2 \mathrm{d}A_2 - \int_{A_1} \boldsymbol{v}_1 \rho_1 v_1 \mathrm{d}A_1$$

Momentum equation in steady total flow:

$$\sum \boldsymbol{F} = \int_{A_2} \boldsymbol{v}_2 \rho_2 v_2 \mathrm{d}A_2 - \int_{A_1} \boldsymbol{v}_1 \rho_1 v_1 \mathrm{d}A_1$$

Defining: $\int_A \rho v^2 \mathrm{d}A = \beta \rho A v^2$, so

$$\beta_2 \rho_2 A_2 \boldsymbol{v}_2^2 - \beta_1 \rho_1 A_1 \boldsymbol{v}_1^2 = \sum \boldsymbol{F} \quad \text{(generally, assuming } \beta = 1.0\text{)}$$

Incompressible momentum equation in steady total flows:

$$\rho Q(\boldsymbol{v}_2 - \boldsymbol{v}_1) = \sum \boldsymbol{F}$$

$x$, $y$, $z$ directions:
$$\begin{cases} \rho Q(u_2 - u_1) = \sum F_x \\ \rho Q(v_2 - v_1) = \sum F_y \\ \rho Q(w_2 - w_1) = \sum F_z \end{cases} \qquad (4-27)$$

The sum of all forces on fluid in the control volume. This external forces include:

(1) The mass forces on fluid in the control volume;

(2) The surface forces on the control volume surface (dynamic pressure and shear stress);

(3) The total forces on fluid exerted by the surrounding boundaries.

### 4.4.2 The application of momentum equation

The limitation of momentum equation:

(1) Ideal fluid, incompressible steady flow;

(2) Two cross sections must be gradually varied flow sections;

(3) Discharge doesn't change along flow.

Steps to solve the problem:

(1) Select the control body;

(2) Select a coordinate system;

(3) Draw calculation diagrams;

(4) List the momentum equation.

### 4.4.3 Deduction

Correction equation:

$$\sum (\beta \rho q_v \boldsymbol{v})_{\text{out}} - \sum (\beta \rho q_v \boldsymbol{v})_{\text{in}} = \sum \boldsymbol{F} \qquad (4-28)$$

**Example 4 – 4** A nozzle that discharges a 10 cm diameter fluid jet into the air is on the right end of a horizontal 4 cm-diameter (see Fig. 4 – 12). The relative density of the fluid is 0.85, $p_{M1} = 7.0 \times 10^5$ Pa, $d_1 = 10$ cm, $d_2 = 4$ cm, What is the force exerted by fluid on the bolt $S$ (neglect friction)?

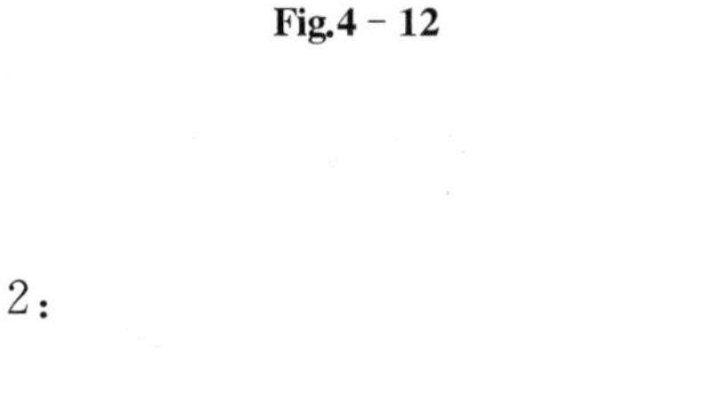

**Solution:** Select control volume 1122:

Fig. 4 – 12

$$v_1 = v_2 \frac{A_2}{A_1} = v_2 \left(\frac{d_2}{d_1}\right)^2$$

Bernoulli equation between cross sections 1 – 1 and 2 – 2:

$$\frac{p_1}{\rho g} + \frac{v_1^2}{2g} = \frac{v_2^2}{2g}$$

so

$$p_1 = \rho g \left(\frac{v_2^2}{2g} - \frac{v_1^2}{2g}\right) = \rho \frac{v_2^2}{2g}\left[1 - \left(\frac{d_2}{d_1}\right)^4\right]$$

$$v_2 = \sqrt{\frac{2p_1}{\rho\left[1 - \left(\frac{d_2}{d_1}\right)^4\right]}} = \sqrt{\frac{2 \times 7 \times 10^5}{0.85 \times 1\,000 \times \left[1 - \left(\frac{4}{10}\right)^4\right]}} = 41.1 \text{ m/s}$$

so:
$$v_1 = v_2 \left(\frac{d_2}{d_1}\right)^4 = 41.1 \times \left(\frac{4}{10}\right)^2 = 6.58 \text{ m/s}$$

$$Q = v_1 \pi \left(\frac{d_1}{2}\right)^2 = 6.58 \times \pi \times \left(\frac{0.1}{2}\right)^2 = 0.051\ 6 \text{ m}^3/\text{s}$$

Momentum equation in $x$-direction:

$$R = p_1 A_1 - \rho Q(v_2 - v_1)$$
$$= 7 \times 10^5 \times \frac{\pi}{4} \times 0.1^2 - 0.85 \times 1\ 000 \times 0.051\ 6 \times (41.1 - 6.58) = 3\ 982 \text{ N}$$

# Exercises

1. Given a varying section of an oil pipeline, with diameters at the two ends being $d_1 = 200$ mm, $d_2 = 60$ mm and the density of oil being 860 kg/m³. Known cross-section 1 - 1 velocity $v_1 = 2$ m/s, try to calculate the velocity at cross-section 2 - 2 and the mass flow rate.

2. As shown in Fig.4 - 13, a pipeline is composed of two pipes with different diameters connected by a gradual transition pipe. It is known that $d_A = 200$ mm, $d_B = 400$ mm, the relative pressure at point $A$, $p_A = 6.86 \times 10^4$ N/m², and the relative pressure at point $B$, $p_B = 3.92 \times 10^4$ N/m²; the sectional average velocity at point $B$, $v_B = 1$ m/s. The elevation difference between points $A$ and $B$, $\Delta z = 1$ m. It is required to determine the direction of flow and calculate the head loss, $h_w$, between these two sections.

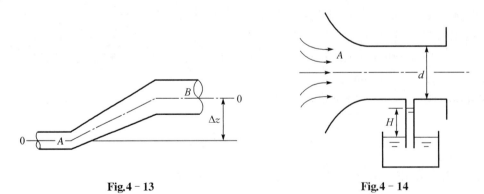

Fig.4 - 13          Fig.4 - 14

3. As shown in Fig.4 - 14, a centrifugal fan draws air from the atmosphere through a collector $A$. A glass tube is connected to a cylindrical pipe with a diameter of $d = 200$ mm, and the lower end of the tube is inserted into a water trough. If the water in the glass tube rises $H = 150$ mm, calculate the volume of air $q_v$ drawn per second. The density of air $\rho = 1.29$ kg/m³.

4. As shown in Fig.4 - 15, given that a pump delivers water, with the pump's shaft power $N = \rho g q_v H = 13.3$ kW. The efficiency $\eta_p = 0.75$. Known $h = 20$ m, and the head loss in

the pipeline $h_w = \dfrac{8v^2}{2g}$. Try to calculate its flow rate and the pump's head $H_p$ ($H_p = h + h_w$), and qualitatively draw its total energy line and the piezometric head line.

5. As shown in Fig. 4 - 16, there is a single-width jet of water with a thickness of 50 mm and velocity $v = 18$ m/s, diagonally impacting a smooth flat plate with a side length $l = 1.2$ m, and the jet is divided into two parts along the surface of the plate. The angle between the plate and the direction of the water flow is $\theta = 30°$. Assuming the friction of water flow, air, and plate is negligible, and the flow is on the same horizontal plane, try to calculate: (1) the flow distribution $q_{v_1}$ and $q_{v_2}$; (2) the impact force of the jet on the plate; (3) if the plate moves at $u = 8$ m/s in the same direction as the water flow, the magnitude of the force exerted on the plate by the water flow.

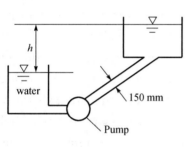

Fig.4 - 15

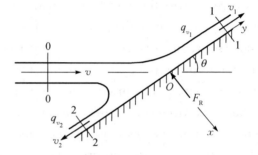

Fig.4 - 16

# Chapter 5  Similarity Theory and Dimensional Analysis

The dimensionless parameters for any particular problem can be determined in two ways. They can be deduced directly from the governing differential equations if these equations are known; this method is illustrated here. However, if the governing differential equations are unknown or the parameter of interest does not appear in the known equations, dimensionless parameters can be determined from dimensional analysis. The advantage of adopting the former strategy is that dimensionless parameters determined from the equations of motion are more readily interpreted and linked to the physical phenomena occurring in the flow. Thus, knowledge of the relevant dimensionless parameters frequently aids the solution process, especially when assumptions and approximations are necessary to reach a solution.

This section describes and presents the primary dimensionless parameters or numbers required in the remainder of the text. Many others not mentioned here are defined and used in the broad realm of fluid mechanics.

## 5.1  Dimension

### 5.1.1  Basic concept

#### 1. Units and dimension

Various physical quantities was constituted by two factors. One of them is dimension such as L and M, another is physical units. A physical dimension is unique, but the unit has multiple representations. Adopting dim $q$ as dimension of physical quantity $q$, dimension of velocity $v$ is:

$$\dim v = LT^{-1}$$

#### 2. Fundamental dimensions and derived dimensions

The dimensions which are independent of each other and can't express to each other is called fundamental dimensions. The dimensions can be deduced by fundamental dimensions are called derived dimensions. For example, $q$ is a physical quantity of derived dimensions, and

$$\dim q = L^{\alpha} T^{\beta} M^{\gamma} \tag{5-1}$$

(1) If $\alpha \neq 0$, $\beta = \gamma = 0$, area $A$ and volume $V$ are $\dim A = L^2$, $\dim V = L^3$, respectively.

(2) If $\alpha \neq 0$, $\beta \neq 0$, $\gamma = 0$, velocity $v$ and acceleration $a$ are $\dim v = LT^{-1}$, $\dim a = LT^{-2}$, respectively.

(3) If $\alpha \neq 0$, $\beta \neq 0$, $\gamma \neq 0$, force $F$ and pressure $p$ are $\dim F = MLT^{-2}$, $\dim p = ML^{-1}L^{-2}$, respectively.

### 3. Quantities of dimension one

If the physical dimension in the equation (5-1) is $\alpha = \beta = \gamma = 0$, the physical dimension is called quantities of dimension one. Such as angle, $\pi$ and strain rate, $\dim \theta = 1$, $\dim \pi = 1$, $\dim \varepsilon = 1$.

Quantities of dimension one has three features. First, the number is independent of the units. Second, quantities of dimension one is purely digital without scale effect. Last, quantities of dimension one can be computed by transcendental functions.

### 5.1.2 Dimensional homogeneity

The fundamental of dimensional analysis is dimensional homogeneity. For example, actual total flow in Chapter 4 is derived by Bernoulli equation.

$$z_1 + \frac{p_1}{\rho g} + \frac{\alpha_1 v_1^2}{2g} = z_2 + \frac{p_2}{\rho g} + \frac{\alpha_2 v_2^2}{2g} + h_w$$

All dimensions in the above equation is 1. The equation of constant total flow is

$$\sum \boldsymbol{F} = \rho q_v \beta_2 \boldsymbol{v}_2 - \rho q_v \beta_1 \boldsymbol{v}_1$$

All dimensions in the above equation is $MLT^{-2}$.

Only the same type of physical quantity can be added and subtracted. It is meaningless to add or subtract different types of physical quantities.

An equation is in the dimensional homogeneity, the form of the equation does not change with the unit. If you use one of the item divided by the entries, you will get a equations constructed by quantities of dimension one. Therefore, the dimensional homogeneity can be used to test a new equation or empirical equation for the correctness and completeness. That is, the dimensional homogeneity can be used to determine the dimensional exponents of physical quantity, or to establish a physical equation.

## 5.2 Dimensional analysis

Dimensional analysis have two kinds of solutions. The Rayleigh method is suitable for simple problem, and the Buckingham $\pi$ method is a universal method.

### 5.2.1 Rayleigh method

Rayleigh method is a dimensional analysis directly from the principle of consistent dimension. It is described by way of example.

**Example 5-1** Solving the expressions of wall shear stress in pipe flow.

**Solution:** The effect of $\tau_0$ have density $\rho$, kinematic viscosity $\mu$, diameter $D$, roughness $\Delta$ and average velocity $v$.

$$\tau_0 = K\rho^a \mu^b D^c \Delta^d v^e$$
$$(ML^{-1}T^{-2}) = (ML^{-3})^a (ML^{-1}T^{-1})^b (L)^c (L)^d (LT^{-1})^e$$

According to dimensional homogeneity,

$$M: 1 = a + b$$
$$L: -1 = -3a - b + c + d + e$$
$$T: -2 = -b - e$$

Combine the three equations,

$$b = 1 - a, \ c = a - d - 1, \ e = 1 + a$$

so
$$\tau_0 = K\rho^a \mu^{1-a} D^{a-d-1} \Delta^d v^{1+a}$$

$$\tau_0 = K\left(\frac{\rho D v}{\mu}\right)^a \left(\frac{\Delta}{D}\right)^d \left(\frac{\mu}{\rho D v}\right)\rho v^2 = K(Re)^{a-1}\left(\frac{\Delta}{D}\right)^d \rho v^2 = f\left(Re, \frac{\Delta}{D}\right)\rho v^2$$

When $f\left(Re, \dfrac{\Delta}{D}\right) = \dfrac{\lambda}{8}$,

then $\tau_0 = \dfrac{\lambda}{8}\rho v^2$.

The step of applying Rayleigh method of dimensional analysis is generally:

(1) Determining physical quantities related to the physical phenomena;

(2) Writing dimensional exponents between the physical quantities in the form of the product;

(3) Using fundamental dimensions to replace each physical dimension;

(4) According to dimensional homogeneity to determine the dimensional exponent of physical quantity.

### 5.2.2 The Buckingham $\pi$ method

**1. $\pi$ method**

The number of independent dimensionless groups of variables needed to correlate the variables in a given process is equal to $n - m$, where $n$ is the number of variables involved and $m$ is the number of fundamental dimensions included in the variables.

$$\varphi(\pi_1, \pi_2, \cdots, \pi_{n-m}) = 0 \qquad (5-2)$$

**2. The solving steps of $\pi$ method**

(1) Finding the relevant physical quantities

$$f(x_1, x_2, \cdots, x_n) = 0$$

(2) Determining the basic quantities. Choosing from a few basic physical quantities as the fundamental dimensions.

(3) Determining the number of $\pi$.

$$\pi_i = x_1^{a_i} x_2^{b_i} x_3^{c_i} \cdot x_i$$

(4) Determining each exponent of quantities of dimension one $\pi$.

(5) Substituting each $\pi$ into equation (5-2).

$$f(\pi_1, \pi_2, \cdots, \pi_{n-m}) = 0$$
$$\pi_4 = f(\pi_1, \pi_2, \cdots, \pi_{n-m})$$

### 3. The basic principles of selecting

(1) The basic quantities should correspond to the fundamental dimensions.

(2) When selecting basic quantities, you should choose important physical quantities.

**Example 5-2**

(1) $f(d, \rho, \mu, v, l, \Delta, \Delta p) = 0$

(2) Choosing $\rho, d, v$ as basic quantities, $m = 3$.

(3) $N(\pi) = n - m = 7 - 3 = 4$

$$\pi_1 = d^{a_1} v^{b_1} \rho^{c_1} \cdot \mu, \quad \pi_2 = d^{a_2} v^{b_2} \rho^{c_2} \cdot l, \quad \pi_3 = d^{a_3} v^{b_3} \rho^{c_3} \cdot \Delta, \quad \pi_4 = d^{a_4} v^{b_4} \rho^{c_4} \cdot \Delta p$$

(4) $\dim \pi_1 = \dim(d^{a_1} v^{b_1} \rho^{c_1} \cdot \mu)$

$$M^0 L^0 T^0 = (L)^{a_1} (LT^{-1})^{b_1} (ML^{-3})^{c_1} \cdot (ML^{-1}T^{-1})$$

$$M: 0 = c_1 + 1$$
$$L: 0 = a_1 + b_1 - 3c_1 - 1$$
$$T: 0 = -b_1 - 1$$

$$a_1 = -1, \; b_1 = -1, \; c_1 = -1, \; \pi_1 = \frac{\mu}{dv\rho} = \frac{1}{Re}$$

Similarly: $\pi_2 = \dfrac{l}{d}, \; \pi_3 = \dfrac{\Delta}{d}, \; \mu_4 = \dfrac{\Delta p}{\rho v^2}$

(5) $f_1\left(\dfrac{l}{d}, \dfrac{1}{Re}, \dfrac{\Delta}{d}, \dfrac{\Delta p}{\rho v^2}\right) = 0$

or

$$\frac{\Delta p}{\rho v^2} = f_2\left(\frac{l}{d}, Re, \frac{\Delta}{d}\right)$$

$$\Delta p = f_3\left(Re, \frac{\Delta}{d}\right) \frac{l}{d} \rho v^2$$

$$\Delta h = \frac{\Delta p}{\rho g} = f_4\left(Re, \frac{\Delta}{d}\right) \frac{l}{d} \frac{v^2}{2g}$$

Defining: $\lambda = f_4\left(Re, \dfrac{\Delta}{d}\right)$, then

$$\Delta h = \lambda \frac{l}{d} \frac{v^2}{2g}$$

## 5.3 Similarity theory basis of fluid motion

The dimensionless parameters obtained from the equations of fluid motion set the conditions under which scale model testing with small models will be proven useful for predicting the performance of larger devices. In particular, two flow fields are considered to be dynamically similar when their dimensionless parameters match, and their geometries are scale similar; that is, any length scale in the first flow field may be mapped to its counterpart in the second flow field by multiplication with a single scale ratio. When two flows are dynamically similar, analysis, simulations, or measurements from one flow field can be directly applicable to the other when the scale ratio is accounted for. Moreover, use of standard dimensionless parameters typically reduces the parameters that must be varied in an experiment or calculation, and greatly facilitates the comparison of measured or computed results with prior work conducted under potentially different conditions.

At any time, all the parameters of the model and prototype are in the same ratio throughout the entire flow field: geometric similitude, motion similitude and dynamic similitude.

### 5.3.1 Similar conditions

#### 1. Geometrical similarity

The model is an exact geometric replica of the prototype. All the linear dimensions of the model and prototype are in the same ratio.

$$\lambda_l = \frac{l_p}{l_m} \tag{5-3}$$

where $\lambda_l$ is scale ratio between model and prototype, $l_p$ is the length of the prototype, $l_m$ is the length of the model.

The results of geometrically similitude lead to corresponding ratio of area and ratio of volume maintained a certain ratio.

Ratio of area: $\lambda_A = \dfrac{A_p}{A_m} = \dfrac{l_p^2}{l_m^2} = \lambda_l^2$

Ratio of volume: $\lambda_V = \dfrac{V_p}{V_m} = \dfrac{l_p^3}{l_m^3} = \lambda_l^3$

#### 2. Kinematic similarity

Velocity of the model and the prototype are in the same ratio throughout the entire flow field. Such as, $u_p$ is the velocity of the prototype, $u_m$ is the velocity of the model, the scale ratio between model and prototype is

$$\lambda_u = \frac{u_t}{u_m}$$

So the ratio of velocity is

$$\lambda_v = \frac{v_p}{v_m} = \frac{\left(\frac{l}{t}\right)_p}{\left(\frac{l}{t}\right)_m} = \frac{\lambda_l}{\lambda_t}$$

and the ratio of acceleration is

$$\lambda_a = \frac{a_p}{a_m} = \frac{\left(\frac{v}{t}\right)_p}{\left(\frac{v}{t}\right)_m} = \frac{\lambda_v}{\lambda_t} = \frac{\lambda_l}{\lambda_t^2}$$

### 3. Dynamic similarity

All the forces that act on corresponding masses in the model and the prototype are in the same ratio throughout the entire flow field.

The ratio of force is $\lambda_F = \dfrac{F_p}{F_m}$ ($F_p$ is one of the force of the prototype, $F_m$ is one of the force of the model).

### 4. Relationship

Geometric similitude, motion similitude and dynamic similitude are interrelated as a whole. Geometric similitude is basic and the most obvious requirement, dynamic similitude is the condition, and motion similitude is the result of another two.

## 5.3.2 Derivation of similitude principles

According to the similitude principles, similarity criterion number is dimensionless parameters with dynamic similitude. The starting point is the Navier-Stokes momentum equation simplified for incompressible flow. The model is

$$\left[ f - \frac{1}{\rho} \nabla p + \nu \nabla^2 u = \frac{\partial u}{\partial t} + (u \cdot \nabla) u \right]_m \qquad (5-4)$$

The prototype is

$$\left[ f - \frac{1}{\rho} \nabla p + \nu \nabla^2 u = \frac{\partial u}{\partial t} + (u \cdot \nabla) u \right]_p \qquad (5-5)$$

If the model is geometric similitude with the prototype, and the boundary and initial conditions as well. The mass force is assumed only gravity, the ratio is

$$\frac{f_p}{f_m} = \frac{g_p}{g_m} = \lambda_g$$

$$\lambda_g f_m - \left(\frac{\lambda_p}{\lambda_\rho \lambda_l}\right) \left(\frac{1}{\rho} \nabla p\right)_m + \left(\frac{\lambda_\nu \lambda_u}{\lambda_l^2}\right) (\nu \nabla^2 u)_m$$

$$= \left(\frac{\lambda_u}{\lambda_t}\right) \left(\frac{\partial u}{\partial t}\right)_m + \left(\frac{\lambda_u^2}{\lambda_l}\right) [(u \cdot \nabla) u]_m \qquad (5-6)$$

The equation (5-4) and equation (5-5) can be rendered dimensionless. Equation (5-4) divided by $(u \cdot \nabla)u$, we get

$$\left[\frac{f}{(u \cdot \nabla)u} - \frac{1}{\rho}\frac{\nabla p}{(u \cdot \nabla)u} + \frac{\nu \nabla^2 u}{(u \cdot \nabla)u} = \frac{\partial u/\partial t}{(u \cdot \nabla)u} + 1\right]_m \tag{5-7}$$

Equation (5-6) divided by $\left(\frac{\lambda_u^2}{\lambda_l}\right)[(u \cdot \nabla)u]$, we get

$$\frac{\lambda_g \lambda_l}{\lambda_u^2}\left[\frac{f}{(u \cdot \nabla)u}\right] - \frac{\lambda_p}{\lambda_\rho \lambda_u^2}\left[\frac{1}{\rho}\frac{\nabla p}{(u \cdot \nabla)u}\right] + \frac{\lambda_\nu}{\lambda_l \lambda_u}\left[\frac{\nu \nabla^2 u}{(u \cdot \nabla)u}\right] = \frac{\lambda_l}{\lambda_t \lambda_u}\left[\frac{\partial u/\partial t}{(u \cdot \nabla)u}\right] + 1 \tag{5-8}$$

Equation (5-7) equals to equation (5-8).

$$\frac{\lambda_g \lambda_l}{\lambda_u^2} = \frac{\lambda_p}{\lambda_\rho \lambda_u^2} = \frac{\lambda_\nu}{\lambda_l \lambda_u} = \frac{\lambda_l}{\lambda_t \lambda_u} = 1 \tag{5-9}$$

Using the average velocity replace to instantaneous velocity.

$$\frac{v_p^2}{g_p l_p} = \frac{v_m^2}{g_m l_m}, \quad \frac{p_p}{\rho_p v_p^2} = \frac{p_m}{\rho_m v_m^2}, \quad \frac{v_p l_p}{\nu_p} = \frac{v_m l_m}{\nu_m}, \quad \frac{l_p}{v_p t_p} = \frac{l_m}{v_m t_m}$$

or

$$Fr_p = Fr_m, \quad Eu_p = Eu_m, \quad Re_p = Re_m, \quad St_p = St_m \tag{5-10}$$

where $Fr = \frac{v^2}{gl}$ is Froude number, $Eu = \frac{p}{\rho v^2}$ is Euler number, $Re = \frac{vl}{\nu}$ is Reynolds number, $St = \frac{l}{vt}$ is Strouhal number. Therefore Froude number, Euler number, Reynolds number and Strouhal number is called similarity criterion number.

### 5.3.3 The physical meaning of similarity criterion number

When model is tested, it is important to identify the major forces affecting fluid movement and select similarity criterion number to model designing. Therefore, it is necessary to understand the physical meaning of similarity criterion number.

#### 1. The physical of N-S equation

Forces on the fluid particle were gravity, pressure and viscosity. N-S equation (3-30) was

$$f - \frac{1}{\rho}\nabla p + \nu \nabla^2 u = \frac{\partial u}{\partial t} + (u \cdot \nabla)u \tag{3-30}$$

$\langle 1 \rangle \quad \langle 2 \rangle \quad \langle 3 \rangle \quad \langle 4 \rangle \quad \langle 5 \rangle$

Obviously, item $\langle 1 \rangle$ is gravity, item $\langle 2 \rangle$ is pressure, item $\langle 3 \rangle$ is viscosity. Items $\langle 4 \rangle$ and $\langle 5 \rangle$ are Euler acceleration. According to the equation, items $\langle 4 \rangle$ and $\langle 5 \rangle$ also are acceleration inertia forces.

Attention:

(1) Impact of these forces on fluid movement is strong or weak. In a similar model, we need to analyze which forces play a leading role in flow.

(2) A variety of forces except inertia force are attempted to change the state of motion. Therefore, the flow is changed by the result of interaction between inertia force and other forces.

### 2. The physical meaning of similarity criterion number

Items of equation (5 - 8) are gravity, pressure, viscosity and inertia force. Froude number is characterized to compared the gravity and the inertia force. Reynolds number is characterized to compared the inertia force and the viscosity. Strouhal number is characterized to compared the time-varying acceleration inertia force and the migration acceleration viscosity.

When two viscous incompressible fluid are completely dynamic similar, their $Fr$, $Eu$, $Re$, $St$ are corresponding equal.

## 5.3.4 The principle of dynamic similarity

Complete dynamic similarity requires each similarity criterion number equally and correspondingly. For example, if two flow are equal to Froude number and Reynolds number, then

$$(Re)_p = (Re)_m$$

or

$$\frac{\lambda_v \cdot \lambda_l}{\lambda_\nu} = 1 \quad (\lambda_\nu = 1)$$

Then

$$(Fr)_p = (Fr)_m$$

or

$$\frac{\lambda_v}{\lambda_g^{1/2} \cdot \lambda_t^{1/2}} = 1, \lambda_g = 1$$

$$\lambda_v = \lambda_t^{1/2}, \text{ thus } \lambda_t^{1/2} = \lambda_t^{-1}$$

Generally, only select one force which play a decisive role on the flow to equal similarity criterion number of two flow.

If the flow is mainly urged by gravity, the Froude number is equal. $(Fr)_p = (Fr)_m$.

If the flow is mainly urged by viscosity, the Reynolds number is equal. $(Re)_p = (Re)_m$.

## 5.3.5 Inference—similarity criterion number in compressible fluid

### 1. Cauchy number

The characterization of Cauchy number is the ratio of inertia force and the elastic force. Cauchy number is $Ca = \frac{\rho v^2}{K}$.

$$\left(\frac{f}{f_E}\right)_p = \left(\frac{f}{f_E}\right)_m$$

Arrange: $\left(\dfrac{\rho v^2}{K}\right)_p = \left(\dfrac{\rho v^2}{K}\right)_m$  or  $(Ca)_p = (Ca)_m$

2. Mach number

Similitude principle of flow acted by elastic force. $Ma = \dfrac{v}{c}$ is Mach number.

$$\dfrac{v_p}{c_p} = \dfrac{v_m}{c_m} \quad \text{or} \quad (Ma)_p = (Ma)_m$$

## 5.4 Model experiment

### 5.4.1 Similarity criterion number of model

Models and prototypes at the same fluid medium is difficult to be completely similar, generally only approximate similarity. Therefore, it is important to identify the main force in study, then select similar guidelines which corresponds to the main force.

### 5.4.2 The designing of model

Model design generally perform the following steps.

(1) Determine the scale of length;
(2) Design the geometry of model;
(3) Select model guidelines;
(4) Calculate the similarity scale of physical quantities.

① Reynolds guidelines

The ratio of velocity: $\dfrac{v_p l_p}{\nu_p} = \dfrac{v_m l_m}{\nu_m}$

If $\nu_p = \nu_m$, $\lambda_v = \dfrac{v_p}{v_m} = \dfrac{l_m}{l_p} = \lambda_l^{-1}$.

② Froude guidelines

The ratio of velocity: $\dfrac{v_p}{\sqrt{g_p l_p}} = \dfrac{v_m}{\sqrt{g_m l_m}}$

If $g_p = g_m$, $\lambda_v = \dfrac{v_p}{v_m} = \sqrt{\dfrac{l_p}{l_m}} = \lambda_l^{1/2}$.

(5) Prorated model flow rate, velocity, etc., and determine the boundary conditions. Part of the flow parameters must be given by the prototype. Then converted to values of model. Such as $q_{v_m} = \dfrac{q_{v_p}}{\lambda_{q_v}}$.

### 5.4.3 Experimental data processing of model

Experimental data are divided into two categories, one are dimensionless quantities, another are quantities of dimension one. Since the quantities of dimension one, there is no need to convert, because models and prototypes are similar and their values are corresponding equal. For dimensionless quantities, model values need to be converted to prototype values by the scale of similar.

## Exercises

1. Provide examples of water flows that are primarily affected by gravity and viscosity.

2. Try proof $\tau = \mu \dfrac{du}{dy}$, $z + \dfrac{p}{\rho g} + \dfrac{u_2}{2g} = H$, the dimension is harmonious.

3. The dimensional analysis method combines the following physical quantities into a quantity of dimension one:

(1) $\tau$, $v$, $\rho$, where $\tau$ is the shear stress, $v$ is the velocity;

(2) $\Delta p$, $v$, $p$, $g$;

(3) $F$, $\rho$, $l$, $v$, where $F$ is the force, $v$ is the velocity, and $l$ is a certain linear length;

(4) $v$, $l$, $p$, $\sigma$, where $\sigma$ is the surface tension coefficient.

4. The rectangular thin-walled weir shown in Fig.5-1 is related to the flow rate $q_v$ of the rectangular weir, the water head $H$ on the weir, the weir width $b$, and the gravitational acceleration $g$, as observed through experiments. Use the Rayleigh method to determine the structural form of the weir flow rate.

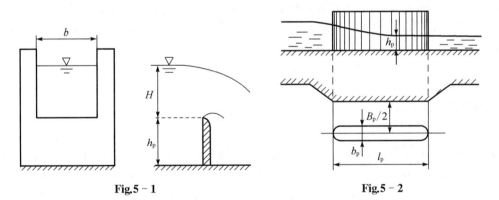

Fig.5-1            Fig.5-2

5. As shown in Fig.5-2, a bridge pier has a length of $l_p = 24$ m, a pier width of $b_p = 4.3$ m, a water depth of $h_p = 8.2$ m, an average velocity under the bridge $v_p = 2.3$ m/s, and a distance of $B = 90$ m between the two abutments. Taking $\lambda_l = 50$ to design hydraulic model experiments with $l = 50$, try to determine the geometric dimensions and experimental flow rate of the model.

# Chapter 6  Fluid Resistance and Energy Loss

The difference between actual fluid and ideal fluid is that the actual fluid is viscous, this is reflected for the head loss in the Bernoulli equation of the actual fluid in Chapter 4. The head loss is closely related to both the physical properties and the boundary conditions of the fluid.

This chapter describes two forms of head loss(frictional head loss and minor loss), two flow patterns (laminar flow and turbulent flow) and the calculation equation of water head loss in two kinds of flowing environment (pressure pipe flow and open channel flow) in two flow patterns and so on. Finally, the boundary layer and the drag are briefly introduced.

In this chapter, we should grasp the concepts of the two flow patterns and Reynolds number, master the characteristics and discrimination methods of two flow patterns, understand the movement law of circular pipe laminar flow, understand features of pipe turbulent flow, concept of mixing length and velocity distribution of the turbulent flow; Understand the regulation of coefficient of frictional loss in pipes, master the computational method of both frictional head loss and minor loss in pipes; Understand the concept of boundary layer, separation of boundary layer and drag.

## 6.1  Introduction

Viscous fluid is bound to produce energy loss in movement, in the total Bernoulli equation, the mechanical energy of the unit weight fluid is represented by $h_w$, called the head loss. Viscosity is the root cause of head loss, there is no head loss in ideal fluid flow because of the absence of viscosity. Fluids in this and subsequent chapters, if there are no special instructions, are all viscous fluids.

### 6.1.1  Flowed friction and head loss categories

According to the differences of flow boundary condition and mechanism that produce water head loss, the head loss can be divided into the frictional head loss and minor loss.

**1. Frictional head loss**

When fluid is uniform flow, due to the viscosity, the flow velocity on the cross section is not uniform. There is relative motion between two adjacent layers, which produces shear stress between adjacent layers and between flow layer and boundary (friction), forming flow resistance. The flow resistance generated by the uniform flow is called the friction drag force or friction resistance. Head loss caused by friction work is called frictional head loss, represented by $h_f$. Frictional head loss is distributed uniformly along the flow process and size is proportional to the length of the process. In the flow section where shape, size and direction of the side wall are unchanged, such as long straight channel and equal diameter pressure pipe, head loss is frictional head loss.

### 2. Minor loss

Fluid in a local area, such as elbows, sudden expansion and contraction and valves etc. (Fig.6-1), the change of the velocity distribution caused by the sharp change of the solid boundary, even makes the mainstream out of the boundary, forming the vortex zone called local resistance. Head loss caused by local resistance work is called minor loss, represented by $h_j$. Minor loss often occurs in a suddenly changed flow cross section, flow axis bends sharply or there is a suddenly changed boundary shape or local obstacle. The formation of the area of vortex is the main cause of minor loss. Minor loss is accomplished in a flow path, even a very long section of length. But for convenience, we often treat it as a concentrated head in fluid mechanics of a section to handle.

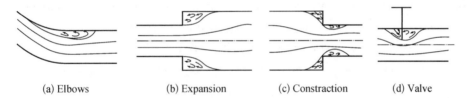

(a) Elbows　　(b) Expansion　　(c) Constraction　　(d) Valve

**Fig.6-1　Minor loss**

## 6.1.2　Calculation equation of head loss

Energy loss in the form of heat energy dissipation, cannot be reversed for other forms of mechanical energy. If the whole distance of the flow water is consist of flow sections with different boundaries and there are several minor loss, the head loss of the whole distance equals the sum of both frictional head loss and minor loss, i.e.

$$h_w = \sum h_f + \sum h_j \qquad (6-1)$$

It is Venturi pipe flow in Fig.6-2 and outlet section 4 connected the atmosphere. $H$ in the figure is the whole head line and $H_p$ is hydraulic grade line. Minor loss included the inlet head loss $h_{j1}$, Venturi flow meter loss $h_{j2}$ and valve loss $h_{j3}$. Frictional head loss included $h_{f1,2}$, $h_{f2,3}$ and $h_{f3,4}$, the whole head loss is

$$h_w = h_{j1} + h_{j2} + h_{j3} + h_{f1,2} + h_{f2,3} + h_{f3,4}$$

Mechanical energy losses of gas pipe flow are still described by Bernoulli equation of gas in Chapter 4, expressed by express loss $p_w$.

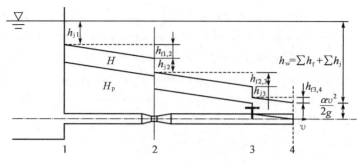

Fig.6 – 2  Head loss

### 1. Calculation equation of frictional head loss

Experiment shows, the frictional head loss of circular pipe flow $h_f$ is proportional to the flow length $l$ and is inversely proportional to the diameter $d$, and the calculation equation is

$$h_f = \lambda \frac{l}{d} \frac{v^2}{2g} \tag{6-2}$$

where $v$ is the average velocity of cross section, and $\lambda$ is frictional head loss coefficient.

The calculation equation above is called Darcy-Weisbach equation, also shorted for Darcy equation or D-W equation. It is a semi-empirical relationship. The same form can also be obtained by dimensional analysis, so it's structure of equation is reasonable. $\lambda$ in the equation is related to both flow characteristics and roughness of pipe wall. The analysis and calculation of $\lambda$ is the primary content of this chapter.

For the cross section of non-circular tube, frictional head loss can be expressed as follows:

$$h_f = \lambda \frac{l}{4R} \frac{v^2}{2g} \tag{6-3}$$

where $R$ is hydraulic radius of cross-section, whose value is

$$R = \frac{A}{\chi} \tag{6-4}$$

where $A$ is area of cross section; $\chi$ is the perimeter of interface where fluid contacts with solid, called wetted perimeter.

For circular tube,

$$R = \frac{\pi d^2}{4\pi d} = \frac{d}{4}$$

### 2. Calculation equation of minor loss

$$h_j = \zeta \frac{v^2}{2g} \tag{6-5}$$

where $\zeta$ is minor loss coefficient, which is determined by experiment (see Section 6.7).

## 6.1.3 Relationship between frictional head loss and shear stress—fundamental equations of uniform flow

As illustrated in Fig.6 – 3, in the constant uniform flow of the tube, taking the axis as the symmetry line and $l$ for length, $r$ for radius, two ends 1 - 1, 2 - 2 for water cross section, all of which are consisted of a fluid column. And make a force analysis of it. Supposed the central pressure of the upstream and downstream section are separately $p_1$ and $p_2$, position heights are $z_1$ and $z_2$, the angle between the direction of gravity and the axis is $\alpha$. External force, acting on the column, along the flow direction, included pressure, shear of tube wall and gravity, and the force balance equation is

$$p_1 A' - p_2 A' + \rho g A' l \cos \alpha - \tau \chi' l = 0$$

where $\tau$ is shear stress; $\chi'$ is wetted perimeter of the cylinder; $A'$ is area of section of the cylinder.

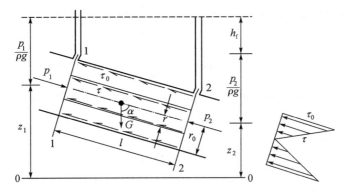

Fig.6 - 3  Uniform flow of circular pipe

Divided each monomial by $\rho g A'$, $l \cos \alpha = z_1 - z_2$:

$$\left(z_1 + \frac{p_1}{\rho g}\right) - \left(z_2 + \frac{p_2}{\rho g}\right) = \frac{\tau \chi' l}{\rho g A'}$$

According to the Bernoulli equation of total flow in Section 1.1 and 2.2,

$$\left(z_1 + \frac{p_1}{\rho g}\right) - \left(z_2 + \frac{p_2}{\rho g}\right) = h_f$$

so

$$h_f = \frac{\tau l}{\rho g R'}$$

or

$$\tau = \rho g R' \frac{h_f}{l} = \rho g R' J \qquad (6-6)$$

Assuming $r_0$ for the radius of the circular pipe, $\tau_0$ for boundary shear, when $r = r_0$, $R' = R$, then

$$h_f = \frac{\tau_0 l}{\rho g R} \qquad (6-7)$$

or

$$\tau_0 = \rho g R J \qquad (6-8)$$

Equation (6-7) and (6-8) shows the relationship between head loss and shear stress. For uniform flow in open channel, we can also get the same conclusion with similar method above. Yet the distribution of the wall shear stress is not as uniform as that of the axial symmetrical pipe flow, $\tau_0$ in the equation is average shear stress.

The derivation process above is not related to the flow characteristics and it is applicative on the condition of uniform flow. Therefore, equation (6-8) is called uniform flow fundamental equations.

### 6.1.4  Friction velocity and shear stress distribution of circular tube

**1. Relationship between frictional head loss coefficient of uniform flow and wall shear stress**

Change the Darcy equation (6-2) into $J = \dfrac{\lambda}{4R}\dfrac{v^2}{2g}$, put it into the fundamental equations of uniform flow (6-8), and we get the relationship between frictional head loss coefficient of uniform flow and wall shear stress $\tau_0$. The equation is

$$\tau_0 = \frac{\lambda}{8}\rho v^2 \tag{6-9}$$

**2. Friction velocity**

Define $v_* = \sqrt{\dfrac{\tau_0}{\rho}}$, with velocity dimension, which is called friction velocity. According to the equation (6-9), we have

$$v_* = v\sqrt{\frac{\lambda}{8}} \tag{6-10}$$

The concept of the friction velocity is repeatedly cited in the later content.

**3. Shear stress distribution of circular tube**

The ratio of equation (6-6) to equation (6-8) is

$$\frac{\tau}{\tau_0} = \frac{r}{r_0} \tag{6-11}$$

As illustrated in Fig.6-3, the shear stress of circular pipe constant uniform flow on the section of water is linear. $\tau = 0$ at the tube axis and $\tau_0$ reached maximum at the pipe wall.

## 6.2  Laminar and turbulent flow

After long-term observation and practice, as early as the beginning of nineteenth century, scientists have found that the flow of fluids has different flow patterns. In different flow patterns, there are different relationships between head loss and flow velocity when the fluid flow. But until 1883, the existence of two kinds of flow pattern and the relationship between head loss and flow velocity were demonstrated by experiments which were conducted by an English physicist Osborne Reynolds.

### 6.2.1 Reynolds experiment, laminar and turbulent flow

Devices for Reynolds experiment are illustrated in Fig. 6 – 4. The side of the water tank is installed with a horizontal glass tube bell with a bell mouth and there is a pipe valve at the downstream of the pipe section to regulate the volume of flow. Needle shaped tubes are installed near the bell mouth to inject color to the water. And there is a overflow plate installed on the tank to keep the water surface of the tank unchangeable and the head of the tube constant. There are also 2 piezometric pipes on section 1—2 of the glass tube to determination of head loss in section 1—2.

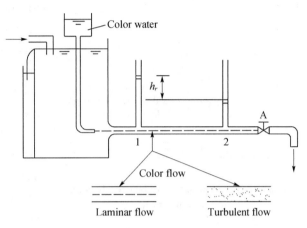

Fig.6 – 4  **Devices for Reynolds experiment**

In experiments, the container is filled with water and the liquid level is kept stable. Flow in pipe is steady flow. Open the valve A slightly and the water will flow at a very small speed.

Water with color will present a cluster of straight line with distinct boundaries and they do not mix with each other. This phenomenon indicates that the fluid particles are not mixed with each other and make an orderly movement. The flow state is called laminar flow. If you continue to turn the valve up, the flow velocity in the tube will increase gradually. When the low velocity reaches a certain value, the water with color begins to vibrate and bend. The line is gradually thickened and gradually broken at some flow section. Finally, then the velocity reached the $v'_c$, the color water is broken thoroughly and spread to the whole pipe rapidly. Then all water in pipe is colored uniformly. This phenomenon shows that the fluid particle is not laminar flow, but the random motion of the fluid particles. This local velocity, pressure and other physical quantities in the time and space pulsed in irregular fluid movement are known as the turbulent flow. The average flow velocity when a laminar flow turned into a turbulent flow is known as the upper critical velocity, represented by $v'_c$. When the experiment is carried out in the opposite direction, that is, open the valve to the maximum first, to ensure that the flow in the pipe is in a fully developed turbulent state, and then turn down the valve gradually. When the velocity decreased to a value that is different from $v'_c$, the uniform color came back to a cluster of straight line, which shows that the flow in the circular tube is changed from turbulence back to laminar flow. The average flow velocity in pipe when turbulence flow is changing into laminar flow is called lower critical velocity, represented by $v_c$, which is smaller than $v'_c$.

## 6.2.2 Relationship between linear head loss and velocity

The water head loss at different velocity $v$ was measured on the logarithmic coordinate paper, and the relationship curve of $v$ and $h_f$ was formed, as illustrated in Fig.6-5.

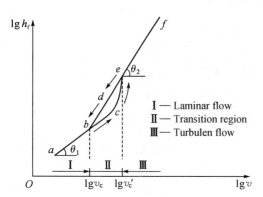

Fig.6-5 Relationship curve between frictional head loss and velocity

If the velocity goes from slow to fast in the experiment, then the experimental points fall in the line $abcef$. Laminar flow can be maintained until point $c$ and then changed into turbulence. And the position of $c$ is not stable. If the velocity goes from fast to slow in the experiment, then the experimental points fall in the line $fedba$, which does not coincide with line $bce$. Turbulence flow can be maintained until point $b$ and then changed into laminar. In interval $bce$, the laminar flow can be destroyed by any accidental reasons and turned into turbulence and won't come back to the previous state. This interval is called transition region. $ab$ is a straight line while $ef$ is an approximate straight line and both of them can be expressed by the following equation:

$$\lg h_f = \lg k + m \lg v$$

that is

$$h_f = kv^m \tag{6-12}$$

Experimental results indicate, when $v < v_c$, laminar flow is linear $ab$ and $\theta_1 = 45°$, $m = 1$, that is all test points are distributed in a straight line with the angle of 45 degrees in the horizontal axis. Frictional head loss is proportional to the first power of velocity, that is $h_f \propto v$. While $v > v_c'$, turbulence flow is linear $ef$ and $\theta_2 = 60° \sim 63°$, $m = 1.75 \sim 2.0$, that is linear head loss is proportional to the 1.75~2.0 power of velocity, also $h_f \propto v^{1.75 \sim 2.0}$.

## 6.2.3 Discrimination of laminar and turbulent flow

The results of Reynolds experiment show that there is close relationship among critical flow velocity, fluid density $\rho$, dynamic viscosity coefficient $\mu$ and pipe diameter $d$. Liquid flow pattern can be used in the following dimensions of dimensional one—Reynolds number, to distinguish:

Upper critical Reynolds number

$$Re_c' = \frac{\rho v_c' d}{\mu} = \frac{v_c' d}{\nu}$$

Lower critical Reynolds number

$$Re_c = \frac{\rho v_c d}{\mu} = \frac{v_c d}{\nu}$$

A large number of experiments prove that lower critical Reynolds number in round pipe pressure flow $Re_c \approx 2\ 300$, which is a especially stable value and almost has nothing to do with external disturbance; While upper critical Reynolds number $Re'_c > Re_c$, which is a unstable value and even reaches 12 000~20 000. These are mainly related to the stability of the fluid before entering the pipeline and the disturbance of the outside world. Any small disturbance will make the laminar flow into turbulent flow because the flow is extremely unstable between upper and lower critical Reynolds number. And disturbance always exists in actual engineering. Therefore, liquid flow, between upper and lower critical Reynolds number, can be regarded as turbulence flow in the practical application. So it is the lower critical Reynolds number that is the discriminant criteria we should use, to distinguish fluid flow state.

In a circular tube

$$Re_c = \frac{v_c d}{\nu} = 2\ 300 \qquad (6-13)$$

If $Re < Re_c$, laminar flow; while $Re > Re_c$, turbulence flow.

For open channel flow and the flow in natural river

$$Re_c = \frac{v_c R}{\nu} = 575 \qquad (6-14)$$

If $Re < Re_c$, laminar flow; while $Re > Re_c$, turbulence flow.

**Example 6-1** Some tap water pipe section, $d = 0.1$ m, $v = 1.0$ m/s. Water temperature 10 ℃, (1) Judge the flow pattern in pipe. (2) Find out maximum flow velocity that keeps the flow laminar.

**Solution:** (1) When the water temperature is 10 ℃, the coefficient of viscosity of water is

$$\nu = \frac{0.017\ 75 \times 10^{-4}}{1 + 0.033\ 7t + 0.000\ 221t^2} = \frac{0.017\ 75}{1.359\ 1} \times 10^{-4} = 1.31 \times 10^{-6} \text{ m}^2/\text{s}$$

so

$$Re = \frac{vd}{\nu} = \frac{1 \times 0.1}{1.31 \times 10^{-6}} = 76\ 336 > Re_c = 2\ 300$$

Flow in the circular pipe is laminar flow.

(2) $$Re_c = \frac{v_c d}{\nu}$$

$$v_c = \frac{\nu Re_c}{d} = \frac{1.31 \times 10^{-6} \times 2\ 300}{0.1} = 0.03 \text{ m/s}$$

So, to keep the flow laminar, the maximal flow velocity is 0.03 m/s.

**Example 6-2** Flow of rectangular open channel in an experiment, bottom width $b = 0.2$ m, depth of water $h = 0.1$ m, flow velocity $v = 0.12$ m/s, water temperature is 20 ℃, try to judge the flow pattern.

**Solution:** Calculate the movement elements of water according to the known conditions:

$$A = bh = 0.2 \times 0.1 = 0.02 \text{ m}^2$$
$$\chi = b + 2h = 0.2 + 2 \times 0.1 = 0.4 \text{ m}$$
$$R = \frac{A}{\chi} = \frac{0.02}{0.4} = 0.05 \text{ m}$$

When the water temperature is 20 ℃,

$$\nu = 1.0 \times 10^{-6} \text{ m}^2/\text{s}$$

then

$$Re = \frac{vR}{\nu} = \frac{0.12 \times 0.05}{1.0 \times 10^{-6}} = 6\,000 > Re_c = 575$$

The flow pattern is turbulence flow.

## 6.3 Laminar flow in a circular pipe

Laminar flow is a simple case study of fluid movement and it is also one of several kinds of flow field, which can be concluded that the flow velocity distribution and the loss of water head. This chapter will show some relevant expressions.

### 6.3.1 Flow velocity distribution

Take an uniform flow section of a circular pipe laminar flow, as illustrated in the Fig.6-6 and analyze the flow velocity distribution of circular pipe laminar flow. In laminar flow, shear stress satisfies the Newton internal friction law (1-10), considering the relationship between $y$ and $r$ is $y = r_0 - r$, then $dy = -dr$. So

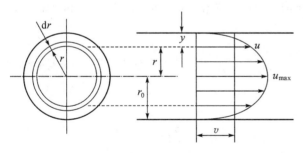

**Fig.6-6  Flow velocity distribution**

$$\tau = \mu \frac{du}{dy} = -\mu \frac{du}{dr} \quad (6-15)$$

Taking equation (6-6) $\tau = \rho g R'J$, where $R' = \frac{r}{2}$, into equation (6-15) and making separation of variables, we get

$$du = -\frac{\rho g J}{2\mu} r\, dr$$

Due to the same $J$ in uniform flow, the indefinite integral of the equation above is

$$u = -\frac{\rho g J}{4\mu} r^2 + C \quad (6-16)$$

At the tube wall, fluid is attached to the wall, satisfying no sliding condition, that is, when $r = r_0$, $u = 0$, so

$$C = \frac{\rho g J}{4\mu} r_0^2$$

Substituted into equation (6 - 16)

$$u = \frac{\rho g J r_0^2}{4\mu} \left(1 - \frac{r^2}{r_0^2}\right) \qquad (6-17)$$

Equation (6 - 17) indicates that the flow velocity distribution of flow section of circular pipe presents a rotational parabolic distribution, which is one of important features of laminar flow.

According to the Fig.6 - 6, maximum flow velocity in circular pipe flow is at the tube axis, according to equation (6 - 17)

$$u_{max} = \frac{\rho g J r_0^2}{4\mu} \qquad (6-18)$$

Therefore, equation (6 - 17) is also

$$u = u_{max}\left(1 - \frac{r^2}{r_0^2}\right) \qquad (6-19)$$

Take a ring section with a wide of $dr$, element area $dA = 2\pi r dr$. According to the definition of mean flow velocity of cross section, circular pipe laminar flow average velocity is:

$$v = \frac{q_v}{A} = \frac{\int_A u\,dA}{A} = \frac{1}{\pi r_0^2}\int_0^{r_0} \frac{\rho g J}{4\mu}(r_0^2 - r^2) \cdot 2\pi r\,dr = \frac{\rho g J}{8\mu} r_0^2 \qquad (6-20)$$

The equation above is Hagen-Poiseuille equation.

Compare equation (6 - 18) and equation (6 - 20), we get

$$v = \frac{1}{2} u_{max} \qquad (6-21)$$

That is, the average velocity of the laminar flow in circular pipe is half of the maximum velocity. Compared to turbulence flow in circular pipe, the velocity of turbulence flow in circular pipe is nonuniform, and the kinetic-energy correction factor is:

$$\alpha = \frac{\int_A u^3\,dA}{Av^3} = 2.0$$

The momentum correction factor is

$$\beta = \frac{\int_A u^2\,dA}{Av^2} = \frac{4}{3}$$

### 6.3.2 Calculation of the linear head loss of laminar flow in circular pipe

Frictional head loss of laminar flow in circular pipe can be obtained by equation (6 – 20)

$$J = \frac{h_f}{l} = \frac{8\mu v}{\rho g r_0^2} = \frac{32\mu v}{\rho g d^2}$$

Then we get

$$h_f = \frac{32\mu v l}{\rho g d^2} \qquad (6-22)$$

The equation above indicates that the frictional head loss is in proportion to the average velocity of cross section in circular pipe of laminar flow and has nothing to do with the roughness of pipe wall.

Besides, equation (6 – 22) can also be rewritten into the form of Darcy equation

$$h_f = \frac{32\mu v l}{\rho g d^2} = \frac{64}{\underbrace{\frac{vd}{\nu}}} \frac{l}{d} \frac{v^2}{2g} = \lambda \frac{l}{d} \frac{v^2}{2g} \qquad (6-23)$$

where $\lambda = \dfrac{64}{Re}$.

**Example 6 – 3** Oil, $\rho = 850$ kg/m³, $\nu = 0.18 \times 10^{-4}$ m²/s, moving in a pipe of which diameter $d = 0.1$ m, at a velocity $v = 0.063\ 5$ m/s and make laminar flow. Try to find out: (1) maximum flow velocity at the center of the tube; (2) the velocity at $r = 20$ mm from the center of the tube; (3) frictional head loss coefficient $\lambda$; (4) pipe wall shear stress $\tau_0$ and head loss of every kilometer.

**Solution:** (1) Maximum flow velocity at the center of the tube

$$u_{max} = 2v = 2 \times 0.063\ 5 = 0.127 \text{ m/s}$$

(2) The velocity, 20 mm from the center of the tube

$$u = u_{max}\left[1 - \left(\frac{r}{r_0}\right)^2\right] = 0.127 \times \left[1 - \left(\frac{0.02}{0.05}\right)^2\right] = 0.107 \text{ m/s}$$

(3) Frictional head loss coefficient

Figure out $Re$ first

$$Re = \frac{vd}{\nu} = \frac{0.063\ 5 \times 0.1}{0.18 \times 10^{-4}} = 353 \quad \text{(laminar flow)}$$

so

$$\lambda = \frac{64}{Re} = \frac{64}{353} = 0.18$$

(4) Shear stress and head loss of every kilometer

$$\tau_0 = \frac{\lambda}{8}\rho v^2 = \frac{0.18}{8} \times 850 \times 0.063\ 5^2 = 0.077 \text{ N/m}^2$$

$$h_f = \lambda \frac{l}{d} \frac{v^2}{2g} = 0.18 \times \frac{1\ 000}{0.1} \times \frac{0.063\ 5^2}{2 \times 9.8} = 0.37 \text{ m}$$

## 6.4 Turbulent flow

The flow problems in nature and engineering are mostly turbulent, so the study of turbulent flow has a wide range of significance.

Different from the laminar flow, the turbulence is a disorder of the mutual doping of fluid particles. The speed of each point and the pressure, concentration and other physical quantities related to the speed, are all random fluctuations (called turbulent pulsation) in both time and space. For more than a century, the theory of turbulence in the study of turbulence can be said to be fruitful. But the turbulent flow is much more complex than that of the laminar flow, and the theoretical analysis is more difficult, so the study of turbulence is still in the ascendant. Below from the concept of turbulence, research methods and the main features, etc., to do a brief introduction.

### 6.4.1 Turbulent structure—the viscous sublayer and turbulent core area

#### 1. Turbulent structure

According to the theoretical analysis and experimental observation, turbulent flow and laminar flow have no sliding condition on the wall. In turbulence, there is a extremely thin layer flow close to the solid boundary. Due to the effect of fluid viscosity and the limit of solid boundary, the doping between particles is eliminated so the flow pattern appears to be laminar flow, which is called viscous sublayer.

Outside the viscous sublayer, the fluid particles are doped. The fluctuation of flow rate and its related physical quantity become obvious and it is called turbulent area. And this area is often called core region of turbulent flow.

There is a thin layer of transition layer between the viscous and turbulent core regions, which is very thin and the boundary is not clear. When doing the analysis of turbulent flow, the whole section is divided into two sections, viscous sublayer and core region of turbulent flow to discuss (see Fig.6 – 7).

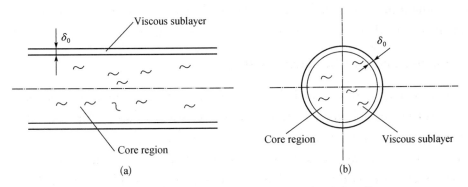

**Fig.6 – 7 Viscous sublayer and core region of turbulent flow**

### 2. Viscous sublayer

In the laminar flow, the fluid velocity on the wall is zero. With the increase of distance from the wall, the fluid velocity increases from zero to a finite value. The velocity is approximately linear distribution and its value of gradient is extremely large. While the thickness of viscous sublayer is $\delta_0$ often less than 1 mm and reduces with the increase of Reynolds number.

In viscous sublayer, the shear stress obeys the law of internal friction of Newton. Take the wall shear stress $\tau = \tau_0$, then

$$\tau_0 = \mu \frac{du}{dy}$$

Integral of the equation above

$$u = \frac{\tau_0}{\mu} y + C$$

According to the boundary conditions, on the surface of tube wall, $y=0$, $u=0$, and we get integral constant $C=0$, so

$$u = \frac{\tau_0}{\mu} y \qquad (6-24)$$

or $\mu = \rho \nu$, $v_* = \sqrt{\dfrac{\tau_0}{\rho}}$ substituted into the equation above, transform and we can get

$$\frac{u}{v_*} = \frac{v_*}{\nu} y \qquad (6-25)$$

The equation above showed the relationship between the viscous sublayer and shear stress.

We can understand that, from the content behind, although the viscous sublayer is very thin, it has a great influence on both the velocity distribution and flow resistance of turbulent.

### 3. Core region of turbulent flow

The turbulent motion in this section is mainly aimed at the fluid movement in the core area of the turbulent flow and will be introduced in the following.

## 6.4.2 Time averaging of turbulent motion

Studying the motion law of turbulence by ignoring random properties of fluid and time averaging of turbulent motion elements, is an effective way to study turbulent motion.

The velocity of a fluid particle when it is moving to a certain point is called the instantaneous velocity, represented by $u_x$. It is known that the instantaneous velocity of the fluid particle is changed with time. Using the method of sample mean and deviation, we can consider that the instantaneous velocity is composed of mean velocity and fluctuating velocity. For an enough long period of time $T$, the average value of the instantaneous flow velocity is

$$\bar{u}_x = \frac{1}{T}\int_0^T u_x(t)\,dt \qquad (6-26)$$

where $\bar{u}_x$ is called mean flow velocity of time, shorted for time average velocity.

The difference between the instantaneous velocity and the mean velocity is called the fluctuating velocity $u'_x$, that is

$$u_x = \bar{u}_x + u'_x$$

Likewise, speed in $y$, $z$ coordinate direction $u_y$, $u_z$ and instantaneous pressure $p$ all can be regarded as two parts consist of the average values of time and fluctuating:

$$u_y = \bar{u}_y + u'_y$$
$$u_z = \bar{u}_z + u'_z$$
$$p = \bar{p} + p'$$

The average value of each physical quantity of pulsation is always equal to zero.

The average value of fluctuating velocity $u'_x$ is

$$\overline{u'_x} = \frac{1}{T}\int_0^T u'_x\,dt = \frac{1}{T}\int_0^T (u_x - \bar{u}_x)\,dt = \frac{1}{T}(\bar{u}_x - \bar{u}_x) = 0$$

But the mean square value of each pulse is not equal to zero. i.e.

$$\overline{u'^2_x} = \frac{1}{T}\int_0^T (u_x - \bar{u}_x)^2\,dt \neq 0$$

Besides, the time mean value of the product of the two pulsation value is not equal to zero neither, that is, $\overline{u'_x u'_y}$ and $\overline{u'_y u'_z}$ both are not equal to zero.

In the study of fluid motion regularity, we often use the mean square root of fluctuating velocity to express the size of pulsing amplitude.

$$N = \frac{\sqrt{\frac{1}{3}(\overline{u'^2_x} + \overline{u'^2_y} + \overline{u'^2_z})}}{\bar{u}} \qquad (6-27)$$

where $N$ is called intensity of turbulent flow.

When the concept of time mean value was introduced, turbulent motion can be regarded as a superposition of the time mean value and the pulsation value and then we can study them separately. Therefore, in engineering practice, calculation of the general flow can be calculated by time average value. And constant flow in the turbulent flow is also the time averaged steady flow.

### 6.4.3 Reynolds stress in turbulent flow

According to the theory of turbulence, the stresses in the turbulent motion include the compressive stress and viscous stress under the action of gravity. Additional stress in turbulent flow is also included. Taking the plane uniform flow as an example, it can be expressed as

$$\tau = \mu \frac{du_x}{dy} - \rho \overline{u'_x u'_y} \qquad (6-28)$$

where $-\rho \overline{u'_x u'_y}$ is called additional stress in turbulent flow, represented by $\tau'$, and the "$-$" is to make additional shear stress positive because of the contrary sign between $u'_x$ and $u'_y$. For 3-dimensional fluid motion, the NS equation is done by the time averaging operation and we get $-\rho \overline{u'_x u'_y}$, $-\rho \overline{u'_x u'_z}$, $-\rho \overline{u'_y u'_z}$, three additional stress component and $\rho \overline{u'^2_x}$, $\rho \overline{u'^2_y}$, $\rho \overline{u'^2_z}$, three additional stress positive component. These additional stress are called Reynolds stress.

The proportion of viscous and additional shear stress is different with the condition of the turbulent flow. Experiments show that, for the fully developed turbulent flow, the viscous stress is mainly in the thin viscous layer at the near wall. While the additional shear stress is much greater than the viscous stress in turbulent core area so viscous shear stress can be neglected.

### 6.4.4 The Prandtl mixing length semi-empirical theory of turbulent flow

Constructing the constitutive relation of Reynolds stress and mean velocity of flow (turbulence model), the flow velocity and stress distribution in turbulent flow can be solved by the method of theoretical explanation or computer numerical method, which is the core problem of the theory of turbulence. There has been a lot of research results in this area, of which Prandtl mixing length theory is the most classical and simple turbulence model that is of certain practicality.

Prandtl L, a German scholar, proposed a hypothesis of turbulent mixing length, using the concept of molecular free path. He thinks that the velocity of turbulent flow is proportional to the product of the mean velocity gradient and a certain length (the mixing length), that is

$$u'_x \propto L_1 \frac{d \bar{u}_x}{dy}, \quad u'_y \propto L_2 \frac{d \bar{u}_x}{dy}$$

Prandtl called $L_1$ and $L_2$ mixing length and used $L^2$ to express the product of proportionality constant. So additional stress in turbulent flow $\tau'$ is

$$\tau' = -\rho \overline{u'_x u'_y} = \rho L^2 \left(\frac{d \bar{u}_x}{dy}\right)^2 \qquad (6-29)$$

where $L$ is still called mixing length, which is determined by experiment.

### 6.4.5 Velocity distribution of turbulent flow

From above we can know, the velocity distribution of the turbulent flow is mainly the velocity distribution in the core area of the turbulent flow. In the core area of the turbulent flow, compared to the additional stress of the turbulent flow, the viscous shear stress can be neglected. Therefore, laminar shear stress is determined by the additional shear stress of turbulent

$$\tau = \tau' = \rho L^2 \left(\frac{d\bar{u}_x}{dy}\right)^2$$

For circular pipe, by equation (6 - 11), we get $\tau = \frac{\tau_0 r}{r_0}$; And according to the experimental results we can also get $L = \kappa y \sqrt{\frac{r}{r_0}}$, and the equation above can be expressed as

$$\tau_0 = \rho \kappa^2 y^2 \left(\frac{d\bar{u}_x}{dy}\right)^2$$

Then

$$v_* = \sqrt{\frac{\tau_0}{\rho}} = \kappa y \frac{d\bar{u}_x}{dy}$$

Separate the variables and make an integral we get

$$\frac{\bar{u}_x}{v_*} = \frac{1}{\kappa} \ln y + C \qquad (6-30)$$

The equation above is the turbulent flow velocity distribution derived by the Prandtl mixing length theory, which is along the normal direction at the flow section of circular tube. And it is of universal meaning and can be extended to any pipe surface flow, channel flow and other flow, etc., included. But the constants $\kappa$, $C$ should be determined by experiment according to specific practice.

In the following discussion, for the sake of simplicity, the line, representing average velocity, is omitted.

## 6.5  Linear head losses of turbulent flow in circular pipes

In this chapter, we mainly introduce the characteristics of turbulent flow in circular tube and the method of determining the frictional resistance coefficient.

### 6.5.1  Nikuradse experiment and resistance partition of turbulent flow

In 1933, a German force scientist and engineer, Nikuradse, carried out the experiments that study the distribution of the loss coefficient; and cross section velocity along the circular pipe and showed the resistance distribution diagram, which is of great significance. Here are a brief introduction to this classic experiment and its main results.

## 1. Nikuradse experiment

Nikuradse thought that there are two factors influencing the resistance factor

$$\lambda = f(Re, \Delta/d)$$

where $Re$ is the pipe flow Reynolds number; $\Delta$ is wall roughness height of the protrusion, called absolute roughness; $d$ is the diameter; $\Delta/d$ is the relative roughness.

With the artificial roughness method, Nikuradse tightly paste the screened uniform sand on the tube wall surface, and used sand particle diameter to represent the absolute roughness. He made six long straight and same diameter artificial rough pipes, but with different relative roughness. He made flow experiments respectively with these pipes. In the experiments, he measured velocity $v$, head loss $h_f$ (piezometric head difference) in length $l$ flow section, and calculated $Re$ and $\lambda$ with equation $Re = \dfrac{vd}{\nu}$ and $\lambda = \dfrac{d}{l}\dfrac{2g}{v^2}h_f$. Then, he plotted the $\lg(100\lambda) - \lg Re$ graph (see Fig.6-8), which is the famous Nikuradse diagram, with the series data.

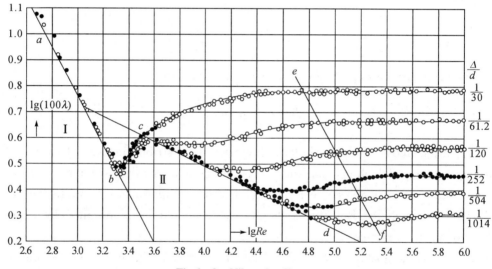

Fig.6-8  Nikuradse diagram

## 2. Turbulent resistance partition

According to the change characteristics of $\lambda$, the Nikuradse curve is divided into 5 resistance regions.

(1) Laminar flow region, $ab$ line, $\lg Re < 3.36$, $Re < 2300$. The experimental points of the 6 kinds of pipe flow all fall on the same line. $\lambda$ has nothing to do with the relative roughness $\dfrac{\Delta}{d}$, and $\lambda = \dfrac{64}{Re}$ can be derived from the graph. The experimental results are consistent with the theoretical results.

(2) The transition region from laminar to turbulent, $bc$ line, $\lg Re = 3.36 \sim 3.6$, $Re = 2300 \sim 4000$. $\lambda$ relates to $Re$ only. Because the flow regime is not stable, the experimental results are quite scattered.

(3) Smooth region of turbulent flow, $cd$ line, $\lg Re > 3.6$, $Re > 4000$. The flow pattern

is turbulent, and six kinds of pipe flow all have the experimental point to fall on the same line. In the hydraulically smooth region, $\lambda$ has nothing to do with $\frac{\Delta}{d}$ and relates to $Re$ only. We can see, from the figure, the smooth flow of the water region is smaller in the tube having larger $\frac{\Delta}{d}$; and the smooth flow of the water region is larger in the tube having smaller $\frac{\Delta}{d}$. For example, the pipe flow with $\frac{\Delta}{d} = \frac{1}{120}$ and the flow with $Re = 4\,000 \sim 12\,600$ are in the hydraulically smooth region, while the pipe flow with $\frac{\Delta}{d} = \frac{1}{1\,014}$ and the flow with $Re = 4\,000 \sim 70\,000$ are in the hydraulically smooth region.

(4) Turbulent transition region, the region between lines $cd$ and $ef$. Six kinds of pipe flow with different relative roughnesses have of six different slope curves, which shows that $\lambda$ in turbulent transition region relates to $\frac{\Delta}{d}$, $Re$.

(5) Turbulent rough region (also known as hydraulic rough region). The experiment curves of six kinds of pipe flow have six different horizontal lines, which shows $\lambda$ in turbulent rough region only relates to $\frac{\Delta}{d}$ and has nothing to do with $Re$. For example, when the flow is in the turbulent rough region, the $\lambda$ of the pipe flow with $\frac{\Delta}{d} = \frac{1}{120}$ is 0.035, regardless of the value of $Re$. In the turbulent rough region, the $\lambda$ of a certain pipeline $\left(\frac{\Delta}{d}\text{ is a certain value}\right)$ is constant. Because the frictional head loss is proportional to the square of flow velocity by equation $h_f = \lambda \frac{l}{d} \frac{v^2}{2g}$, turbulent rough region is also known as drag square region.

When the pipe flow is in smooth flow region, the flow is called the hydraulically smooth pipe. When the pipe flow is in rough flow region, the pipe flow is called hydraulically rough pipe. The hydraulically smooth pipe and hydraulically rough pipe are names of the pipeline flow characteristics. For a certain pipe, the flow is a hydraulically smooth pipe at low Reynolds number, while it becomes hydraulically rough pipe at high Reynolds number.

**Corollary** In the similar experiment, if the prototype and the model are in a rough area, it can automatically achieve resistance similar (equal to $\lambda$) only in geometric similarity $\left(\text{equal to }\frac{\Delta}{d}\right)$, without equaling to $Re$. That is to say, as long as the model is designed according to the Froude criterion, the model can automatically realize the viscous force similar, without satisfying the Reynolds similarity criterion. Therefore, square resistance region criterion is also called automatic model region in the model of similarity.

### 3. The causes analysis of turbulent flow resistance changes

The turbulence is divided into smooth region, transition region and rough region. The changes of $\lambda$ in different districts are different. The reason can be analyzed by the variation of the viscous sublayer of Fig.6-9. Fig.6-9a shows the flow situation that the thickness $\delta_0$ of viscous sublayer is bigger than the pipe wall roughness, which rough projections is completely concealed in the viscous sublayer and has almost no influence on the turbulent core region. Therefore, $\lambda$ is only related to $Re$, and has nothing to do with $\frac{\Delta}{d}$, which forms the flow resistance characteristics of turbulent smooth region. Fig. 6-9b shows the flow situation $\delta_0$ is close to $\Delta$, which roughness has influence on the turbulent core region. Therefore, $\lambda$ is related to both $\frac{\Delta}{d}$ and $Re$, which forms the flow resistance characteristics of transition region of turbulent flow. Fig.6-9c shows the flow situation $\delta_0$ is far less than $\Delta$, which coarse prominency almost completely goes into turbulent core. In this situation, the change of $Re$ has little influence on viscous sublayer and the degree of turbulent dynamic. Besides, $\lambda$ is only related to $\frac{\Delta}{d}$ and has nothing to do with $Re$, which forms the flow resistance characteristics of the rough region of turbulent flow.

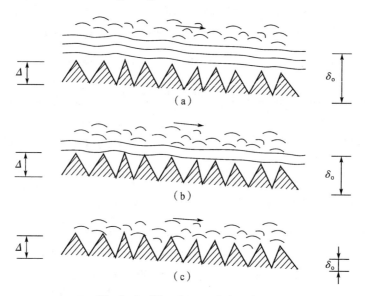

Fig.6-9  The change of viscosity

## 6.5.2  Velocity distribution of turbulent flow in circular tube

### 1. Smooth region of turbulent flow

It has been introduced the velocity of viscous flow at the bottom has a linear distribution

$$u = \frac{\tau_0}{\mu} y \quad (y \leqslant \delta_0)$$

In the core area of the turbulent flow, by the equation (6-30)

$$\frac{u}{v_*} = \frac{1}{k} \ln y + C$$

Referencing boundary conditions $y = \delta_0$, $u = u_b$, we determine the integral constant $C = \dfrac{u_b}{v_*} - \dfrac{1}{k}\ln \delta_0$ of above equation and consider $\delta_0 = \dfrac{\mu}{\tau_0}u_b = \dfrac{u_b}{v_*^2}v$. Then we can get

$$\frac{u}{v_*} = \frac{1}{k}\ln \frac{yv_*}{v} + C_1$$

So $C_1 = \dfrac{u_b}{v_*} - \dfrac{1}{k}\ln \dfrac{u_b}{v_*}$. Then, we confirm $k = 0.4$, $C_1 = 5.5$ by Nikuradse experiment and get the equation of the flow velocity distribution on the cross flow section of the hydraulic smooth circular tube, i.e.

$$\frac{u}{v_*} = 5.75 \lg \frac{yv_*}{v} + 5.5 \quad (y > \delta_0) \tag{6-31}$$

### 2. Rough region of turbulent flow

Distribution of flow velocity in turbulent pipe flow in a circular tube is (derivation omitted)

$$\frac{u}{v_*} = 5.75 \lg \frac{y}{\Delta} + 8.48 \quad \text{(core region)} \tag{6-32}$$

Equation (6-31) and equation (6-32) show that the flow velocity distribution of the turbulent flow in a circular tube conforms to the logarithm distribution law. Compared with parabolic distribution of laminar flow of circular tube, the flow velocity distribution on the cross-section of turbulent flow of circular tube is more uniform, showing as Fig.6-10.

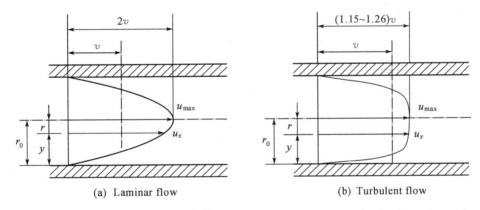

(a) Laminar flow  (b) Turbulent flow

Fig.6-10  **Comparison of flow velocity distribution in cross flow section of circular tube**

The above equation is based on the Prandtl mixing length theory and their undetermined constants are determined by the experiment, which were called the semi-empirical equation.

### 3. The exponential equation for the velocity distribution of the turbulent flow in a circular tube

In order to facilitate the engineering calculation, a more simple equation, the index distribution of flow velocity of turbulent flow in circular tube, was proposed.

$$\frac{u}{u_{max}} = \left(\frac{y}{r_0}\right)^n \tag{6-33}$$

where $u_{max}$ is the maximum velocity at the tube axis; $r_0$ is the radius of the tube; $n$ is the exponent, which can be calculated by the following empirical equation.

$$\frac{1}{n} = -2.8\lg\left(\frac{0.034\ 55}{Re^{0.1}} - 0.007\ 875\right) \quad (6-34)$$

where $4 \times 10^3 \leqslant Re < 2 \times 10^6$; When $Re \geqslant 2 \times 10^6$, $n = \frac{1}{10}$.

The $n$ value in equation (6-33) can also be taken as a constant $n = \frac{1}{7}$, which called the law of $\frac{1}{7}$ party of the velocity distribution.

### 4. Several parameters of turbulent pipe flow

According to equation (6-33), the ratio of average velocity and the tube center maximum velocity is $\frac{v}{v_{max}} = 0.79 \sim 0.87$, and increases with the increase of $Re$; the kinetic energy correction factor of flow section is $\alpha = 1.077 \sim 1.031$, and decreases with the increase of $Re$; the momentum correction factor of flow section is $\beta = 1.027 \sim 1.011$, and decreases with the increase of $Re$.

## 6.5.3 Semi-empirical equation for the linear resistance factor of circular tube turbulent motion

The mean velocity of cross section is calculated by the semi-empirical equation (6-31) of velocity distribution and equation (6-32). Then according to $v_s = v\sqrt{\frac{\lambda}{8}}$, we adjust the constant with experimental data. Finally, we can get Nikuradse semi-empirical equation of along the way resistance factor corresponding resistance region.

Smooth region of turbulent flow: $\quad \frac{1}{\sqrt{\lambda}} = -2\lg\left(\frac{2.51}{Re\sqrt{\lambda}}\right) \quad (6-35)$

Rough region of turbulent flow: $\quad \frac{1}{\sqrt{\lambda}} = -2\lg\left(\frac{\Delta}{3.7d}\right) \quad (6-36)$

## 6.5.4 Industrial pipelines, Moody diagram, and Barr equation

### 1. Industrial pipelines and Colebrook-White equation

The actual application pipeline without manual roughing processing called industrial pipeline. The manual rough pipelines have much different with industrial pipelines in the rough. Connecting the two different rough forms is a real factor of how to use Nikuradse semi-empirical equations in industrial pipelines.

Colebrook made a comparative experiment of industrial pipe resistance loss and Nikuradse experiment. He measured the $\lambda$ of industrial pipeline in the drag square zone. Comparing with Nikuradse curve, he defined the rough protrusion height $\Delta$ of manual rough tube with the same diameter and the same $\lambda$ in turbulent drag square zone as equivalent roughness of industrial pipe.

The equivalent roughnesses of common industrial pipes are as follows. Such as the $\Delta$ of welded steel pipe is 0.06~1.0 mm, the $\Delta$ of cast iron pipe is 0.3~1.2 mm, the $\Delta$ of cement pipe is 0.5 mm, the $\Delta$ of concrete pipe is 0.3~3.0 mm. The Nikuradse semi-empirical equation can be applied in industrial pipelines with equivalent roughness.

In 1939, Colebrook and White merged equation (6-35) and equation (6-36) together and got the empirical equation of $\lambda$ in turbulent transition region.

$$\frac{1}{\sqrt{\lambda}} = -2\lg\left(\frac{2.51}{Re\sqrt{\lambda}} + \frac{\Delta}{3.7d}\right) \tag{6-37}$$

The equation is called Colebrook-White equation.

Equation (6-37) is consistent with the experimental results of industrial pipe flow, and can be applied to the smooth region, the transition region and the rough region of turbulent flow. It's a general equation for calculating the factor of the turbulent flow along the path. Therefore, it is widely used in the calculation of industrial pipe flow.

### 2. Moody diagram

Equation (6-37) is an implicit equation for solving $\lambda$ value, which is not easy to calculate. In 1944, American engineer Moody drew a logarithmic curve of the equation, known as Moody diagram, as shown in Fig.6-11. In the diagram, we can directly find the $\lambda$ value by $\frac{\Delta}{d}$ and $Re$. In the age without computer, Moody diagram was very practical and one of the classic charts of Engineering fluid mechanics. Comparing Moody diagram and Nikuradse diagram, we also find industrial pipe flow and the flow region of industrial pipe can be divided into laminar region, critical region (laminar turbulent transition region), turbulence smooth region, transition region of the turbulent flow and rough region of turbulent flow.

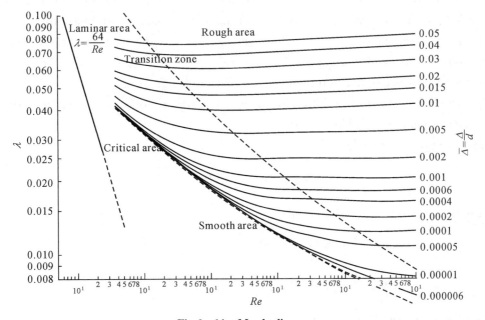

**Fig.6-11 Moody diagram**

$\lambda$ has the same changing law, only a large difference in turbulent transition zone. Because it is a gradual process to break through the non-uniform rough into the turbulent core area of the industrial pipeline, which is different from the artificial roughness with uniform particle size.

### 3. Barr equation

Except for the semi-empirical relationship referred above, there are a lot of empirical relationships, such as Bula Huges equation, Seffrin loose equation and so on. In which the equation proposed by Barr in 1977 is:

$$\frac{1}{\sqrt{\lambda}} = -2\lg\left(\frac{\Delta}{3.7d} + \frac{5.128\,6}{Re^{0.89}}\right) \tag{6-38}$$

The above is Barr equation, the applicable range of the equation is the same as that of equation (6-37), which is also the general experience equation of resistance area. Compared with equation (6-37), the maximum error is only 1%. As it is the explicit equation of $\lambda$, it is convenient to calculating and programming. For example, Moody diagram referred before is based on Barr equation, it is exactly alike with the traditional Moody diagram. With all the equations solving $\lambda$, priority being recommended as the Barr equation.

## 6.5.5 Discrimination of turbulent resistance region

The application of the discrimination of the turbulent resistance region in two aspects is as follows.

First is based on turbulent resistance region, select the appropriate equation to calculate the value of $\lambda$; The second is based on general equations, although the turbulent flow resistance region is not required to be determined in advance, it is necessary to determine the turbulent resistance area of the pipeline when the flow similarity simulation is carried out. Therefore, there is practical significance to distinguish the turbulence resistance region.

The semi-empirical equation for the discrimination of the turbulent resistance region derived from the artificial rough pipe flow can refer to the reference books. The calculation results show that the existing semi-empirical equation has great error in the use of industrial pipeline turbulent resistance region, pipe flow resistance as follows obtained by fitting the Moody discriminant is with high accuracy.

Smooth region of circular tube

$$\frac{\Delta}{d} \leqslant 0.000\,8 \text{ and } 4\,000 < Re \leqslant 10\left(\frac{\Delta}{d}\right)^{-1} \tag{6-39}$$

Turbulent transition zone of circular tube

$$\text{when } \frac{\Delta}{d} > 0.000\,8,\ 4\,000 < Re \leqslant 576.12\left(\frac{\Delta}{d}\right)^{-1.119}$$

or

$$\text{when } \frac{\Delta}{d} \leqslant 0.000\,8,\ 10\left(\frac{\Delta}{d}\right)^{-1} < Re \leqslant 576.12\left(\frac{\Delta}{d}\right)^{-1.119} \tag{6-40}$$

Circular tube turbulent rough region

$$Re > 576.12\left(\frac{\Delta}{d}\right)^{-1.119} \qquad (6-41)$$

## 6.6 The linear head loss of the turbulent flow in a non circular tube

### 6.6.1 Pressure pipe flow in non circular pipe

If $d_e = 4R$ is the equivalent diameter of a non circular tube, $R$ is the hydraulic radius, it can be got from the equation (6-3) that:

$$h_f = \lambda \frac{l}{d_e} \frac{v^2}{2g} \qquad (6-42)$$

Applicable scope: Rectangular tubes with $a$ and $b$ on short and long side applies to the turbulence with $\frac{a}{b} < 8$; The major axes were respectively $d_1$ and $d_2$ in the oval tube which applies to the turbulence with $\frac{d_1}{d_2} < 3$.

### 6.6.2 Channel flow

The 575 is the critical Reynolds number of the open channel flow, when $Re > 575$, open channel flow is turbulent flow. Laminar open channel flow has little practical value, so we focus on the study of turbulent channel flow.

French engineer Chézy presents empirical equation of average velocity of uniform flow in open channel in 1775,

$$v = C\sqrt{RJ} \qquad (6-43)$$

The above equation is Chézy equation. The $C$ is Chézy coefficient.

$$h_f = \frac{lv^2}{RC^2} = \frac{8g}{C^2} \frac{l}{4R} \frac{v^2}{2g} \qquad (6-44)$$

Compared with the equation (6-3), it is apparent that $\lambda = \frac{8g}{C^2}$, that is

$$C = \sqrt{8g/\lambda} \qquad (6-45)$$

Dimension of $C$ is $m^{0.5}/s$.

$C$ is usually determined by empirical equation. The most famous empirical equation is proposed by Irish engineer Manning in 1890.

$$C = \frac{1}{n} R^{1/6} \qquad (6-46)$$

The above is Manning equation, in which $n$ is roughness coefficient, a comprehensive coefficient to measure the influence of side wall roughness. In general, it is not listed in the unit. The value is got by measuring open channel or pipe flow turbulence rough region. For example, the value of $n$ of cast iron pipe and steel pipe is $0.011 \sim 0.013$; value of $n$ of concrete wall is $0.011 \sim 0.014$. From the above we can know, the use condition of the equation (6-46) is the rough region of turbulent flow, when $n < 0.02$, $R > 0.5$ m, the calculated results are in accord with the actual situation. Conditions of use are the same when equation (6-43) is introduced into the equation (6-46) calculation. When eliminating unit of $n$, the equation (6-43) and the equation (6-46) are not in agreement with the unit. Therefore, the unit of the length in accordance with the provisions can only choose meters while can not choose cm or other units. The Chézy coefficient is only related to the $Re$ and sidewall roughness, has nothing to do with the shape of cross flow. Equation (6-44) and equation (6-46) can also be applied to the area of resistance of the pipeline.

**Example 6-4** A water pipe, long $l = 500$ m, diameter $d = 0.2$ m, pipe wall roughness height $\Delta = 0.1$ mm, water temperature $t = 10$ ℃. How much is the head loss along the way if transport flow rate $q_v = 10 \times 10^{-3}$ m³/s?

**Solution:**

$$v = \frac{q_v}{\frac{1}{4}\pi d^2} = \frac{10 \times 10^{-3}}{\frac{1}{4} \times 3.14 \times 0.2^2} = 0.318\ 3 \text{ m/s}$$

Cause
$$t = 10℃, \nu = 1.31 \times 10^{-6} \text{ m}^2/\text{s}$$

$$Re = \frac{vd}{\nu} = \frac{0.318\ 3 \times 0.2}{1.31 \times 10^{-6}} = 48\ 595 > 2\ 300$$

So there is turbulent flow in pipe.

According to Barr equation

$$\frac{1}{\sqrt{\lambda}} = -2\lg\left(\frac{\Delta}{3.7d} + \frac{5.128\ 6}{Re^{0.89}}\right)$$

$$\frac{\Delta}{d} = 0.000\ 5, Re = 48\ 595, \text{ so } \lambda = 0.022\ 7$$

$$h_f = \lambda \frac{l}{d} \frac{v^2}{2g} = 0.022\ 7 \times \frac{500}{0.2} \times \frac{0.318\ 3^2}{2 \times 9.8} = 0.293 \text{ m}$$

**Example 6-5** A new water pipeline, pipe diameter $d = 0.4$ m, tube length $l = 100$ m, roughness coefficient $n = 0.011$, linear loss $h_f = 0.4$ m, water flow is turbulent rough region. How much is the flow rate?

**Solution:** Area of cross section of pipeline

$$A = \frac{\pi}{4}d^2 = \frac{3.14}{4} \times 0.4^2 = 0.126 \text{ m}^2$$

Hydraulic radius

$$R = \frac{d}{4} = 0.1 \text{ m}$$

Chézy coefficient

$$C = \frac{1}{n}R^{1/6} = \frac{1}{0.011} \times (0.1)^{1/6} = 61.94 \text{ m}^{1/2}/\text{s}$$

So the flow rate is

$$q_v = vA = AC\sqrt{RJ} = 0.126 \times 61.94 \times \sqrt{0.1 \times \frac{0.4}{100}} = 0.156 \text{ m}^3/\text{s}$$

## 6.7 Local head loss

From the equation $h_j = \zeta \dfrac{v^2}{2g}$, the head loss can be calculated, mainly to determine the local head loss factor $\zeta$. The effect of flow Reynolds number on $\zeta$ is very small, it is negligible, under normal circumstances, $\zeta$ is determined only by the boundary of the flow channel. Its value is determined by the trial, expect that of the sudden-expansion tube.

### 6.7.1 Sudden-expansion tube

Sudden-expansion tube's $\zeta$ can be determined by three equations of the total flow.

Assuming that the fluid flow is a constant turbulent flow. As shown in Fig.6-12, section 1-1 and 2-2 are the control sections. The shear stress of the fluid in the pipe wall acting on the flow section 1-2 is neglected. Experiment shows, the distribution of pressure of section 1-1 is consistent with the

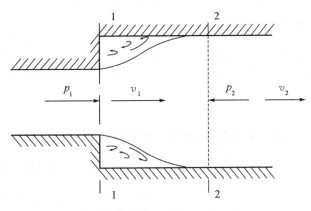

Fig.6-12  Sudden-expansion tube

distribution of static pressure. From the momentum equation (4-30) (taking $\beta_1 \approx \beta_2 \approx 1.0$), it can be deduced that

$$\sum F = p_1 A_1 - p_2 A_2 = \rho q_v (v_2 - v_1)$$

Apply momentum equation to section 1-1 and section 2-2, now $h_w = h_j$, then

$$\frac{p_1}{\rho g} + \frac{v_1^2}{2g} = \frac{p_2}{\rho g} + \frac{v_2^2}{2g} + h_j$$

From the two equations above, we get

$$h_j = \frac{(v_1 - v_2)^2}{2g}$$

From the continuous equation, we know that $v_2 = \dfrac{v_1 A_1}{A_2}$ or $v_1 = \dfrac{v_2 A_2}{A_1}$, so

$$h_j = \left(1 - \frac{A_1}{A_2}\right)^2 \frac{v_1^2}{2g} = \zeta_1 \frac{v_1^2}{2g} \tag{6-47}$$

or

$$h_j = \left(\frac{A_2}{A_1} - 1\right)^2 \frac{v_2^2}{2g} = \zeta_2 \frac{v_2^2}{2g} \tag{6-48}$$

where $\zeta_1$ and $\zeta_2$ respectively correspond to the average velocity of the two sections of the sudden expansion.

### 6.7.2 Sudden shrink tube

Fig.6-13 shows the flow in sudden shrink tube, from mainstream section 1-1, transiting to necked-down section c-c, expanding to section 2-2, local losses are mainly caused by the expansion, its value calculated according to empirical equation.

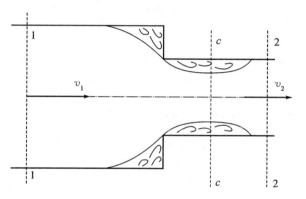

Fig.6-13 Sudden-shrink tube

$$h_j = \zeta \frac{v_2^2}{2g}, \quad \zeta = 0.5\left(1 - \frac{A_2}{A_1}\right) \tag{6-49}$$

### 6.7.3 Pipeline outlet and inlet

(1) Outlet loss coefficient $\zeta$. When the pipe suddenly expands into the pool (or container) (see Fig.6-14), $\dfrac{A_1}{A_2} \approx 0$, from equation (6-47) we know

$$h_j = \zeta \frac{v^2}{2g}, \quad \zeta = 1.0$$

(2) Inlet loss coefficient $\zeta$, equals to $A_1 \to \infty$ suddenly shrink to $A_2$, we have

$$h_j = \zeta \frac{v^2}{2g}$$

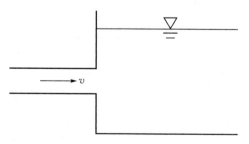

Fig.6-14 Pipe outlet

where $\zeta$ decreases with the increase of the degree of inlet shape close to the streamline (see Fig.6-15).

**Example 6-6** In a sudden-expansion tube, in which flow speed $v_1$ changes into $v_2$, if

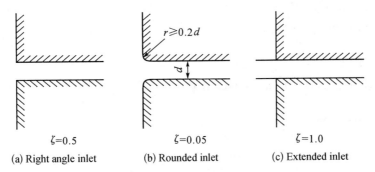

(a) Right angle inlet $\zeta=0.5$  
(b) Rounded inlet $\zeta=0.05$  
(c) Extended inlet $\zeta=1.0$

Fig.6-15  Pipe entrance

the middle plus a medium thickness of the tube section to form twice sudden expansion (see Fig.6-16), to omit the mutual interference of local resistance, using superposition method to calculate:

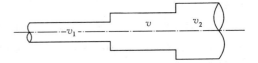

Fig.6-16  Twice sudden expansion tube

(1) When the total local head loss is minimum, how much is the velocity in the intermediate pipe;

(2) Calculate the total local head loss, and then compare with value of once-expansion.

**Solution:** (1) Local head loss at two sudden enlargement

$$h_j = h_{j1} + h_{j2} = \frac{(v_1-v)^2}{2g} + \frac{(v-v_2)^2}{2g}$$

The velocity in the middle tube is $v$. When the total local head loss is the minimum

$$\frac{dh_j}{dv} = 0$$

That is

$$\frac{dh_j}{dv} = -\frac{2(v_1-v)}{2g} + \frac{2(v-v_2)}{2g} = 0$$

That is

$$v = \frac{v_1+v_2}{2}$$

(2) Total local loss

$$h_j = \frac{\left(v_1 - \frac{v_1+v_2}{2}\right)^2}{2g} + \frac{\left(\frac{v_1+v_2}{2} - v_2\right)^2}{2g} = \frac{(v_1-v_2)^2}{4g}$$

Local head loss in a sudden enlargement $h_j = \frac{(v_1-v_2)^2}{2g}$, the total local head loss was 0.5 times when the sudden expansion of the two time.

To be sure, local head loss superposition principle can only be applied to a partial loss with interval long enough. If the interval is very close, mutual interference will happen. Total loss of head is about 0.5~3 times of that of a single normal local loss.

## 6.8 The concept of boundary layer

### 6.8.1 Boundary layer

As shown in Fig.6-17, when velocity of uniform flows is $U_0$ through the leading edge of the plate surface, due to the viscous effect, a layer of fluid particles close to the plate adhesion on the plate surface, and speed down to zero. A fluid in the outer layer will be blocked by this layer of fluid, the velocity decreases also. The farther the distance from the wall is, the smaller the decrease of the flow velocity is. When the distance from the plate surface to a certain distance, the velocity will be close to the original velocity $U_0$. Due to the influence of viscosity, a flow velocity distribution is not uniform between the plate surface and the non disturbed liquid flow, velocity is gradient, there is also large shear stress. This adhesive which is not be ignored near the wall of the thin layer is known as the boundary layer. Distance from the flat surface along the normal to the flow velocity $u_x = 0.99U_0$ is the thickness of boundary layer, denoted by $\delta$. Boundary layer thickness $\delta$ increased along the flow of water, because the influence of the boundary is gradually extending to the flow region along with the length of the boundary. Using the concept of boundary layer, the solution of the flow field can be divided into two zones. One is the flow in the boundary layer, the effects of fluid viscosity must be included in this layer. But because the boundary layer is thin, the N-S equation can be simplified, and the approximate solution can be obtained by using the momentum equation. The second is outside the boundary layer flow, velocity gradient is zero, there is no internal friction force. Thus, it can be considered as the flow of an ideal fluid, so it can be solved as potential flow.

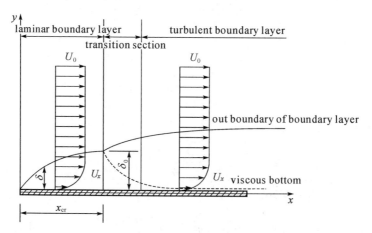

Fig.6-17  Flat-plate flow

As shown in Fig.6-17, the flow in the boundary layer of a flat plate is in a state of laminar flow, and the thickness of the boundary layer increases. After a transition period, the laminar boundary layer will be transformed into a turbulent boundary layer. So, the Reynolds number in the flat plate boundary layer is

$$Re_x = \frac{U_0 x}{\nu} \tag{6-50}$$

The distance of plate end farther away, the greater the Reynolds number. When the Reynolds number reaches a certain critical value, the flow from laminar flow to turbulent flow. This transition from laminar to turbulent flow is called the turning point. Now $x = x_{cr}$, its corresponding Reynolds number

$$Re_{cr} = \frac{U_0 x_{cr}}{\nu} \tag{6-51}$$

called critical Reynolds number, and the size of the value is related to the fluctuation degree of the flow, the fluctuation is strong, and the $Re_{cr}$ is small.

The range of Reynolds number for the smooth plate boundary layer is $3 \times 10^5 < Re_{cr} < 3 \times 10^6$.

The experimental results show that the thickness of the plate boundary layer can be calculated by the equation.

Laminar boundary layer

$$\delta = \frac{5x}{Re_x^{1/2}} \tag{6-52}$$

Turbulent boundary layer

$$\delta = \frac{0.377x}{Re_x^{1/5}} \tag{6-53}$$

Within the turbulent boundary layer, in the turbulent boundary layer, the most close to the flat plate, there is a thin layer, a large flow velocity gradient, viscous shear stress is still the main role. Turbulent shear stress can be ignored. The flow pattern is still laminar flow, this layer is the aforementioned viscous bottom.

**Deduction** According to the theory of flat plate boundary layer, the distribution of flow velocity in the inlet section of the circular pipe is changed along the course. As shown in Fig.6-18, fluid from the tank through a smooth circular inlet flow into the pipe, the velocity of flow is almost uniform throughout the entire water cross section. However, with the development of the boundary layer along the flow direction, the velocity of flow in the

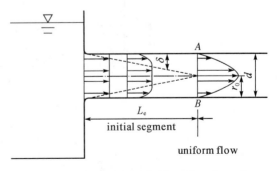

Fig.6-18 Development of boundary layer in the initial segment of the inlet

vicinity of the boundary gradually decreases, in the center of the pipe gradually increased to the maximum, along the flow velocity distribution is no longer change. From the inlet to the pipe center flow rate reaches the maximum, that is, the development of boundary layer thickness to the center of the circular pipe between the section of the pipeline called the starting section. The full experience of the beginning of turbulent flow is $L_e = (50 \sim 100)d$, but it is generally believed that when $L_e < (20 \sim 40)d$, the velocity is close to uniform.

### 6.8.2 Separation of boundary layer

As shown above, when the water flow along the wall, the boundary layer will be generated, and increased along the flow direction. In this process, the phenomena of boundary layer and over current wall detachment may be produced, this phenomenon is called the separation of boundary layer.

For flat-plate flow, when the pressure gradient is zero, one that is $\frac{dp}{dx} = 0$, no matter how long it is, there will be no separation, and the boundary layer will only increase continuously along the direction of flow. However, when the boundary is spread along the direction of the flow, as shown in Fig. 6-19, the pressure gradient is positive. Hence $\frac{dp}{dx} > 0$, the boundary layer thickness increases rapidly, which leads to the boundary layer separation. The kinetic energy of water flow in the boundary layer, on the one hand, is to be converted into the potential energy of pressure which increases gradually. And also consumed in the energy loss along the way, which causes the fluid flow in the boundary layer to stop. The upstream flow is forced to move away from the solid side wall, and separation is thus produced. At the separation point $C$, there is

$$\left(\frac{\partial u}{\partial y}\right)_{y=0} = 0$$

$$\tau_0 = \mu \left(\frac{\partial u}{\partial y}\right)_{y=0} = 0$$

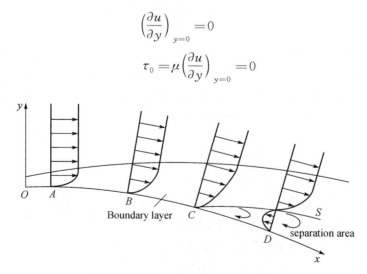

Fig.6-19  **Boundary layer separation**

Since the separation point $C$, there are formation of reflux (or vortex) at the lower reaches of the wall. Usually the downstream boundary separation line and surrounded by objects called wake flow. Wake flow will make the energy loss increases, and the reduction of pressure, so that the flow around the object before and after the formation of a larger pressure drop resistance. Moreover, reflux will also cause the foundation scouring brush and sediment accumulation. The strong turbulent vortex splitting may also induce the fluid around the random vibration of structure damage. And the bigger the tail, the more serious the consequences. The main way to wake decreases is to make the fluid body as far as possible changes in fluid flow.

The concept of the boundary layer was first proposed by Prandtl in 1904. The boundary layer theory is of great significance in the history of the development of fluid mechanics, especially in the research of aviation, ship, and fluid machinery. There is always canal flow in water conservancy project. In addition to the import part, almost all of the flow area is the boundary layer flow, and thus, it is no longer divided between the boundary layer and the outer region. However, it is still needed to apply the concept of boundary layer in the analysis of the phenomena and the loss of flow and resistance of the dam.

## 6.9  Drag force

### 6.9.1  The characteristic of around flow

Fluid discussed earlier in the solid boundary, such as pipes, open channels of flow resistance and head loss is within the so-called flow problem. The flow resistance of water flow around an object is an outflow problem. Such as water supply and drainage engineering, water conservancy project in a variety of gate piers, railway and highway bridge piers and transport and defense industry in a variety of body and aircraft flow around the problem, both belong to the flow around the problem.

When the fluid and submerged in the fluid solid for the relative motion, the effect of fluid solid, according to the direction can be divided into two components: One is the component parallel to the flow direction of the role of the object $F_D$, known as drag, including the frictional resistance caused by the viscous friction in the boundary layer and the resistance caused by the boundary layer separation. The second is perpendicular to the flow direction of the force acting on a body of $F_L$, known as lift. The force can only occur in an asymmetric fluid (see Fig.6 - 20).

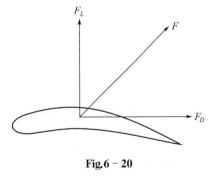

Fig.6 - 20

Fig.6 - 21a shows the streamline flow, the flow resistance is mainly friction force. Fig.6 - 21b shows a cylindrical fluid, the flow resistance is mainly the body resistance. When the same flow area is compared, the flow resistance of the streamline is 1/10 of the cylinder. So, the flow resistance is mainly caused by the shape of the body.

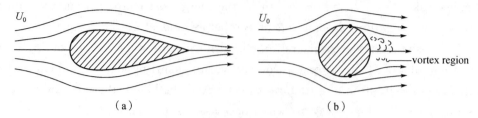

Fig.6 - 21  Flow around a circular cylinder

Flow around the fluid can also make the fluid vibration. When $Re = \dfrac{U_0 d}{\nu}$ reaches a certain value, the fluid around the occurrence of boundary layer separation, formed around the fluid on both sides of the alternate shedding vortex, was brought to the downstream, arranged in two rows, called the Karman vortex. This periodic vortex can subject the fluid alternating lateral force and resulting the transverse vibration of the fluid. If the frequency of vortex flow and vibration frequency, resonance occurs, causing damage to buildings. Such as vibration, wind in the wire rack, all derived from this.

### 6.9.2  Flow drag force calculation

In 1726, Newton proposed the calculation equation of the flow resistance $F_D$:

$$F_D = C_D \rho \dfrac{U_0^2}{2} A \tag{6-54}$$

where $C_D$ is the flow resistance factor, its value is mainly determined by the flow around the body and the Reynolds number $Re$; $C_D$ is available on the chart; $\rho$ is fluid density; $U_0$ is relative speed; $A$ is projected area.

The above equation can be applied to the flow resistance of various body shapes. For a small sphere, the flow resistance factor can be determined by the following equation:

$$C_D = \begin{cases} \dfrac{24}{Re} & (Re < 1.0) \\ \dfrac{13}{\sqrt{Re}} & (1.0 < Re < 10^3) \\ 0.44 & (10^3 < Re < 2 \times 10^5) \end{cases} \tag{6-55}$$

### 6.9.3  Ball settling velocity

We use flow drag force to discuss $d$ diameter ball subsidence phenomenon in fluid.

When the ball began to sink in the fluid, due to the acceleration of gravity and buoyancy difference, ball sinking speed increases gradually. At the same time, the flow resistance is also increased. When the gravity, buoyancy and viscous flow resistance balance and the ball starts at a uniform speed sinking, the velocity known for settling velocity, referred to as the settling velocity.

When $Re = \dfrac{U_0 d}{\nu} < 1$, flow drag force can be obtained by equation (6 - 54) and equation (6 - 55), among them $U_0 = u_0$,

$$F_D = 3\pi\mu d u_0 \quad (Re < 1.0) \tag{6-56}$$

$U_0$ is the ball sinking speed.

According to the principle of force balance, the sum of the sphere buoyancy and the resistance of the ball should be equal to the weight of the ball. That is

$$\frac{1}{6}\pi d^3 \cdot \rho g + 3\pi\mu d u_0 = \frac{1}{6}\pi d^3 \cdot \rho_s g$$

Solving can get the limit velocity of fluid in the sinking of the ball

$$u_0 = \frac{d^2}{18\mu}(\rho_s g - \rho g) \tag{6-57}$$

The $\rho$ and $\rho_s$ in the equation are the density of the fluid and the sphere.

In the analysis of the above equation can be used in the design of the sedimentation tank, dredging and sewage treatment etc.

# Exercises

1. At a certain point in the open channel flow, a laser velocimeter (a highly sensitive flow velocity meter) was used to measure the velocity. Measurements were taken every 0.5 seconds, and the results are shown in the table below. Calculate: (1) turbulence intensity and (2) turbulent shear stress.

| Flow velocity | 1 | 2 | 3 | 4 | 5 | 6 | 7 | 8 | 9 | 10 |
|---|---|---|---|---|---|---|---|---|---|---|
| $u_x$ / (m · s$^{-1}$) | 1.88 | 2.05 | 2.34 | 2.30 | 2.17 | 1.74 | 1.91 | 1.91 | 1.98 | 2.19 |
| $u_y$ / (m · s$^{-1}$) | 0.10 | −0.06 | −0.21 | 0.19 | 0.12 | 0.18 | 0.21 | 0.06 | −0.04 | −0.10 |

2. Prove, using equation (6 - 30), that in a wide rectangular cross-section channel as shown in Fig.6 - 22, the velocity at a certain water depth $y' = 0.63h$ is equal to the average velocity of that cross section. $\left(\text{Hint: } \int_0^h \ln\dfrac{y}{h} \mathrm{d}\left(\dfrac{y}{h}\right) = -1\right)$

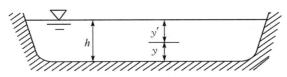

**Fig.6 - 22**

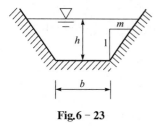

Fig.6 - 23

3. In a trapezoidal cross-section soil channel as shown in Fig. 6 - 23, uniform flow is occurring. Given the bottom width $b=2$ m, side slope factor $m=1.5$, water depth $h=1.5$ m, bottom slope $i=0.000\,4$ m, and roughness coefficient of the soil $n=0.022\,5$. Calculate: (1) the velocity of flow in the channel $v$; (2) the flow rate in the channel $q_v$.

4. In a prismatic concrete channel with uniform flow, as shown in Fig. 6 - 24 the hydraulic slope $J=1/800$ is given and the flow rate $q_v=7.05$ is in m³/s. Calculate: (1) the average friction velocity along the wetted perimeter; (2) the roughness coefficient $n$.

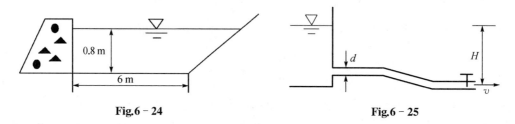

Fig.6 - 24                     Fig.6 - 25

5. In the water intake pipe shown in Fig.6 - 25, with a diameter of $d=500$ mm and a total length of $l=1\,200$ m, the resistance factors along the pipe are given $\lambda=0.015$, as well as the flow rate $q_v=0.2$ m³/s and the local resistance factors at the inlet $\zeta_c=0.5$, each bend $\zeta_b=0.8$, and the gate $\zeta_f=0.2$. Calculate the required effective head $H$.

# Chapter 7  Orifice Outflow and Nozzle Flow

The phenomenon of the liquid flow out of a orifice, which is on the side of a chamber, is called orifice flow. When a pipe, the length of which is about 3~4 times of the orifice's dimension, is linked to the orifice, this pipe is called a cylindrical nozzle. And it is called nozzle efflux when the liquid flows through the nozzle, and then form a full pipe flow at the outlet. The orifice outflow and the pipe nozzle flow are both common phenomenon in engineering. The water intaking, sluice hole, and orifice plate flow rate measurement equipment in water supply and sewerage works are orifices. The phenomenon such as the water flow through a pressure culvert under the roadbed, and the drain pipes in the dams are relevant with pipe nozzle, while the nozzle is also applied in fire branches and the hydraulic giants used in the possess of hydromechanization. This chapter will analyses the principle and methods of the hydraulic calculation about orifice outflow and pipe nozzle flow, applied the theory introduced before.

We demand that you should grasp the essential conception, equations, the application in engineering, as well as the ways to do the hydraulic calculation of non steady orifice outflow and pipe nozzle flow.

## 7.1  Steady orifice outflow

### 7.1.1  The steady orifice outflow in the sharp edged orifice

When there is a sharp edge on the orifice, the wall of hole and the water flowing through the orifice contact only at a contour, which means that the thickness of the orifice will not affect the outflow. This kind of orifice is called sharp edged orifice, the shape of which can be round, rectangle, triangle, or elliptocytosis. Here we mainly discuss the round shape.

The orifices can be classified as big orifice and small orifice according to the dimension $d$ (or the height $e$), and the head height above the poid of the orifice: it's small orifice when $\frac{d}{H} < 0.1$; it's big orifice when $\frac{d}{H} \geqslant 0.1$. For the small orifice, we can assume that the velocity, as well as the head height, of each mass point on the vena contraction of the orifice is equal, because the dimension $d$ is much less than the head height $H$.

## 1. Free steady orifice outflow

It's called free outflow when the water flow from a orifice into the atmosphere, as is shown in Fig.7 – 1. When the shape of the orifice is round, the water level of the water tank is constant, the water stream flow towards the orifice in many direction from the head water, and the streamline can't be break angle because of the effect of inertia, but only in smooth, continuous curve, so that the streamline in the orifice section can't be parallel. And the stream will continue drawing back after flowing out of the orifice until it reached the distance of $\dfrac{d}{2}$ away from the the orifice section, when the streamline will be parallel. This section is called vena constracta, as the c-c section in Fig.7 – 1. The specific value of the area of the vena constracta and the orifice section is shrinkage coefficient, which is substituted by $\varepsilon$. $\varepsilon$ is related to the shape, the border situation, and the position of the orifice. When the distance from each side of the orifice to the side wall is greater than 3 times of the dimension (or length of the side) of the orifice, the shrink of the water stream can't be effected by the side wall, which is called the perfect contraction. The shrinkage coefficient of the orifice has been found in the experiments,

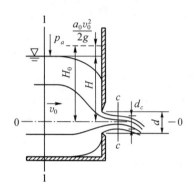

Fig.7 – 1  Free steady orifice outflow

$$\varepsilon = \frac{A_c}{A} = \left(\frac{d_c}{d}\right)^2 = 0.64.$$

Choose the horizontal plane which the poid of the orifice is on as the base level 0 – 0, as is shown in Fig.7 – 1, then get the energy equation according to the section 1 – 1 and the vena constracta c-c which conform to the conditions of gradually varied flow:

$$H + \frac{p_a}{\rho g} + \frac{\alpha_0 v_0^2}{2g} = 0 + \frac{p_c}{\rho g} + \frac{\alpha_c v_c^2}{2g} + h_w \qquad (7-1)$$

When it is the free outflow at the orifice, the pressure at the vena constracta will be atmospheric pressure, that is $p_c = p_a$. On the other hand, the tiny loss of the head height inside the water tank can be neglected, so the head loss $h_w$ is just the local head loss at the orifice.

$$h_w = h_j = \zeta_0 \frac{v_c^2}{2g}$$

The $v_c$ of the equation means the average velocity at the vena constracta.

Assume $H_0 = H + \dfrac{\alpha_0 v_0^2}{2g}$, which is called the head height of the free flow at the orifice, and $\alpha_c = 1.0$, so that the equation (7 – 1) can be written as:

$$H_0 = (1 + \zeta_0) \frac{v_c^2}{2g} \qquad (7-2)$$

So that we can get the equation to calculate the average velocity $v_c$ at the vena constracta.

$$v_c = \frac{1}{\sqrt{1+\zeta_0}}\sqrt{2gH_0} = \varphi\sqrt{2gH_0} \qquad (7-3)$$

where $H_0$ represents the acting head; $\zeta_0$ means the local resistance factor as the water flow through the orifice; $\varphi$ is the velocity factor, $\varphi = \frac{1}{\sqrt{1+\zeta_0}}$. If we neglect the head loss, then $\zeta_0 = 0$, $\varphi = 1$. That is the velocity factor $\varphi$ is the ratio of the actual velocity $v_c$ to the ideal fluid velocity $\sqrt{2gH_0}$. According to the experimental data, the velocity factor of the small orifice $\varphi = 0.97 \sim 0.98$. So we can get the local resistance factor, that is

$$\zeta_0 = \frac{1}{\varphi^2} - 1 = \frac{1}{0.97^2} - 1 = 0.06$$

Assuming the area of the orifice section as $A$, the equation to calculate the discharge of free outflow comes out of the orifice

$$q_v = A_c v_c = \varepsilon A \varphi \sqrt{2gH_0} = \mu A \sqrt{2gH_0} \qquad (7-4)$$

The $\mu$ in the equation means the flow rate factor at the orifice, $\mu = \varepsilon\varphi$. For the thin round small orifice, $\mu = \varepsilon\varphi = 0.60 \sim 0.62$. If the tank at the upstream of the orifice is big enough, $v_0 \approx 0$, so $H_0 = H$, and the equation of velocity and flow rate could be improved to

$$v_c = \varphi\sqrt{2gH} \qquad (7-5)$$

$$q_v = \varepsilon A \varphi\sqrt{2gH} = \mu A \sqrt{2gH} \qquad (7-6)$$

For the non-perfect contraction, as the hole 1 and hole 2 in Fig.7 - 2, the contraction will be decided by experiments.

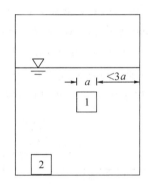

Fig.7 - 2  Non-perfect contraction

### 2. Constant submerged orifice outflow

The water flowing from the orifice and then coming into another part of the ford is called submerged orifice, as shown in Fig. 7 - 3. Because of the effect of inertia, when the stream comes to the orifice, it will contract at first and then expand, just like the free outflow. Assuming the section of uniform flow upstream and downstream as 1 - 1 and 2 - 2, the vena constracta of the outflow at orifice is $c$-$c$. And the $H_0$ in the picture is

$$H_0 = \left(H_1 + \frac{\alpha_1 v_1^2}{2g}\right) - \left(H_2 + \frac{\alpha_2 v_2^2}{2g}\right) \qquad (7-7)$$

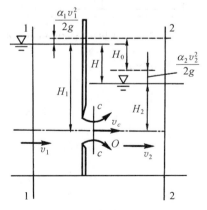

Fig.7 - 3  Constant submerged orifice outflow

which is the acting head of the submerged outflow at the orifice, obviously

$$H_0 = h_w = \sum h_j = (\zeta_0 + \zeta_s)\frac{v_c^2}{2g} \tag{7-8}$$

where $\zeta_0$ is the local resistance factor when the water stream reaches the orifice; $\zeta_s$ is the local resistance factor when the water stream comes out of the orifice and suddenly begins to expand, when $A_2 \gg A_C$, $\zeta_s \approx 1$. Substituting them into the equation above, after some arrangement, we can get the equation of the velocity and the flow rate of the submerged outflow as follows:

$$v_c = \frac{1}{\sqrt{1+\zeta_0}}\sqrt{2gH_0} = \varphi\sqrt{2gH_0} \tag{7-9}$$

$$q_v = \varepsilon A\varphi\sqrt{2gH_0} = \mu A\sqrt{2gH_0} \tag{7-10}$$

If the tank on the downstream of the orifice is big enough, $v_1 \approx v_2 \approx 0$, so $H_0 = H$, and then the equation can be improved to

$$v_c = \varphi\sqrt{2gH} \tag{7-11}$$

$$q_v = \varepsilon A\varphi\sqrt{2gH} = \mu A\sqrt{2gH} \tag{7-12}$$

The $\mu$ is the flow rate factor of the submerged orifice outflow, the value of which could be equal with the flow factor of free outflow, that is $\mu = 0.60 \sim 0.62$.

Comparing equation (7-6) and equation (7-12), we notice that the forms of them are exactly the same, so that their flow rate factors is also equivalent. But we must notice that the acting head $H$ at the orifice is the height difference between the affluent level and the center of the orifice when it's free outflow. In addition, every spot on the section of submerged orifice has the same acting head, so the velocity and the flow rate of the submerged orifice outflow have nothing to do with the depth under the water, and the size of the orifice.

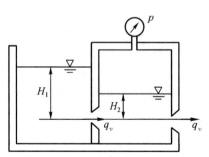

Fig.7-4 Double mouth water tank

**Example 7-1** There are two orifices that is same on the water tank, $H_1 = 6$ m, $H_2 = 2$ m, as is shown in Fig.7-4. Try to figure out the table pressure of the airtight container.

**Solution:**

$$q_v = \mu A\sqrt{2g\left[(H_1 - H_2) - \frac{p}{\rho g}\right]} = \mu A\sqrt{2g\left(\frac{p}{\rho g} + H_2\right)}$$

$$H_1 - H_2 - \frac{p}{\rho g} = \frac{p}{\rho g} + H_2$$

$$\frac{p}{\rho g} = \frac{1}{2}(H_1 - 2H_2)$$

So

$$p = \frac{\rho g}{2}(H_1 - 2H_2) = \frac{9\,800}{2} \times (6 - 2\times 2) = 9\,800 \text{ Pa}$$

## 7.1.2 Free outflow from the big thin-wall orifice

It has been proven in the experiments that the flow rate equation (7 - 6) are suitable for the big orifices. The $H$ in the equation should be the head height of the center of the big orifice ($H_C$), as is shown in Fig.7 - 5. However, because the shrinkage coefficient ε for the big orifices are greater, so that the flow rate factor $\mu$ should be greater too. In most of the irrigation works, we can treat the gate holes as big orifices, and the value of the flow rate factor is depend on the contracting situation, $\mu = 0.66 \sim 0.9$.

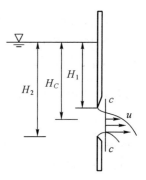

Fig.7 - 5 Free outflow frow big thin-wall orifice

## 7.2 Nozzle outflow

### 7.2.1 The steady outflow of cylindrical outer nozzle

#### 1. Free outflow

After the water flowing into the nozzle, at the edge of the entrance, the water will form a contraction, separation with wall, and forming a vortex region. Then streams gradually expanded to the entire cross-section by the contraction section, as can be seen in Fig.7 - 6.

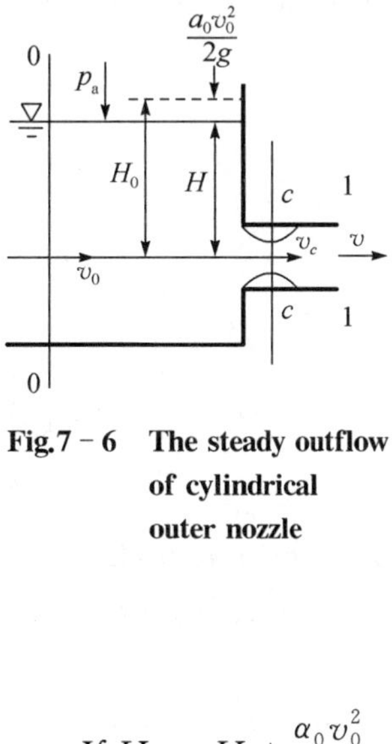

Water level of tank remains unchanged, and the water pressure is atmospheric pressure. Nozzle efflux is free outflow. List the energy equation of section 0 - 0 and 1 - 1.

$$H + \frac{\alpha_0 v_0^2}{2g} = \frac{\alpha v^2}{2g} + h_w \quad (7-13)$$

Fig.7 - 6 The steady outflow of cylindrical outer nozzle

Because the length of nozzle is very short, so the head loss along the way can be negligible.

$$h_w = \sum h_j = \zeta_n \frac{v^2}{2g}$$

If $H_0 = H + \frac{\alpha_0 v_0^2}{2g}$, the equation of velocity and flow rate is obtained as follows:

$$v = \frac{1}{\sqrt{\alpha + \zeta_n}} \sqrt{2gH_0} = \varphi_n \sqrt{2gH_0} \quad (7-14)$$

$$q_v = Av = A\varphi_n \sqrt{2gH_0} = \mu_n A \sqrt{2gH_0} \quad (7-15)$$

where $\zeta_n$ is the local resistance factor of the orifice; $\varphi_n$ is the volocity factor of the orifice outflow, $\varphi_n = \frac{1}{\sqrt{\alpha + \zeta_n}}$; $\mu_n = \varphi_n$.

For the right angle sharp edge nozzle, $\varphi_n = 0.82$.

## 2. Submerge outflow

Submerged outflow analysis shows that Submerged nozzle flow has the same equations with equations (7 - 14) and (7 - 15).

### 7.2.2 Vacuum of cylindrical outer nozzle at sharp-edged rectangular imports

After adding the nozzle orifice outside, increasing the resistance, but flow rate has actually increased. Actually, the vena constratra have vacuum, the flow velocity is greater than the velocity of the orifices. Taking $\alpha_c = \alpha = 1$, then

$$\frac{p_c}{\rho g} + \frac{v_c^2}{2g} = \frac{p_a}{\rho g} + \frac{v^2}{2g} + h_f \quad (7-16)$$

The continuity equation can be obtained that $v_c = \frac{A}{A_c} v = \frac{1}{\varepsilon} v$, partial loss occurs mainly in the expansion of the flow.

$$h_j = \zeta_{se} \frac{v^2}{2g} \quad (7-17)$$

$$\frac{p_c}{\rho g} = \frac{p_a}{\rho g} - \frac{v^2}{\varepsilon^2 2g} + \frac{v^2}{2g} + \zeta_{se} \frac{v^2}{2g}$$

For a cylindrical outer nozzle, $v = \varphi_n \sqrt{2gH_0}$; the local resistance factor consider the sudden expansion,

$$\zeta_{se} = \left(\frac{A}{A_c} - 1\right)^2 = \left(\frac{1}{\varepsilon} - 1\right)^2 \quad (7-18)$$

$$\frac{p_c}{\rho g} = \frac{p_a}{\rho g} - \left[\frac{1}{\varepsilon^2} - 1 - \left(\frac{1}{\varepsilon} - 1\right)^2\right]\varphi_n^2 H_0$$

As cylindrical outer nozzle,

$$\varepsilon = 0.64, \quad \varphi_n = 0.82$$

$$\frac{p_c}{\rho g} = \frac{p_a}{\rho g} - 0.75 H_0 \quad (7-19)$$

As equation (7 - 19) shows that the vacuum of cylindrical outer nozzle increase 0.75 times. Therefore, in the same diameter, and the same flow head, the flow rate of the outer nozzle is larger than the aperture.

The normal operating conditions of right angle import cylindrical outer nozzle are:
(1) the flow head is $H_0 \leqslant 9$ m;
(2) the length of nozzle is $l = (3 \sim 4)d$;
(3) the nozzle maintain full tube flow.

### 7.2.3 The classification of Nozzle

(1) Streamlined nozzle (Fig.7 - 7b).
(2) Conically contractive nozzle (Fig.7 - 7c).
(3) Conically expanded nozzle (Fig.7 - 7d).
All the basic equations of the nozzle flow are the same as equations (7 - 14) and (7 - 15).

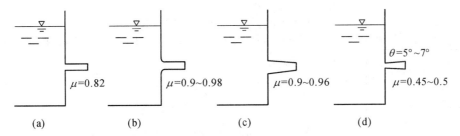

(a)　(b)　(c)　(d)

Fig.7-7　The classification of nozzle

## 7.3 Unsteady outflow of orifice and nozzle

### 7.3.1 Time calculation of container drainage in unsteady outflow of orifice and nozzle

Fig.7-8 shows a cylindrical containers which area is $A_\Omega$. Fluid through the wall outflow freely. If liquid in the container are not added with the flow process, orifice head will gradually decrease. At some time, the head orifice is $H$, the volume of fluid flowing through the orifice with small time is:

$$q_v dt = \mu A \sqrt{2gH}\, dt$$

At the same time, the liquid of container reduces $-dH$ in the small hours. Liquid volume changes $-A_\Omega dH$, its value is equal to the volume of outflow.

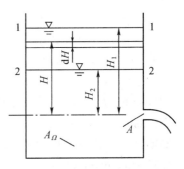

Fig.7-8　Unsteady orifice free outflow

$$-A_\Omega dH = \mu A \sqrt{2gH}\, dt$$

$$dt = -\frac{A_\Omega dH}{\mu A \sqrt{2gH}}$$

The head by a set opening sermon $H_1$, $H_2$, and integration

$$t = \int_{t_1}^{t_2} dt = \int_{H_1}^{H_2} \frac{-A_\Omega}{\mu A \sqrt{2gH}} dH = \frac{2A_\Omega}{\mu A \sqrt{2g}}(\sqrt{H_1} - \sqrt{H_2})$$

The emptying time of container is $t$ ($H_2 = 0$),

$$t = \frac{2V}{\mu A \sqrt{2gH_1}} = \frac{2V}{q_{v\max}} \qquad (7-20)$$

The above analysis also applies to the outflow nozzle. The above analysis also applies to the outflow nozzle.

### 7.3.2 Time calculation of container outflow in submerged orifice

Fig.7-9 shows that water level of upstream fixed in cylindrical container, and water level of downstream changes. The fluid through the wall to the cylindrical container with gradually filling,

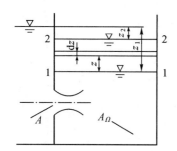

**Fig.7 - 9  Drowned out unsteady flow orifice**

which is container drainage in submerged orifice.

The equation can be obtained according to a similar method in the front.

$$t = \frac{2V}{\mu A \sqrt{2gz}} = \frac{2V}{q_{v\max}} \qquad (7-21)$$

The equation shows that in the case of variable head, the time of cylindrical containers filled with a cross-section of the orifice flow is two times more than the starting level at a constant case.

## Exercises

1. What is the difference between the flow rate calculation equations of small orifice free flow and submerged flow?

2. Upstream or downstream water level constant water tank, the water depth in the cabinet for $H$ and $h$. Three thin-walled openings of equal diameter 1, 2 and 3 are located at different locations on the partition, submerged in water, and are the perfect contraction. Ask: Is the flow rate of the three ports equal? Why? What happens if there is no water in the downstream tank?

3. Thin wall sharp edge circular orifice, diameter $d = 10$ mm, acting head $H = 2$ m, regardless of approach velocity. It is found that the diameter of water at the vena constrata is $d_c = 8$ mm, and the amount of water flowing out through the orifice is 0.01 m³ within 32 s. Try to find the shrinkage coefficient $\varepsilon$, the velocity factor $\varphi$, the flow rate factor $\mu$, and calculate the local resistance factor of the orifice $\zeta_0$.

4. As shown in Fig.7 - 10, a drainage pipe is set in the concrete dam body, the pipe length $l = 4$ m, acting head $H = 6$ m, and the flow rate $q_v = 10$ m³/s. Try to determine the diameter of the pipe $d$, and find the vacuum degree at the contracted section of the pipe. Desirable flow rate factor $\mu_n = 0.82$.

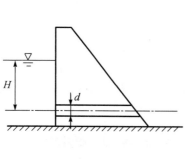

**Fig.7 - 10**

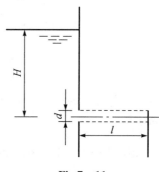

**Fig.7 - 11**

5. As shown in Fig.7 - 11, there is a small orifice on the side wall of the thin-walled

vessel, its diameter $d=20$ mm, the water depth above the center line of the orifice $H=5$ m, and the outflow velocity $v_c$ and flow rate $q_v$ at the orifice are measured. If a short tube with length $l=8d$ is connected externally at the orifice, the loss coefficient at the inlet of the short tube $\zeta=0.5$, and the loss coefficient along the path $\lambda=0.02$, try to find the outflow velocity $v$ and flow rate $q_v$ of the short tube.

# Chapter 8  Fluid Motion in Pressure Pipe

Fluid flow along pipeline as the hydraulic process is referred to as pressurized pipe flow. Pressurized pipe flow consists the major part in water conveyance systems for production and domestic water supply. Hydraulic performance calculation meets the need of practical engineering issue, such as pump pipeline in industrial, agricultural and municipal water system, diversion pipes or pressure drain pipes for reservoir in water distribution system and hydropower station. This chapter introduces aforementioned fundamental equations: continuity equation as well as momentum utilized for hydraulic performance calculation of fluid flow in piping mentioned above. To simplify calculation, pipes are classified as long pipe or short pipe, according to the corresponding percentage of local head loss and frictional head loss in total head loss. Long pipe refers to fluid flow in the pipe where frictional head loss is dominant while local head loss can be ignored. Short pipe refers to fluid flow in the pipe where both local head loss and frictional head loss make major part in total head loss and are not to be ignored in calculation. Hydraulic calculation of incompressible steady flow in basic pipes is firstly introduced in this chapter, examples of complex pipes including pipe in series and parallel, frictional head loss along the uniform slicing pipe as well as hydraulic calculation of incompressible steady flow is then introduced.

It's recommended that readers grasp hydraulic calculation methods of short pipes (siphon, pump suction pipe, pressure culvert), long pipes (pipe in series and parallel) and the ways to draw water head line of them.

## 8.1  Hydraulic calculation of equal diameter short pipe

Suction pipe, siphon, inverted siphon, drain pipe inside dam, culvert on railway for water pump are usually calculated as short pipe.

### 8.1.1  Fundamental equation

For a free outflow short pipe, neither frictional head loss nor local head loss of it is ignored. According to Bernoulli equation, both total head line $H$ and piezometric tube head line $H_p$ are qualitatively shown in Fig.8 − 1. Apparently, total frictional head loss of fluid flow in pipe is the summation of that in every flow segment, expressed as $h_f = \sum \lambda \dfrac{l}{d} \dfrac{v^2}{2g}$, while total local head loss is the summation of that in every flow segment, expressed as $h_j = \sum \zeta \dfrac{v^2}{2g}$. Hence, Bernoulli equation describes section 1-1 and 2-2 of water tank as follows:

$$H_0 = \dfrac{\alpha v^2}{2g} + h_w$$

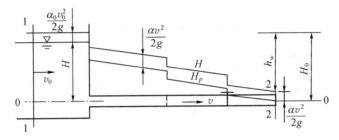

Fig.8-1 Free outflow in short pipe

Assuming $\alpha = 1$, thus

$$H_0 = \left(1 + \sum \lambda \dfrac{l}{d} + \sum \zeta\right) \dfrac{v^2}{2g} \qquad (8-1)$$

where $H_0$ refers to free outflow in short pipe, $H_0 = H + \dfrac{\alpha_0 v_0^2}{2g}$, if $v_0 \approx 0$, the $H_0 = H$; $\sum \zeta$ refers to total local resistance factor.

The equation above is the fundamental equation of hydraulic calculation of free out flow in constant diameter short pipe.

For the submerged discharge of short pipe shown in Fig.8-2. Similarly, fundamental equation of submerged discharge of constant diameter short pipe is acquired as:

$$H_0 = \left(\sum \lambda \dfrac{l}{d} + \sum \zeta\right) \dfrac{v^2}{2g} \qquad (8-2)$$

$\sum \zeta$ is added into head loss factor of the outlet of pipe compared to the case of free outflow, and $\zeta_{outlet} = 1$.

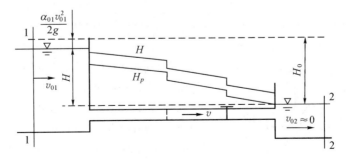

Fig.8-2 Submerged discharge of short pipe

Corollary: For non equal diameter short pipes, it is easy to deduce the equation for free flow as

$$H_0 = \sum_{i=1}^{n}\left(\lambda_i \frac{l_i}{d_i} + \zeta_i\right)\frac{v_i^2}{2g} + \frac{\alpha_n v_n^2}{2g} \quad (8-3)$$

for submerged outflow

$$H_0 = \sum_{i=1}^{n}\left(\lambda_i \frac{l_i}{d_i} + \zeta_i\right)\frac{v_i^2}{2g} \quad (8-4)$$

where $i$ refers to the number of certain segment of pipe, the last one of them is $i = n$.

## 8.1.2 Hydraulic calculation

### 1. Calculation of $\lambda$ with Barr equation

The last term of Barr equation (6-38) is substituted by

$$\frac{5.1286}{Re^{0.89}} = 5.1286\left(\frac{\nu}{vd}\right)^{0.89} = 4.1365\left(\frac{\nu d}{q_v}\right)^{0.89}$$

Thus

$$\frac{1}{\sqrt{\lambda}} = -2\lg\left[\frac{\Delta}{3.7d} + 4.1365\left(\frac{\nu d}{q_v}\right)^{0.89}\right] \quad (8-5)$$

The equation above is available to calculate of smooth region, transition region and rough region.

### 2. Calculation of $C$ with Manning equation

Put $R = \dfrac{d}{4}$ and Manning equation $C = \dfrac{1}{n}R^{1/6}$ into the equation (6-45), then

$$\lambda = \frac{8g}{C^2} = \frac{12.7gn^2}{d^{1/3}} \quad (8-6)$$

Due to the substitution of empirical equation, dimensional homogeneity is not acquired, and units of the terms above are m or s. The equation above is only available for calculation of rough region of turbulent flow.

Most problems of calculation of constant diameter short pipe for solution to discharge capacity, diameter of pipe or water head, the three types of that can be solved by simultaneous equations above or iteration solution see Examples 8-1, 8-2. If incorporating Bernoulli equation, solution to pressure of flow section is also acquired.

**Example 8-1** The reinforced concrete inverted siphon pipe installed on the dam (with a smooth surface), as shown in Fig.8-3, has a length of $l = 200$ m a water level difference of $H = 8$ m between upstream and downstream, and various local loss factors at the inlet $\zeta_A = 0.5$, outlet $\zeta_D = 1.0$, elbow

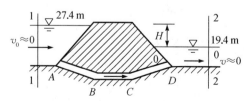

**Fig.8-3** Water supply short pipe

$\zeta_B = \zeta_C = 0.1$, calculate the required pipe diameter using a flow rate of $q_v = 25$ m³/s and $\nu = 1.011 \times 10^{-6}$ m²/s.

**Solution:** $\sum \zeta = \zeta_A + \zeta_B + \zeta_C + \zeta_D = 0.5 + 0.1 + 0.1 + 1.0 = 1.7$

According to the table, the reinforced concrete pipe (with a smoother surface) can be taken as $n = 0.013$, as shown in equations (8-6)

$$\sum \lambda \frac{l}{d} = \frac{12.7 g n^2}{d^{1/3}} \frac{\sum l}{d} = \frac{12.7 \times 9.8 \times 0.013^2 \times 200}{d^{4/3}} = \frac{4.207}{d^{4/3}}$$

The dimension in the above equation is one, so introducing empirical equations results in disharmonious dimensions.

The siphon is submerged for outflow, and since $v_0 \approx 0$, there is $H_0 = H$, which can be obtained from equation (8-2)

$$v = \sqrt{\frac{2gH}{\sum \lambda \frac{l}{d} + \sum \zeta}} = \sqrt{\frac{2 \times 9.8 \times 8}{4.207/d^{4/3} + 1.7}} = \frac{12.52}{\sqrt{4.207/d^{4/3} + 1.7}} \; (\text{m/s})$$

Thus

$$d = \left(\frac{4q_v}{\pi v}\right)^{1/2} = \left(\frac{4 \times 25}{\pi} \times \frac{\sqrt{4.207/d^{4/3} + 1.7}}{12.52}\right)^{1/2} = 1.594 \times \left(\frac{4.207}{d^{4/3}} + 1.7\right)^{1/4} \; (\text{m})$$

Write in the form of an iterative solution as

$$d_{(i+1)} = 1.594 \times \left(\frac{4.207}{d_i^{4/3}} + 1.7\right)^{1/4} \; (\text{m}) \quad (i = 0, 1, 2, \cdots)$$

Assuming the initial value $d_{(0)} = 1$ m, substituting it into the above equation yields $d_{(1)} = 2.485$ m, and iterating back and forth, we can obtain $d_{(2)} = 2.089$ m, $d_{(3)} = 2.144$ m, $d_{(4)} = 2.135$ m and $d_{(5)} = 2.137$ m. Taking the standard pipe diameter $d = 2.25$ m, equation (8-6) is only applicable to turbulent rough areas, so it is necessary to distinguish the resistance zone where the flow is located:

$$Re = \frac{4q_v}{\pi \nu} = \frac{4 \times 25}{\pi \times 2.137 \times 1.011 \times 10^{-6}} = 1.47 \times 10^7$$

Reinforced concrete pipe with a smooth surface, according to the table, it can be taken as $\Delta = 0.001$ m

$$\frac{\Delta}{d} = \frac{0.001}{2.139} = 0.000\,47$$

According to equations (6-41), $Re = 1.47 \times 10^7 > 576.12 \left(\frac{\Delta}{d}\right)^{-1.119} = 3.05 \times 10^6$, It is a turbulent rough zone. Explain the applicable conditions that satisfy equations (8-6).

**Example 8-2** The centrifugal pump shown in Fig.8-4 has a pumping flow rate of $q_v =$

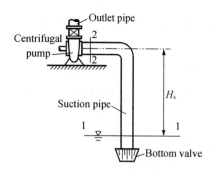

**Fig.8-4 Installation height of centrifugal pump**

306 m³/h, a suction pipe length of $l = 12$ m, a diameter of $d = 0.3$ m, and a loss factor along the way $\lambda = 0.016$, local loss factor: suction port with bottom valve $\zeta_1 = 5.5$, elbow $\zeta_2 = 0.3$. Allow the suction vacuum degree $[h_v] = 6$ m, and calculate the allowable installation height $H_s$ for this water pump.

**Solution:** Using the water surface of the suction tank as the reference plane, formulate the Bernoulli equation for the water surface of the suction tank 1-1 and the inlet section of the water pump 2-2.

$$\frac{p_a}{\rho g} = H_s + \frac{p_2}{\rho g} + \frac{\alpha v^2}{2g} + h_w$$

where

$$v = \frac{4q_v}{\pi d^2} = \frac{4}{\pi \times 0.3^2} \times \frac{360}{3\,600} = 1.20 \text{ m/s}$$

$$h_w = \left(\lambda \frac{l}{d} + \sum \zeta\right)\frac{v^2}{2g} = \left(0.016 \times \frac{12}{0.3} + 5.5 + 0.3\right) \times \frac{1.20^2}{2 \times 9.8} = 0.473 \text{ m}$$

Substituting $\frac{p_a - p_2}{\rho g} = [h_v] = 6$ m and $\alpha = 1$ into the above equation, it can be obtained that

$$H_s = \frac{p_a - p_2}{\rho g} - \frac{v^2}{2g} - h_w = 6 - \frac{1.20^2}{2 \times 9.8} - 0.473 = 5.45 \text{ m}$$

## 8.2 Hydraulic calculation of long pipe

Long pipe are classified as simple one and complex one. That diameter of pipes and flow rate in pipes are constant indicates simple pipes, while else of pipes belong to complex ones.

### 8.2.1 Simple pipe of long pipes

Calculation of simple one of long pipe as one of the simple long pipes shown in Fig.8-5 is the foundation of calculation of complex one of that. Building of Bernoulli equation is based on section 1-1 as well as section 2-2 after setting ground level 0-0. When velocity, head or local resistance are ignored in calculation of long pipe, only frictional head loss is considered, hence

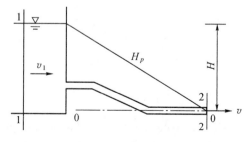

**Fig.8-5 Simple lines**

$$H = h_f$$

Put $h_f = \lambda \dfrac{l}{d} \dfrac{v^2}{2g}$, $v = \dfrac{4q_v}{\pi d^2}$ into the equation above, setting that

$$S = \dfrac{8\lambda}{g\pi^2 d^5} \qquad (8-7)$$

Then

$$H = Slq_v^2 \qquad (8-8)$$

where $S$ refers to resistance, determined by head needed when unit flow rate runs in unit length, with dimension as $[T^2 L^{-6}]$. By coupling equations (8 – 5), (8 – 7) and (8 – 8), mathematical model of calculation of simple pipe is acquired as follows:

$$\begin{cases} H = Slq_v^2 \\ S = \dfrac{8\lambda}{g\pi^2 d^5} \\ \dfrac{1}{\sqrt{\lambda}} = -2\lg\left[\dfrac{\Delta}{3.7d} + 4.1365\left(\dfrac{\nu_d}{q_v}\right)^{0.89}\right] \end{cases} \qquad (8-9)$$

The equation above is available for calculation of various regions in turbulent flow.

For flow in pipe as rough region in turbulent flow, $\lambda = \dfrac{12.7gn^2}{d^{1/3}}$ can be plug into the equation (8 – 7), thus resistance $S$ shall be calculated as:

$$S = \dfrac{10.3n^2}{d^{5.33}} \qquad (8-10)$$

Due to the introduction of empirical equation, dimension homogeneity cannot be acquired, unit of the terms in the equation above is m or s.

**Example 8 – 3** Water is supplied to the factory from the reservoir, as shown in Fig.8 – 6, using cast iron pipes with $n = 0.013$. Given a water consumption of $q_v = 300 \text{ m}^3/\text{h}$, a total pipeline length of 2 500 m, a reservoir water surface elevation of $z_1 = 87$ m, and a factory ground elevation of $z_2 = 42$ m, the service head (also known as free head, which is the pressure head that is still retained at the end of the water supply pipe under a certain flow rate as specified in the water supply design) $H_z = 25$ m. Calculate the diameter $d$ of the water supply pipe (flowing in a turbulent rough area).

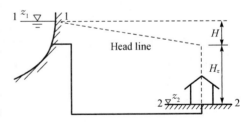

Fig.8 – 6 Factory water supply pipeline

**Solution:**

$$q_v = \dfrac{300}{3\ 600} = 0.083\ 3 \text{ m}^3/\text{s}$$

$$H = z_1 - (z_2 + H_z) = 87 - (42 + 25) = 20 \text{ m}$$

Calculate based on long and simple pipelines. As the flow is a turbulent rough region, according to equations (8-8) and (8-10), there are

$$S = \frac{H}{lq_v^2} = \frac{20}{2\,500 \times (0.083\,3)^2} = 1.153 \text{ s}^2/\text{m}^6$$

$$d = \left(\frac{10.3n^2}{S}\right)^{1/5.33} = \left(\frac{10.3 \times 0.013^2}{1.153}\right)^{1/5.33} = 0.296 \text{ m}$$

Take the standard pipe diameter, $d = 0.3$ m.

If calculated according to equation (8-9), there are

$$d = \left(\frac{8\lambda l q_v^2}{g\pi^2 H}\right)^{1/5}$$

$$\frac{1}{\sqrt{\lambda}} = -2\lg\left[\frac{\Delta}{3.7d} + 4.136\,5\left(\frac{\nu d}{q_v}\right)^{0.89}\right]$$

According to the table, at 20 ℃, the equivalent roughness of cast iron pipes $\Delta = 1.0 \times 10^{-3}$ m, $\nu = 1.01 \times 10^{-6}$ m²/s. In the above two equations only $\lambda$ and $d$ are unknown variables and they can be solved by using an iterative method. Assuming the initial value $d_{(0)} = 0.3$ m, the convergence solution obtained iteratively from the above two equations is $\lambda = 0.027\,7$, $d = 0.288$ m. The results of the two methods are basically consistent.

### 8.2.2 Long pipe in series

As the long pipe in series along the flow rate shown in Fig.8-7, the flow in which may either be constant or not. For pipe in series,

$$\begin{cases} q_{v1} = q'_{v1} + q_{v2} \\ q_{v2} = q'_{v2} + q_{v3} \\ q_{vi} = q'_{vi} + q_{vi+1} \end{cases} \quad (8-11)$$

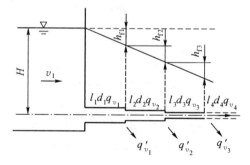

Fig.8-7  Pipe in series

For $k$ segments of longer pipes that are calculated as simple long pipe,

$$H = \sum_{i=1}^{k} h_{fi} = \sum_{i=1}^{k} S_i l_i q_{vi}^2 \quad (8-12)$$

For the flow in the rough region of pipe in series, the mathematical model of hydraulic calculation of which incorporates equation (8-10), equation (8-11) and equation (8-12),

$$\begin{cases} H = \sum_{i=1}^{k} S_i l_i q_{vi}^2 \\ S_i = \frac{10.3 n_i^2}{d_i^{5.33}} \\ q_{vi} = q'_{vi} + q_{vi+1} \end{cases} \quad (8-13)$$

where $i$ refers to number of the segment of pipe from upstream to downstream, $q'_{vi}$ refers branch flow rate of join node between $i$ segment of pie and $(i+1)$ segment of pipe, $q_{vi+1}$ refers to excess flow rate in $(i+1)$ segment of pipe.

The calculation of pipe in series usually aims at head $H$, flow rate $q_v$ or diameter of pipe $d$. Complementary equation is needed in some cases so as to acquire the solution of the terms above, see in Section 8.3.1.

### 8.2.3 Pipe in parallel

System for setting two or more pipes between two nodes is pipe in parellel, as illustrated in Fig.8-8.

The diameter, length and flow rate of each parallel pipes are not necessarily equal. But they all have the same initial head and head of pipe end, that is, the head loss of each pipe is equal represented by specific resistance and flow rate

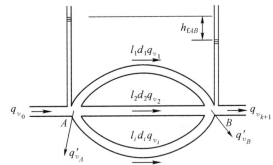

Fig.8-8 Parallel pipeline

$$h_{f1} = h_{f2} = \cdots = h_{fk} \quad (8-14)$$

Represented by specific resistance and flow rate, it is

$$S_1 l_1 q_{v1}^2 = S_2 l_2 q_{v2}^2 = \cdots = S_k l_k q_{vk}^2$$

The flow of any pipe section (express by subscript) is

$$q_{vi} = q_{v1} \left( \frac{S_1 l_1}{S_i l_i} \right)^{0.5} \quad (8-15)$$

Assumed there are $k$ parallel pipe sections, combined node flow rate equilibrium condition, we get

$$A: \sum q_{vA} = 0 \quad \text{or} \quad q_{v0} - (q_{v1} + q_{v2} + \cdots + q_{vk} + q'_{vA}) = 0 \quad (8-16a)$$

$$B: \sum q_{vB} = 0 \quad \text{or} \quad q_{v1} + q_{v2} + \cdots + q_{vk} - (q'_{vB} + q'_{vk+1}) = 0 \quad (8-16b)$$

The "$-$" in the equation above expresses outflow rate from the node.

For a parallel pipe in rough area, whose mathematical model of hydraulic calculation are got by getting equation (8-8), equation (8-10), equation (8-15) and equation (8-16) simultaneous, that is

$$q_{vi} = q_{v1} \left( \frac{S_1 l_1}{S_i l_i} \right)^{0.5}$$

$$q_{v0} = q'_{vA} + q_{v1} \sum_{i=1}^{k} \left( \frac{S_1 l_1}{S_i l_i} \right)^{0.5} \quad (8-17)$$

$$S_i = \frac{10.3n_i^2}{d_i^{5.33}}$$

$$h_{fAB} = S_i l_i q_{vi}^2$$

where $i$ is pipe segment number of pipe in parallel; $q_{v0}$ is flow rate of upstream nodes; $q'_{v0}$ is split flow rate of pipe in parallel in upstream node.

The calculation model can solve flow distribution of parallel pipe section and other hydraulic problems.

**Example 8-4** There is a water supply pipeline system connected in series, with the $D$-end being a free outlet, as shown in Fig.8-9. Given the pipes $AB$ flow rate $q_{v0} = 0.2$ m³/s, length $l_0 = 500$ m, diameter $d_0 = 0.35$ m, parallel pipes $BC$ have $d_1 = 0.25$ m, $d_2 = d_3 = 0.2$ mm, $l_1 = 800$ m, $l_2 = 1000$ m, $l_3 = 600$ m, $l_4 = 300$ m, $d_4 = 0.25$ m, point $B$ flow rate $q'_{v1} = 0.0295$ m³/s, and point $C$ flow rate $q'_{v2} = 0.0705$ m³/s. Calculate the total acting head $H$ of pipeline $AD$, the flow distribution of parallel pipes, and the head loss of section $BC$ (The flow is in the turbulent rough zone, $n = 0.013$).

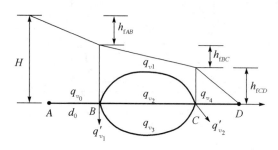

Fig.8-9 Parallel and Series Pipeline Systems

**Solution:** (1) Calculate the flow distribution of parallel pipelines first

$$S_1 = \frac{10.3n_1^2}{d_1^{5.33}} = \frac{10.3 \times 0.013^2}{0.25^{5.33}} = 2.82 \text{ s}^2/\text{m}^6$$

Similarly

$$S_2 = S_3 = 9.25 \text{ s}^2/\text{m}^6$$

Thus

$$q_{v2} = q_{v1}\left(\frac{S_1 l_1}{S_2 l_2}\right)^{0.5} = q_{v1}\left(\frac{2.82 \times 800}{9.25 \times 1000}\right)^{0.5} = 0.494 q_{v1}$$

Similarly

$$q_{v3} = 0.638 q_{v1}$$

Substitute $q_{v0} = 0.2$ m³/s and $q'_{v1} = 0.0295$ m³/s into equation (8-16)

$$q_{v0} - (q'_{v1} + q_{v1} + q_{v2} + q_{v3}) = 0$$
$$(0.2 - 0.0295) - q_{v1}(1 + 0.494 + 0.638) = 0$$

Solving the equation yields

$$q_{v1} = 0.08 \text{ m}^3/\text{s}, \quad q_{v2} = 0.0395 \text{ m}^3/\text{s}, \quad q_{v3} = 0.0510 \text{ m}^3/\text{s}$$

(2) Calculate the acting head $H$ from a series of pipes

$$H = h_{fAB} + h_{fBC} + h_{fCD} = S_0 l_0 q_{v0}^2 + S_1 l_1 q_{v1}^2 + S_4 l_4 q_{v4}^2$$

where $S_0 = \dfrac{10.3n^2}{d_0^{5.33}} = \dfrac{10.3 \times 0.013^2}{0.35^{5.33}} = 0.469$; $S_4 = S_1 = 2.82$; $l_0 = 500\,\text{m}$, $l_1 = 800\,\text{m}$, $l_4 = 300\,\text{m}$, $q_{v_0} = 0.2\,\text{m}^3/\text{s}$, $q_{v_1} = 0.08\,\text{m}^3/\text{s}$, $q_{v_4} = q_{v_0} - q'_{v_1} - q'_{v_2} = (0.2 - 0.0295 - 0.0705) = 0.1\,\text{m}^3/\text{s}$.
Substituting them into the above equation yields

$$h_{\text{f}AB} = 9.38\,\text{m},\ h_{\text{f}BC} = 14.44\,\text{m},\ h_{\text{f}CD} = 6.84\,\text{m},\ H = 30.66\,\text{m}$$

The total acting head is 30.66 m, and the head loss of each branch pipe in the parallel pipeline is 14.44 m.

### 8.2.4 Linear uniformity drain line

The pipe flow rate discussed earlier is constant along the way, the outflow is concentrated at the end of the pipe segment, which is called transmission flow rate. In practical engineering, in the irrigation pipeline, the artificial rainfall pipeline and the filter flushing pipe, besides transmission flow rate in these pipes, there is flow rate lapaxis along the distance $q_{vt}$, called way leak flow rate. Generally speaking, transmission flow rate is inhomogeneous, of which the simplest is the flow rate of very section and length is equal. This is called uniform flow along the way. In order to simplify the calculation, this is often used as a continuous process.

As is illustrated in Fig. 8 – 10, assumed length of uniform flow along the way is $l$, diameter $d$. Therefore, the whole flow rate of the uniform flow along the way is $q_{v_t}$, transmission flow rate at the end of section is $q_{v_z}$. At the point $M$ of distance $x$ from the start point of the discharge point is $A$. Take a micro tube segment $\text{d}x$ and $\text{d}x$ is very small so the flow rate of it ($q_{v_x}$) can be thought to be constant. And it's head loss can be calculated according to simple lines.

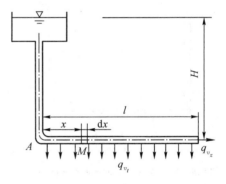

Fig.8 – 10  Uniform flow along the way

The above integral equations

$$q_{v_x} = q_{v_z} + q_{v_t} - \dfrac{q_{v_t}}{l} x$$

$$\text{d}h_{\text{f}} = Sq_{v_x}^2 \cdot \text{d}x = S\left(q_{v_z} + q_{v_t} - \dfrac{q_{v_t}}{l} x\right)^2 \text{d}x$$

$$h_{\text{f}} = \int_0^l \text{d}h_f = \int_0^l S\left(q_{v_z} + q_{v_t} - \dfrac{q_{v_t}}{l} x\right)^2 \text{d}x$$

When pipe section of the rough situation and the diameter of pipe keep the same, and in the area of resistance square, the specific resistance $S$ is a constant, integral equation above

$$h_{\text{f}} = Sl\left(q_{v_z}^2 + q_{v_z} q_{v_t} + \dfrac{1}{3} q_{v_t}^2\right) \tag{8-18}$$

The equation above can be similar to the written

$$h_f = Sl(q_{v_z} + 0.55q_{v_t})^2 = Slq_{vc}^2 \tag{8-19}$$

$q_{v_c}$ in the equation above is called calculated flow rate

$$q_{vc} = q_{v_z} + 0.55q_{v_t} \tag{8-20}$$

Equation (8-19) agreed with the simple pipeline calculation equation (8-8) $h_f = Slq_v^2$, so the calculation of the head loss of the uniform flow along the way can be conducted according to the simple lines.

Under special circumstances of $q_{v_z} = 0$, the head loss of the uniform flow along the way can be got by the equation (8-18)

$$h_f = \frac{1}{3} Slq_{v_t}^2 \tag{8-21}$$

The equation above indicated: when there is only uniform flow rate along the way, the head loss is one-third of that when the same number of transport flow rate went through.

**Example 8-5** The water transmission tower supplied by the water tower is shown in Fig.8-11, consisting of three sections of cast iron pipes, with the middle section being a uniform discharge pipe. Known: $l_1 = 500$ m, $d_1 = 0.2$ m, $l_2 = 150$ m, $d_2 = 0.15$ m, $l_3 = 200$ m, $d_3 = 0.125$ m, node B has a flow rate of $q_v' = 0.01$ m³/s, a discharge flow rate of $q_{v_t} = 0.015$ m³/s, and a transmission flow rate of $q_{v_z} = 0.02$ m³/s. Find the required height of the water tower (acting head).

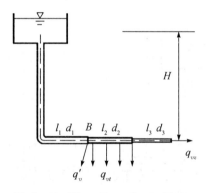

Fig.8-11 Water Tower Supply Pipeline

**Solution:** Convert the discharge flow rate of the second section of the pipeline to the calculated flow rate $q_{vc2}$, and the flow rate of each section of the pipeline is:

$$q_{v1} = q + q_{v_t} + q_{v_z} = 0.01 + 0.015 + 0.02 = 0.045 \text{ m}^3/\text{s}$$

$$q_{vc2} = q_{v_z} + 0.55q_{v_t} = 0.02 + 0.55 \times 0.015 = 0.028 \text{ m}^3/\text{s}$$

$$q_{v3} = q_{v_z} = 0.02 \text{ m}^3/\text{s}$$

The entire pipeline is composed of three pipe sections connected in series, so the acting head is equal to the sum of the head losses of each pipe section. For cast iron pipes, $n = 0.013$, as shown in equations (8-13)

$$H = \sum h_f = 10.3n^2 \left( \frac{l_1 q_{v1}^2}{d_1^{5.33}} + \frac{l_2 q_{vc2}^2}{d_2^{5.33}} + \frac{l_3 q_{v3}^2}{d_3^{5.33}} \right)$$

$$= 10.3 \times 0.013^2 \times \left( \frac{500 \times 0.045^2}{0.2^{5.33}} + \frac{150 \times 0.028^2}{0.15^{5.33}} + \frac{200 \times 0.02^2}{0.125^{5.33}} \right)$$

$$= 23.47 \text{ m}$$

## 8.3 Hydraulic calculation basis of pipe networks

The pipe network is composed of series and parallel pipelines and one example is the town's pipe network. In water supply network, a pipeline connected by a plurality of pipe sections is called a pipeline. Between any two nodes, there is only one pipeline in the network called tree network; Between any two nodes, there is at least two pipelines in the network called tree network. The hydraulic calculation of pipe network must satisfy the conditions of:

(1) Outflow and inflow of flow from either node, and its algebraic sum equals zero—continuity equation.

(2) In the pipe network of any closed pipe, the sum of head loss of the pipe section is equal to zero—Bernoulli equation.

(3) The total flow rate into the pipe network is equal to the sum of all node flow rate.

The tree and ring network are introduced later.

### 8.3.1 Tree networks

The hydraulic calculation of the tree network can be divided into two kinds of new and expanded.

#### 1. Hydraulic calculation of new water supply network

In the design of water supply network, after the completion of the pipe network layout, the length and flow rate of each pipe, node flow rate and elevation are usually told. We should determine the diameter of the pipe and the water pressure of the pipe network. And the method is:

(1) Draw pipe network diagram. Actual pipeline is very massive, it is unnecessary and sometimes impossible to calculate all the pipe lines. Therefore, the actual pipe network can be simplified. Main lines are reserved, while the minor line omitted. But the simplified pipe network should be able to response to the actual use of water.

(2) Number on nodes and pipe segments of pipe network. It is a tree pipe network as shown in Fig.8 – 12, 10 nodes in total. Each pipe segment, number on starting and stopping node encoding, for example "9, 10." Therefore, parameters like flow rate, length of pipes and specific resistance etc. can be distinguished with numbers. $q'_{v5}$ representatives the flow rate of the node 5 and $q_{v9,10}$ representatives the flow rate of the tubulation "9, 10."

(3) Calculate the flow rate of each pipe section according to the connection equation.

$$\sum \mp q_{v_{i,j}} - q'_{v_i} = 0 \qquad (8-22)$$

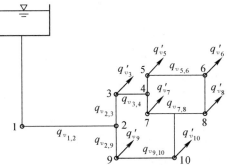

**Fig.8 – 12 Tree pipe network**

(Inflow is positive and outflow negative) Calculate $q_{v_i,j}$ from downstream to upstream according to the $q'_{v_i}$, we can determine flow rate of every tubulation.

(4) Determination of main dry pipe and control point of tree pipe network. We usually choose the point as control point which is far from the source of water, with a high terrain, high head demand and large water demand. Pipeline from water source to control point as main pipe. Line drawn from a main pipe is called branch pipe. The calculation of main pipeline of the new hydraulic pipe network should be the first.

(5) Determination of pipe diameter of main pipeline according to economic velocity. The economic velocity is when the cost is the lowest in the rated water supply flow rate, the pipeline cost is the lowest, and the technology is reasonable. The economic velocity is determined by the professional demand of economy and technology. In economic benefit, we should consider the investment and operation cost of pipeline and pumping station. In technology, water supply network to prevent water hammer caused by high pressure, limit the maximum flow rate is less than 2.5~3.0 m/s. Meanwhile, to avoid the deposition of impurities in water, the minimum flow velocity is 0.6 m/s. The determination of economic velocity is more complicated. In the water supply project, we can use the average economic velocity $v_e$ as the economic velocity of small pipe. When $d = 0.1 \sim 0.4$ m, $v_e = 0.6 \sim 0.9$ m/s; When $d > 0.4$ m, $v_e = 0.9 \sim 1.4$ m/s.

According to the economic velocity, determine the diameter according to the following equation

$$d_{i,j} = \sqrt{\frac{4q_{v_i,j}}{\pi v_e}} \tag{8-23}$$

where $q_{v_i,j}$ is the pipe flow rate of $i$ and $j$.

(6) Calculated the head loss in each pipe section by long tube and the head loss from starting point to control point by pipe in series.

(7) Determine the pump lift or tower height

$$H = \sum h_f + H_z + z_t - z_0 \tag{8-24}$$

where $H$ is the pump lift or tower height from the ground; $\sum h_f$ is total head loss of the pipe system; $H_z$ is service head required for control point; $z_t$ is control point elevation; $z_0$ is elevation of the starting point of the water tower network or pump suction sump elevation.

### 2. Hydraulic calculation of the existing pipe network expansion

The differences between hydraulic calculation of the existing pipe network expansion (or the new network in the branch) and the new network main pipe is that the pressure of starting point is known and we should only determine the diameters. In other words, expansion or extension of the pipe network calculation problem is that the whole head $H$, pipeline layout and flow rate of each pipe section are known and to determine the diameters. The method is:

(1) Calculate the average hydraulic gradient of each branch pipe. When the head and

length of both the starting and ending points are known, we can find out each average hydraulic gradient of each branch pipe $\overline{J}_{k,n}$

$$\overline{J}_{k,n} = \frac{H_k - H_n}{l_{k,n}} \qquad (8-25)$$

where $k$ is number of some starting point; $n$ is the same branch's ending point; $H_k$ is starting head; $H_n$ is ending head; $H_k$, $H_n$ is called water pressure elevation in water supply engineering, relative to the same datum.

(2) Calculate the pipe resistance ratio in branches

$$S_{i,j} = \frac{\overline{J}_{k,n}}{q_{v_i,j}^2} \qquad (8-26)$$

where $q_{v_i,j}$ is the flow rate of tubulation numbered $i$, $j$ in the branches starting and ending points numbered $k$, $n$.

(3) Calculate the corresponding pipe diameter according to each specific resistance. The obtained diameter is not necessarily the standard pipe diameter. We can select some of the standard diameters larger than the calculated diameter in practice, and other less than that to make the total hydraulic performance unchanged.

**Example 8 - 6** Fig.8 - 13 shows a diagram of a newly built water supply pipe network in a development zone. The pump suction sump elevation is $z_0 = 150$ m. The elevations of nodes 2, 4 and 6 are $z_{t2} = 153$ m, $z_{t4} = 155$ m and $z_{t6} = 154.5$ m respectively. The service heads of nodes 4 and 6 are $H_{z4} = 20$ m and $H_{z6} = 12$ m. Other known values are in the calculation table. Please try to calculate the pump lift H and the diameter of each pipe. It is known that pipe $n = 0.013$ (the head loss of the pump suction pipe is ignored).

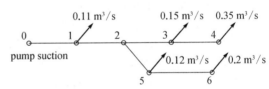

Fig.8 - 13  Example of tree pipe network calculation

**Solution:** (1) The 0 - 1 - 2 - 3 - 4 pipe will be used as the main pipe for main pipe hydraulic calculation, including diameter and head loss of each pipe. Taking pipe section 0 - 1 as an example, the economic velocity is 1.2 m/s. According to experience, the flow is in the rough region of turbulent flow, then

$$d_{0,1} = \sqrt{\frac{4q_{v0,1}}{\pi v_e}} = \sqrt{\frac{4 \times 0.93}{\pi \times 1.2}} = 0.993 \text{ m}$$

The standard pipe diameter is selected as $d_{0,1} = 1$ m and its specific resistance and head loss are calculated.

$$S_{0,1} = \frac{10.3 n_{0,1}^2}{d_{0,1}^{5.33}} = \frac{10.3 \times 0.013^2}{1} = 0.001\ 74 \text{ s}^2/\text{m}^6$$

$$h_{f0,1} = S_{0,1} l_{0,1} q_{v0,1}^2 = 0.001\ 74 \times 1\ 000 \times 0.93^2 = 1.505 \text{ m}$$

The calculation results of the remaining pipe sections are listed in the following table.

The water pump lift $H$ is

$$H = \sum h_f + H_{z4} + z_{t4} - z_0$$
$$= (1.505 + 1.642 + 1.456 + 3.245) + 20 + 155 - 150 = 32.85 \text{ m}$$

| Pipe | Pipe number | Known value | | | Calculated value | | Head loss $h_f/\text{m}$ |
| --- | --- | --- | --- | --- | --- | --- | --- |
| | | Pipe length $l/\text{m}$ | Flow rate $q_v/(\text{m}^3 \cdot \text{s}^{-1})$ | | Pipe diameter $d/\text{mm}$ | Specific resistance $S/(\text{s}^2 \cdot \text{m}^{-6})$ | |
| Main pipe | 0, 1 | 1 000 | 0.93 | | 1 000 | 0.001 74 | 1.505 |
| | 1, 2 | 800 | 0.82 | | 900 | 0.003 05 | 1.642 |
| | 2, 3 | 500 | 0.5 | | 700 | 0.011 65 | 1.456 |
| | 3, 4 | 1 000 | 0.35 | | 600 | 0.026 5 | 3.245 |
| Branch pipe | 2, 5 | 500 | 0.32 | | 450 | 0.117 2 | |
| | 5, 6 | 600 | 0.2 | | 400 | 0.30 | |

(2) The diameter of the branch pipe is calculated. Take branch pipe "2, 5" pipe section as an example:

The work head (water pressure elevation) of node 2 is calculated

$$H_2 = H_{z4} + z_{t4} + h_{f4,3} + h_{f2,3} = 20 + 155 + 3.245 + 1.456 = 179.701 \text{ m}$$

The work head of node 6 is

$$H_6 = H_{z6} + z_{t6} = 12 + 154.5 = 166.5 \text{ m}$$

The average hydraulic gradient of the branch pipe is

$$\overline{J}_{2,6} = \frac{H_2 - H_6}{l_{2,6}} = \frac{179.701 - 166.5}{500 + 600} = 0.012$$

The specific resistance of pipe section "2, 5" is

$$S_{2,5} = \frac{\overline{J}_{2,6}}{q_{v2,5}^2} = \frac{0.012}{0.32^2} = 0.117\,2 \text{ s}^2/\text{m}^6$$

The pipe diameter of pipe section "2, 5" is

$$d_{2,5} = \left(\frac{10.3n^2}{S_{2,5}}\right)^{\frac{1}{5.33}} = \left(\frac{10.3 \times 0.013^2}{0.117\,2}\right)^{\frac{1}{5.33}} = 0.454 \text{ m}$$

The standard pipe diameter is selected as 0.45 m. Similarly, the diameter of pipe section "5, 6" can be calculated, which is listed in the table.

## 8.3.2 Loop pipe networks

Loop system is an extension of the parallel line (as showed in Fig.8 – 14), its main role is to improve the reliability of pipeline network of water supply. According to the solving conditions, the hydraulic calculation of pipe network can be divided into three categories, such as power network equation, one node equation, one pipe equation. Power network equation method is described as followed.

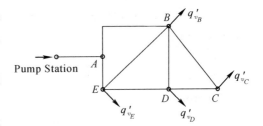

Fig.8 - 14  Ring network

The design of loop pipe network is first according to the engineering requirements and local conditions to fix up the whole pipe network pipeline, to determine the length of each section and leading outward flow rate of each node, then through the annular pipe network hydraulic calculation to determine the flow rate $q_v$, pipe diameter and the relevant head loss, to acquire the required pressure of water supply system. Its pipe diameter can be determined by the economic velocity. According to the characteristic of ring pipe network flow, water flow must meet two conditions that is continuity and energy balance in every closed loop pipe network.

(1) Continuity condition should be satisfied on the node, that is to say, the flow rate of the node to the node should be equal to the flow rate of the node

$$\sum q_{v_i} = 0 \tag{8-27}$$

If set $n_p$ as the number of nodes, $n_p - 1$ independent equations can be established by this. Since the flow balance equation of the $n_p - 1$ node is not an independent equation, it can be derived from $n_p - 1$ equation.

(2) Energy balance. In any closed loop of pipe network, we assume the head loss caused by the flow in the clockwise direction is positive, the water head loss caused by the water flow in the reverse clock direction is negative, thus the algebraic sum of the two should be equal to zero in each ring:

$$\sum h_f = \sum S_i l_i q_{v_i}^2 = 0 \tag{8-28}$$

Above a total of $n_k$ equation, with the equation (8-27), total number is $n_q = n_k + n_p - 1$, to analysis the pipe network diagram, we can know $n_q$ is just the number of pipe network. As Fig.8 - 14 shows, cylinder number $n_k = 3$, number of nodes $n_p = 6$, number of the pipe network number $n_q = 3 + 6 - 1 = 8$. That shows the number of equations is equal to that of the flow rate in each pipe section, it can be solved.

It is difficult to directly solve the flow rate of each pipe section by the above nonlinear equations, we use the asymptotic method to solve it. First of all, it is necessary to estimate the flow distribution in the pipeline, but they often do not meet the conditions of closure, error is inevitable. Therefore, we need to correct the initial assessment of the flow distribution. Assuming that the flow rate of any segment of the ring is $q_{v_i}$, the error is $\Delta q_{v_i}$, the true value of $h_{fi}$ is

$$h_{fi} = S_i l_i (q_{v_i} + \Delta q_{v_i})^2 = S_i l_i (q_{v_i}^2 + 2 q_{v_i} \Delta q_{v_i} + \Delta q_{v_i}^2)$$

Ignoring the two trace $\Delta q_v^2$, and thinking that $\Delta q_v$ is the same for each line, we can get the equation:

$$\sum h_{fi} = \sum S_i l_i q_{vi}^2 + 2\Delta q_v \sum S_i l_i q_{vi} = 0$$

Therefore

$$\Delta q_v = -\frac{\sum S_i l_i q_{vi}^2}{2\sum S_i l_i q_{vi}} = -\frac{\sum h_{fi}}{2\sum S_i l_i \frac{q_{vi}}{q_{vi}^2}} = -\frac{\sum h_{fi}}{2\sum \frac{h_{fi}}{q_{vi}}} \tag{8-29}$$

If the result is positive, which shows the clockwise flow of pipe flow increased by $\Delta q_v$, counter clockwise flow of pipe flow rate reduced by $\Delta q_v$. If the result is negative, then the result is negative.

There are many kinds of hydraulic calculation methods of ring network, the above method is called the head balance method. The calculation steps of the calculation of the annular pipe network using the head balance method are as follows:

(1) Estimate pipeline flow rate, and make the nodes meet the requirements of type (8-27), that is $\sum q_{vi} = 0$.

(2) The pipe diameter is determined by economic velocity $v_e$, $d_{i,j} = \sqrt{\frac{4q_{vi,j}}{\pi v_e}}$, and according to the calculated value, selecting a appropriate standard diameter.

(3) Based on the estimated flow rate to calculate the head loss of each pipeline (only calculate the head loss along the way).

(4) Check loop is satisfied with equation (8-28) or not, that is $\sum h_{wi} = \sum h_{fi} = 0$. If not satisfied, then according to equation (8-29) to calculate the correction flow rate $\Delta q_v$, and correct the initial estimate of the flow rate $q_v$. Repeat the above calculation until the error reaches the required precision.

(5) The head of each node is determined by the calculation of the head loss of each pipe segment. Then get the pump head or the height of the water tower.

### 8.3.3 The characteristics of pump and pipeline system

The actual water consumption of the water supply pipeline system and the design water consumption are often different, and as the pump head changes with the flow rate, the head loss of pipeline system also changes with the flow rate. The problem of coordination and balance between flow rate and water pressure in the pipeline system is existed when the water is pumped directly to the pipeline system. This requires an understanding of the hydraulic characteristics of the pump and the hydraulic characteristics of the piping system. The following is a brief introduction.

#### 1. Hydraulic characteristics of pump

The relationship curves of the pump water flow rate $q_v$ with the head $H$, efficiency $Y$, shaft power $N$, respectively is called the pump performance curve. The $q_v - H$ curve of the pump is called the hydraulic performance curve of the pump. As shown in Fig.8 – 15 is curve of centrifugal pump which commonly used in water supply pipe network. In the design of pipe network, the highest efficiency of pump hydraulic parameters is generally selected as the value of pipe network designation. But the actual water use in the pipe network will be varied in different periods, we can get some useful information by analysis the graph, that is, the smaller the centrifugal pump pump out the water, the higher the pressure, which shows that the water head of pipe network is different with the change of water consumption.

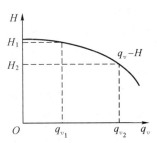

Fig.8 – 15  Centrifugal pump hydraulic performance curve

### 2. Hydraulic characteristics of piping system

The relation curve of flow rate and head loss of pipeline system is called the performance curve of pipeline system. For long pipe systems, the head loss value of a series of flow rate can be calculated by the equation (8 – 8), which can be draw the curve of $q_v - \sum h_f$, which can also be obtained through experiments, as Fig.8 – 16 shows

$$H_g = H_z + (z_1 + z_0)$$

where $z_t$ is the control point elevation; $z_0$ is the height of the water pump suction sump.

From the analysis of figure we can know, for a certain pipeline system, the greater the flow rate, the greater the head loss, then the pump is required to provide the greater the role of head.

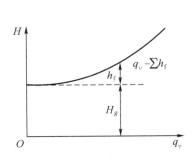

Fig.8 – 16  Line performance curve

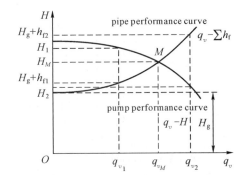

Fig.8 – 17  Hydraulic coupling relationship between pump and pipe

### 3. Hydraulic coupling condition of pump and pipeline system

The hydraulic performance curve of pump and pipeline system is plotted on the same coordinate chart with the same ratio. $q_v - H$ curve and $q_v - \sum h_f$ curve will have a cross point $M$, as Fig.8 – 17 shows. Under the condition of crossing point $M$ in the graph, the head of the pump is just to meet the loss head of the pipeline system, the service head of the control point and the geometry of the water supply, this condition is just the design of pump pipeline system. $q_{vM}$ is the design flow rate, $H_M$ is the design head.

There always be export control valve in the control node of water supply network, which can regulate water consumption. It is assumed that the head loss of the regulating valve is not included in the performance curve of the pipeline, that's to say, the performance curve of the pipeline can not be changed when the valve is adjusted. The pressure change in the pipe when the water quantity is adjusted is analyzed following. When the valve is down, the flow rate of the pipe reduces, pump head and pump outlet pressure increases, rather the loss head of the pipeline is reduced, which makes the remaining head of the control point in the end of the pipe improved; if the pipeline flow rate increases, the results is opposite.

Based on the above analysis, for water supply network, thinks of the great change of the actual flow rate and expects the pressure of the pipe network to be stable, the pump's hydraulic performance curve be relatively flat will satisfy the requirements.

## 8.4 Water hammer in a pressure pipe

What we discuss above is constant flow which has nothing to do with time $t$. In this section we will discuss the unsteady flow of a pressure pipe which is related to $s$ and $t$.

### 8.4.1 The phenomenon of water hammer

#### 1. The propagation process of water hammer

As Fig. 8 - 18 shows, a pressure pipeline is connected with a reservoir, the valve is used to adjust the flow rate at the end of the pipe. The length of the pipeline is $l$ and diameter $d$. In the normal condition the valve opening remains unchanged, and the flow in the pipeline is constant. The flow rate, the velocity and the pressure of the valve is $q_v$, $v_0$ and $p_0$, respectively.

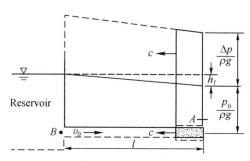

**Fig.8 - 18 Water hammer**

When the valve is shut down, water hammer boost $\Delta p$ will be generated in the valve and spread in the pipeline with speed $c$. A series of processes of water hammer wave propagation can be divided into four stages. Due to the phenomenon of water hammer, elastic force and inertia force play a major role, head loss and flow velocity are negligible when analysis the water hammer phenomenon which does not affect the cycle characteristics of water hammer wave propagation.

(1) $0 < t \leqslant \dfrac{l}{c}$, stage of boost back wave (Fig.8 - 19a).

$\dfrac{l}{c}$ stands for the time that the water hammer wave passes through the full length of the pipeline $l$. $0 < t \leqslant \dfrac{l}{c}$ shows a stage of water hammer wave from the front of the valve to import $B$.

When the valve $A$ is closed, the tiny water layer, which is close to the valve in the pipe, is $dl$, stops flowing. Its velocity reduced from $v_0$ to 0. The fluid is compressed by the inertia force, which makes the pressure increases from $p_0$ to $p_0 + \Delta p$. Under the action of the pressure increased suddenly, the small water layer is compressed and the density increases. Tube wall will expansion simultaneously. The phenomenon above is called propagation phenomena of water hammer wave.

This kind of water hammer wave is called the pressure wave.

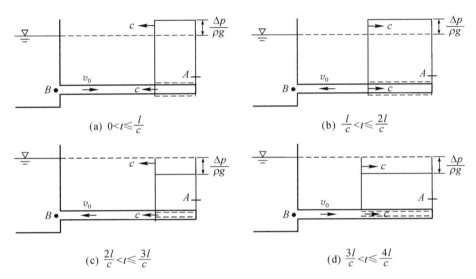

Fig.8 – 19  Pressure water hammer in pipeline

(2) $\dfrac{l}{c} < t \leqslant \dfrac{2l}{c}$, stage of reduced pressure wave (Fig.8 – 19b)

Because the reservoir water level will not change because of water hammer in the pipe, at the time $t = \dfrac{l}{c}$, the pressure at the left side of the inlet section of the pipeline is maintained $p_0$ at a constant current of $B$, the pressure for after water hammer impaction is $p_0 + \Delta p$, this indicates that the water body is not balanced. Because of the pressure difference, the water in the pipe network flows from the static state to the direction of the reservoir with the velocity of $v_0$. So the water body of $B$ point is restored to its original state first and its pressure decreases from $p_0 + \Delta p$ to $p_0$. Then the water and the pipe wall have been restored to its original state. Pressure is also reduced to $p_0$. Which is also a water hammer that is transmitted by the pipeline inlet to the valve. It's called reduced pressure wave.

Due to consistency of the compressibility of water and elasticity of the pipe wall, it is also concluded that the reflection wave velocity is $c$, that's to say, when $t = \dfrac{2l}{c}$, reflected wave reaches the downstream valve, usually this time is called water hammer constructive, we use $T_r$ to show it, $T_r = \dfrac{2l}{c}$. The period of time, from $t=0$ to $t=\dfrac{2l}{c}$, is called the head of a phase.

(3) $\dfrac{2l}{c} < t \leqslant \dfrac{3l}{c}$, stage of pressure relief (Fig.8 - 19c).

At the end of the first phase, all parameters have been restored. Due to the inertia effect, water layer adjacent to the valve $A$, flow towards the reservoir with speed $v_0$ consistence. As the valve is fully closed, the water can not be restored completely, so the water layer close to the valve has a tendency stay away from the valve. Due to the requirement of continuity, the flow ceases, velocity decreased from $v_0$ to 0, and the pressure decreased from $p_0$ to $\Delta p$, which causes the water body to expand, the density reduces, the tube wall contraction. This phenomenon since the valve layer by layer has been passed to the pipe inlet in the same way, which means a buck reflected wave was generated at the valve. At the time $t = \dfrac{3l}{c}$, the whole water body is in a static state, and the wall of the pipe is contracted, the pressure decreased from $p_0$ to $\Delta p$.

(4) $\dfrac{3l}{c} < t \leqslant \dfrac{4l}{c}$, stage of boost wave (Fig.8 - 19d).

At the time $t = \dfrac{3l}{c}$, the whole water body is in a state of static, decompression and expansion. The pressure at the left side of the pipeline $B$ points is $p_0$, while the right pressure is $p_0 - \Delta p$, the force is not balanced. Under the action of pressure difference $\Delta p$, water flows to the valve with a velocity of $v_0$, the expanded water body is compressed and the pressure recovered to $p_0$, the contracted tube wall recovers immediately. At the time $t = \dfrac{4l}{c}$, wave spread to the valve. The flow rate, pressure and wall of the whole pipe are restored to the state before the water hammer, the velocity of the whole tube is $v_0$, the pressure is $p_0$.

The period of time, from $t = \dfrac{2l}{c}$ to $t = \dfrac{4l}{c}$, is called the water hammer.

To sum up, quick opening and closing of valve is an important reason of water hammer. Water hammer pressure is not possible in the whole tube at the same time due to the compression of the water and the flexibility of the pipe wall, this is a process, and which makes water hammer pressure is limited.

### 2. Direct water hammer and indirect water hammer

It is assumed that the valve is suddenly closed simple occurred in pressure pipeline water hammer above discussion. In fact, opening and closing of the valve is not instantaneous, this needs a period of time as $t_c$. In the process of closing the valve, the pressure at the valve is rising. Which can be seen as the valve continuously produce a water hammer pressure wave, and spread to the reservoir, when the first pressure wave reached the reservoir after a period of $t=\dfrac{l}{c}$, a reduced pressure wave reflected back immediately. When the pressure relief meets with other pressure wave, the pressure in the pipe must be reduced. As a result, the water hammer is divided into direct water hammer and indirect water hammer.

Under normal circumstances, when the pipe valve suddenly closed, the velocity in the pipe is decreased rapidly, water hammer with obviously increased, we call it positive water hammer, which can cause water pipe burst. While for the pipeline valve open quickly, the flow rate in the pipe increases rapidly, the pressure is significantly reduced, this water hammer is called a negative water hammer. Negative water hammer can cause hollow and cavitation in the pipe, even can cause pipeline depression.

## 8.4.2 Calculation of water hammer

### 1. Propagation velocity of water hammer wave

$$c=\dfrac{c_0}{\sqrt{1+\dfrac{K_0}{K}\dfrac{D}{\delta}}}=\dfrac{1\,435}{\sqrt{1+\dfrac{K_0}{K}\dfrac{D}{\delta}}}\,(\text{m/s}) \qquad (8-30)$$

where $c_0$ is the speed of sound waves in water, $c_0 = 1\,435$ m/s, when the water temperature is about 5 ℃ and the pressure is between 1~25 atm; $K_0$ is the bulk modulus of water, $K_0 = 2.06 \times 10^3$ MPa, when the water temperature is about 5 ℃, $K$ is the elastic modulus of the pipe; $D$ is the pipe diameter, m; $\delta$ is the pipe wall thickness, m.

### 2. Calculation of water hammer pressure

Set valve closed instantaneously, means $v=0$, the calculation equation of the maximum value of direct water hammer pressure is

$$\Delta p = \rho c v_0 \qquad (8-31)$$

or

$$\dfrac{\Delta p}{\rho g} = \dfrac{cv_0}{g} \qquad (8-32)$$

Indirect water hammer interaction due to the positive water hammer wave and reflected wave, it is hard to calculate, approximate equation is

$$\Delta p = \rho c v_0 \dfrac{T_r}{T_z} \qquad (8-33)$$

or

$$\dfrac{\Delta p}{\rho g} = \dfrac{cv_0}{g}\dfrac{T_r}{T_z} = \dfrac{2l}{T_z} \qquad (8-34)$$

where $T_z$ says the time of valve closed; $v_0$ indicates the mean velocity before water hammer; $T_r = \dfrac{2l}{c}$ indicates the water hammer wave length.

### 8.4.3 Prevention of water hammer damage

(1) Set air room or install water hammer eliminating valve.

(2) Set the pressure regulating tower or tank.

(3) Extend the valve closing time, shorten the length of the pipeline, reduce the velocity in the tube, which are useful methods.

# Exercises

1. As shown in Fig.8-20, the water level difference between the two water tanks is $H=8$. If two parallel pipes with the same elevation are arranged between the two water tanks, one of which has a diameter of $d_1 = 50$ mm, another pipeline diameter $d_2 = 100$ mm. The length of the two pipes is equal, i.e. $l_1 = l_2 = 30$ m.

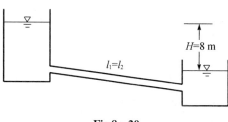

**Fig.8-20**

(1) Try to find the flow rate through each pipeline.

(2) Change to arrange a pipeline with the same length and total required flow rate, and calculate the diameter of the pipeline. Let all local head loss factors in each pipeline be $\sum \zeta = 0.5$, the resistance factors along the pipeline are all $\lambda = 0.02$.

2. A pipeline is divided into two branch pipes $A$ and $B$ at a certain point, and then re merged. The $A$ pipe is galvanized, $n=0.011$, 1 500 m long, and 150 mm in diameter. $B$ pipe is a cast iron pipe, $n=0.013$, with a diameter of 200 mm. Request that the flow rate through pipes $A$ and $B$ be equal, and try to determine the length of pipe $B$.

3. As shown in Fig. 8-21, two water pools $A$ and $B$ converge through pipes $AE$ and $BE$ at point $E$, and then supply water to point $C$ through the $EC$ pipeline, where it flows into the atmosphere. The water level of pool $A$ is 36 m, pool $B$ is 40 m, and the elevation of water supply point $C$ is 10 m. It is required to calculate the flow rate through each pipe.

The data for each pipeline is shown in the table below.

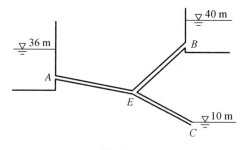

**Fig.8-21**

| Pipeline number | Length/m | Diameter/mm | λ |
| --- | --- | --- | --- |
| AE | 300 | 300 | 0.024 |
| BE | 450 | 375 | 0.02 |
| CE | 60 | 450 | 0.024 |

4. The tree network is shown in Fig.8 – 22. It is known that the ground elevation of the water tower is $z_A = 15$ m, the elevations of the pipe network terminal points $C$ and $D$ are $z_C = 20$ m, and $z_D = 15$ m. All service heads $H_z$ are 5 m. $q'_{vC} = 0.02$ m³/s, $q'_{vD} = 0.0075$ m³/s, $l_1 = 800$ m, $l_2 = 700$ m, $l_3 = 500$ m, $n = 0.013$. Please try to design the height of the water tower and the pipe diameters of sections $AB$, $BC$ and $BD$.

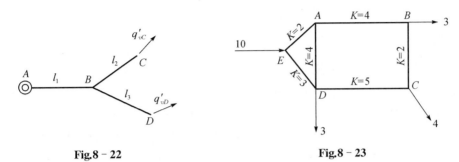

Fig.8 – 22    Fig.8 – 23

5. As shown in Fig.8 – 23, there are quadrilateral pipe network $ABCD$ and triangular pipe network $ADE$, and the two pipe networks are connected by $AD$. 10 units of flow rate will flow into point $E$, and 3, 4, and 3 units of flow rate will flow out of points $B$, $C$, and $D$ respectively. Other information about the pipeline is marked in the figure. Please try to find the flow rate through each pipe section, and indicate the direction of the flow rate in the figure ($K = Sl$, $h_f = Kq_v^2$).

# Chapter 9 Open channel flow

Artificial channels, natural rivers, and pipelines that are not filled with water are collectively referred to as open channels. Open channel flow is a flow with a free surface, where each point on the free surface is affected by the local atmospheric pressure, and its relative pressure is zero, so it is also known as an pressure-free flow. Unlike pressurized pipe flow, gravity is the main driving force of open channel flow, while pressure is the main driving force of pressurized pipe flow. Open channel water flow is divided into constant flow and unsteady flow based on whether its hydraulic elements change over time. The constant flow in open channels is divided into uniform flow and non-uniform flow based on whether the streamline is parallel to a straight line. The theory of open channel flow will provide scientific basis for the design and operation control of water conveyance, drainage, and irrigation channels.

This chapter mainly introduces the formation conditions and characteristics of steady uniform flow in open channels, as well as the hydraulic calculation of uniform flow in open channels, including channel bottom slope, allowable velocity, hydraulic optimal section, hydraulic optimal section conditions, and basic equation for uniform flow in open channels; Characteristics, hydraulic elements, filling degree, and calculation of uniform flow force in non pressurized circular pipes. Then discuss the three flow patterns of steady non-uniform flow in open channels: slow flow, critical flow, and jet flow, as well as the discrimination of the three flow patterns; The concepts of water jump and water drop, as well as the analysis of non-uniform gradient flow surface curves in prismatic channels. The teaching focus is on the basic concepts and hydraulic calculations of steady uniform flow in open channels, as well as the concepts of three flow states of steady non-uniform flow in open channels; The teaching difficulty lies in the analysis of the non-uniform gradient flow surface curve of prismatic channels.

## 9.1 Classification of open channel

### 9.1.1 Prismatic channel and non-prismatic channel

Open channel can be divided into prismatic and non-prismatic. Sectional shape and size remains constant called prismatic channel. The area of cross section only changes with depth. Sectional shapes and sizes changed constantly called non-prismatic channel. The area of cross section not only changes with depth, but also changes with position.

There are various forms of open channel section. The sectional of natural rivers is generally irregular in shape, but the shape of the artificial channels sectional are generally rectangular, trapezoidal, circular, etc, as show in Fig.9 – 1.

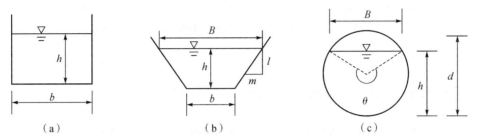

Fig.9 – 1  Sectional shape of artificial channels

### 9.1.2  Downhill slope, horizontal slope and adverse slope

The bottom of open channel generally slopes downward slightly. On the longitudinal section, channel bottom will be a straight line called channel bottom line. The elevation difference between channel bottom line and flow along the process in unit length is called bottom slope or longitudinal slope of the channel, that is

$$i = \frac{\Delta z}{l'} = \sin\theta \qquad (9-1)$$

where $\theta$ is the angle between channel bottom and a horizontal line, as shown in Fig.9 – 2.

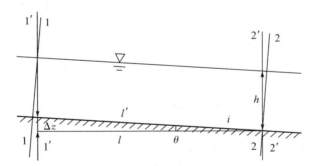

Fig.9 – 2  Channel bottom slope

In pratical engineering, bottom slope of open channel is small. So $\sin\theta \approx \tan\theta$,

$$i = \tan\theta = \frac{\Delta z}{l} \qquad (9-2)$$

As shown in Fig.9 – 3, depending on the value of channel bottom slope, the channel is usually divided into downhill slope, horizontal slope and adverse slope. Channel along the way reduce called downhill slope ($i>0$). Channel bottom is horizontal channels called horizontal slope ($i=0$). Channel bottom along the way increased called adverse slope ($i<0$).

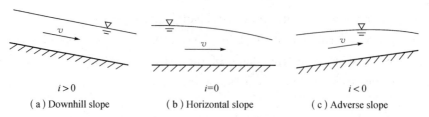

Fig.9 - 3 Downhill slope, horizontal slope and adverse slope

## 9.2 Uniform open channel flow

### 9.2.1 The characteristics and conditions of uniform open channel flow

**1. The characteristics of uniform open channel flow**

Uniform open channel flow has characteristics of both uniform flow and gravity flow. Its streamline is parallel to each other, and all liquid particles move uniformly in a straight line along the same direction. The combined external force in the direction of water flow is zero. From this, it can be inferred that uniform open channel flow has following characteristics:

(1) The elements of movement remains constantly along the process such as depth, velocity distribution, average velocity, flow rate and so on.

(2) The pressure of flow cross-section distributes by the hydrostatic pressure.

(3) Total head line, water surface profile (piezometric head line) and channel bottom line are parallel to each other. So, the hydraulic slope $J$, water surface slope $J_p$ and bottom slope $i$ remain unchanged and equal along the process.

(4) Open channel uniform flow is the flow in which the component of gravity in the flow direction is balanced by the flow resistance.

**2. Forming conditions of uniform open channel flow**

Due to the aforementioned characteristics of uniform open channel flow, its formation requires certain conditions to be met.

(1) It must be a long and straight prismatic channels with sectional shape and size, bottom slope $i$ and roughness $n$ unchanged along the process.

(2) The channel must be a downslope channel ($i>0$). Uniform flow cannot be formed in flat-slope channels or reverse-slope channels.

(3) The water flow must be steady, and flow rate does not change, and there is no water flowing out or merging along the process.

(4) There is no local interference from structures such as gates, dams or waterfalls in the channel.

These conditions cannot meet in natural channels, its sectional slope and size, bottom slope, and roughness will change on the way, so natural channels are usually non-uniform flow.

### 9.2.2 Foundational equation of uniform open channel flow

The basic equation of hydraulic calculation used Chézy equation, that is

$$v = C\sqrt{RJ} = C\sqrt{Ri} \tag{9-3}$$

Based on the continuity equations and Chézy coefficient, flow rate equation of open channel flow is

$$q_v = AC\sqrt{RJ} = AC\sqrt{Ri} = K\sqrt{J} = K\sqrt{i} \tag{9-4}$$

In the equation, $K = AC\sqrt{R}$, called flow rate modulus, has a dimension of flow rate, which represents the flow rate passing through a prismatic channel with a certain cross-sectional shape and size when bottom slope $i$ is equal to 1. Chézy coefficient is determined by the empirical equation of Manning, $C = \dfrac{1}{n}R^{1/6}$.

### 9.2.3  Optimal hydraulic cross section and the allowable velociy

#### 1. Optimal hydraulic cross section

The conveyance capacity of the open channel flow depends on channel bottom slope, roughness, shape and size. Water and drainage capacity of a channel refers a certain uniform flow depth adopted. When designing channels, the bottom slopes is determined by the terrain conditions or other general technical requirements, and roughness mainly depends on the material of channel. From a design perspective, it is hoped that the minimum flow cross-sectional area can be obtained at a certain flow rate to reduce engineering workload and save investment; another way is under certain conditions of flow cross-sectional area, roughness coefficient, and bottom slope, the flow rate through the channel can be maximized. Any cross section form that meets these conditions is called optimal hydraulic cross section. Optimal hydraulic cross section refers to that when the bottom slope, roughness and cross-sectional area is constant, the flow rate is largest in cross section. Geometry area is constant, the minimum wetted perimeter, the largest hydraulic radius of shape is round. In a variety of cross-sectional shape, the hydraulic radius of semicircular cross section is same with circular. So semicircular cross section is optimal hydraulic cross section.

By the Manning equation available:

$$q_V = AC\sqrt{Ri} = A\frac{1}{n}R^{2/3}i^{1/2} = \frac{1}{n}\frac{A^{5/3}}{\chi^{2/3}}i^{1/2} \tag{9-5}$$

The above equation shows that the smaller the wetted perimeter $\chi$ is, the greater the water carrying capacity is when $i$, $n$ and $A$ are constant. When the geometric area is fixed, the wetted perimeter is minimum and the hydraulic radius is maximum. For the open channel, the semicircle section is the most hydraulic, but the emicircle section is not easy to construct and is only used in hydraulic structures such as concrete aqueducts.

In engineering, trapezoidal cross section is used mostly. Its slope coefficient $m$ is determined by the slope stability requirements. According to optimal hydraulic cross section, when the cross-section area is fixed and the wetted perimeter is minimum, the flow rate is maximum.

$$\frac{dA}{dh} = 0 \text{ and } \frac{d\chi}{dh} = 0$$

For trapezoidal open channel, its cross-sectional area is

$$A = h(b + mh)$$

then

$$b = \frac{A}{h} - mh \tag{9-6}$$

Wetted perimeter of trapezoidal open chanal is

$$\chi = b + 2h\sqrt{1+m^2} = \frac{A}{h} - mh + 2h\sqrt{1+m^2} \tag{9-7}$$

Derivation of depth $h$ is

$$\frac{d\chi}{dh} = -\frac{A}{h^2} - m + 2\sqrt{1+m^2} = -\frac{b+mh}{h} - m + 2\sqrt{1+m^2} = 0$$

$$\beta_m = \left(\frac{b}{h}\right)_m = 2(\sqrt{1+m^2} - m) \tag{9-8}$$

The ratio of width to depth in the optimal trapezoidal hydraulic cross section only related with slope coefficient.

The hydraulic radius of the trapezoidal section is

$$R = \frac{A}{\chi} = \frac{h(b+mh)}{b + 2h\sqrt{1+m^2}} \tag{9-9}$$

Simultaneous equations (9 - 8) and (9 - 9), yields

$$R_m = \frac{h}{2}$$

The hydraulic radius of the optimal trapezoidal hydraulic cross section is equal to half of the water depth and is independent of the slope coefficient. For rectangular cross sections, substituting $m = 0$ into equation (9 - 8), yields

$$\beta_m = 2$$

The bottom width $b$ of the optimal rectangular hydraulic cross section is twice the water depth $h$.

It is worth noting that the concept of hydraulic optimal section is only proposed from the perspective of hydraulics, and is not completely equivalent to technical and economic optimization. In engineering practice, it is also necessary to comprehensively consider and compare various conditions such as cost, construction technology, operation requirements and maintenance to select the most economical and reasonable section form.

### 2. Allowable velocity of channel

A well-designed channel should make the design velocity not be too large so that avoiding to suffer erosion. Design velocity can not be too small to prevent siltation. Therefore, when designing channels, in addition to considering the hydraulic optimal conditions and economic factors mentioned above, the average velocity $v$ of the channel cross section should also be kept within the allowable velocity range, that is

$$v_{max} > v > v_{min} \quad \text{or} \quad v_s > v > v_d$$

where $v_{max}$ is the maximum allowable velocity to avoid scouring, referred to as the no-scouring allowable velocity; $v_{min}$ is the minimum allowable velocity to avoid siltation, referred to as the no-siltation allowable velocity.

## 9.2.4 Mathematical model of trapezoidal open channel uniform flow and its solution

Simultaneous equations (9-5), (9-6) and (9-7), mathematical model of trapezoidal open channel uniform flow is

$$q_v = \frac{\sqrt{i}}{n} \frac{[h(b+mh)]^{5/3}}{(b+2h\sqrt{1+m^2})^{2/3}} \tag{9-10a}$$

$$v = \frac{\sqrt{i}}{n} R^{2/3} \quad \text{or} \quad v = \frac{q_v}{A} \tag{9-10b}$$

$$R = \frac{A}{\chi} \tag{9-10c}$$

In the above equations, there are six variables: $q_v$, $n$, $i$, $b$, $h$, and $m$. Generally, the slope coefficient $m$ and roughness coefficient $n$ can be determined based on soil conditions, while the other four variables are determined in advance according to engineering conditions, and then the other variable is solved. In practical engineering, there may be many variables to be solved, which can be difficult to solve. Therefore, the above parameters in the actual project may be the unknown quantity with solving difficulties, we need to build a new computer program to calculate equations.

(1) Checking water capacity of channel $q_v$, or determining channel bottom slope $i$, or calculating channel roughness $n$ are questions that can be solved by other five quantities, and it can be calculated by equation (9-10) directly.

(2) Channel cross-sectional dimensions $b$ and $h$ need to be decided.

① If we know $q_v$, $i$, $n$, $m$, $b$, then to solve the height $h$. Set the $h$ of $h^{5/3}$ in equation (9-10a) to the value to be calculated $h_{j+1}$, and the iterative equation for calculating $h$ is as follows:

$$h_{j+1} = \left(\frac{nq_v}{\sqrt{i}}\right)^{0.6} \frac{(b+2h_j\sqrt{1+m^2})^{0.4}}{b+mh_j} \quad (j=0, 1, 2, \cdots) \tag{9-11}$$

If $q_v$, $i$, $n$, $h$ and $m$ is known, the channel bottom width $b$ needs to be solved. Similarly, the iterative equation for solving $b$ is as follows:

$$b_{j+1} = \left[\frac{1}{h}\left(\frac{nq_v}{\sqrt{i}}\right)^{0.6}(b_j+2h\sqrt{1+m^2})^{0.4} - mh\right] \times b_j^{-0.3} \quad (j=0, 1, 2, \cdots) \tag{9-12}$$

The above equation is a weighted correction from accelerating convergence, and its general equation is $x_{j+1} = x_{j+1}^y \cdot x_j^{1-y}$. If the convergence of iterative solution was up and down around the true solution convergence, the exponent value $y$ is less than 1. If convergence of iterative solution is unilateral asymptotic convergence in the true value, the exponent value $y$ is more than 1. It is significant effect that weighted exponent accelerate the convergence of iterative solution. The iterative equations (9-11) and (9-12) converge quickly. Generally, 3 or 4 iterations are sufficient to meet the engineering accuracy requirements.

② If we know $\beta = b/h$, then to solve the height $h$ and $b$.

The ratio of width to depth in small channel is $\beta = \beta_m = 2(\sqrt{1+m^2} - m)$. The ratio of width to depth in large channel is given according to project conditions. For navigation channels the design should meet special requirements.

Substituting $b = \beta h$ into equation (9-10a), yiealds

$$q_v = \frac{\sqrt{i}}{n} \frac{[h^2(b+mh)]^{5/3}}{n[h(\beta+2h\sqrt{1+m^2})]^{2/3}} \tag{9-13}$$

$$\begin{cases} h = \left(\dfrac{nq_v}{\sqrt{i}}\right)^{0.375} \dfrac{(\beta+2\sqrt{1+m^2})^{0.25}}{(\beta+m)^{0.625}} \\ b = \beta h \end{cases} \tag{9-14}$$

③ $b$ and $h$ are calculated according no-slurry velocity.

The method to solve such problems is to consider $v_{max}$ as the actual average velocity of the designed channel cross section. The corresponding cross-sectional area $A$, hydraulic radius $R$, and wetted perimeter $\chi$ can be determined as follows:

$$A = \frac{q_v}{v_{max}}$$

$$R = \left(\frac{nv_{max}}{i^{1/2}}\right)^{3/2}$$

$$\chi = \frac{A}{R}$$

Applying geometric euations between hydraulic elements on trapezoidal sections:

$$A = h(b+mh)$$

$$\chi = b + 2h\sqrt{1+m^2}$$

Last
$$\begin{cases} h = \dfrac{\chi \pm \sqrt{\chi^2 - 4A(2\sqrt{1+m^2}-m)}}{2(2\sqrt{1+m^2}-m)} \\ b = \chi - 2h\sqrt{1+m^2} \end{cases} \tag{9-15}$$

Equations (9-11)-(9-15) are introduced Manning equation. Application conditions is the flow resistance in the square area.

**Example 9-1** The bottom width of a trapezoidal channel is $b = 3.6$ m, the water depth is $h = 1.2$ m, the slope coefficient is $m = 2.0$, the roughness coefficient is $n = 0.019\,2$, and the bottom slope is $i = 0.000\ 625$. Find the passing flow rate $q_v$.

**Solution:**

Water surface width $B = b + 2mh = 3.6 + 2 \times 2 \times 1.2 = 8.4$ m

Water cross-sectional area $A = (b + mh)h = (3.6 + 2 \times 1.2) \times 1.2 = 7.2$ m$^2$

Wetted perimeter $\chi = b + 2h\sqrt{1+m^2} = 3.6 + 2 \times 1.2\sqrt{1+2^2} = 8.97$ m

Hydraulic radius $R = \dfrac{A}{\chi} = \dfrac{7.2}{8.97} = 0.803$ m

Chézy coefficient $C = \dfrac{1}{n}R^{1/6} = \dfrac{1}{0.019\ 2} \times 0.803^{1/6} = 50$ m$^{1/2}$/s

Flow rate $q_v = AC\sqrt{Ri} = 7.2 \times 50\sqrt{0.803 \times 0.000\ 625} = 8.07$ m$^3$/s

**Example 9 - 2** An earth channel with a trapezoidal cross-section has a flow rate of $q_v = 1$ m$^3$/s, a bottom slope of $i = 0.005$, a slope coefficient of $m = 1.5$, a roughness coefficient of $n = 0.025$, and a non-flushing allowable velocity of $v_{\max} = 1.2$ m/s. Try to design the cross-section dimensions according to the allowable velocity and optimal hydraulic conditions.

**Solution:** (1) Design according to the allowable velocity $v_{\max} = 1.2$ m/s without flushing

$$A = \dfrac{q_v}{v_{\max}} = \dfrac{1}{1.2} = 0.83 \text{ m}^2$$

Since $v_{\max} = \dfrac{1}{n}i^{1/2}A^{2/3}\chi^{-2/3} = \dfrac{1}{0.025} \times 0.005^{1/2} \times 0.83^{2/3}\chi^{-2/3} = 1.2$ m/s

the solution is $\chi \approx 3.0$ m.

From the trapezoidal section condition, we get

$$A = (b + mh)h = bh + 1.5h^2 = 0.83 \text{ m}^2$$

$$\chi = b + 2h\sqrt{1+m^2} = b + 3.61h = 3.0 \text{ m}^2$$

Solve by combining the above two equations

$$b_1 = -0.79 \text{ m}, \quad h_1 = 1.05 \text{ m} \quad \text{(discard)}$$

$$b_2 = 1.63 \text{ m}, \quad h_2 = 0.38 \text{ m}$$

(2) Design according to optimal hydraulic conditions

The optimal hydraulic width-to-depth ratio is

$$\beta = \dfrac{b}{h} = 2(\sqrt{1+m^2} - m) = 0.61$$

then

$$b = 0.61h$$

Since

$$A = (b + mh)h = (0.61h + 1.5h)h = 2.11h^2$$

$$C = \dfrac{1}{n}R^{1/6}$$

$$R = 0.5h$$

therefore

$$q_v = AC\sqrt{Ri} = \dfrac{1}{n}AR^{2/3}i^{1/2} = 3.76h^{8/3} = 1 \text{ m}^3/\text{s}$$

Solve      $h = 0.61$ m, $b = 0.37$ m

Check      $A = 2.11h^2 = 0.79$ m$^2$, $v = \dfrac{q_v}{A} = \dfrac{1}{0.79} = 1.27$ m/s

Therefore, appropriate reinforcement measures need to be taken, otherwise erosion will occur.

## 9.3 Uniform flow in non-pressure circular conduits

### 9.3.1 Hydraulic mathematical model and solving method

Non-pressure pipe refers to a long pipeline with a circular cross section that is not filled with flow. The flow inside the pipeline has a free surface and the surface pressure is atmospheric pressure. The uniform flow in non-pressure pipe belongs to a special cross-sectional form of uniform flow in open channels, and its formation conditions, hydraulic characteristics, and basic equations are the same as those of uniform flow in open channels mentioned above. The mathematical model for hydraulic calculation of uniform flow in a non-pressure circular pipe is still given by combining the cross-sectional geometric relationship and the above-

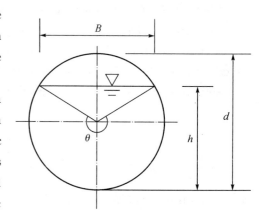

Fig.9 – 4   The inside section of the non-pressure pipe uniform flow

mentioned basic equation for uniform flow in an open channel. The geometric elements of the cross-sectional area of uniform flow in a non-pressure circular pipe are shown in Fig. 9 – 4. The basic quantities are the diameter of the pipe $d$, the water depth $h$, and the filling degree $\alpha$ or the filling angle $\theta$. The filling degree is defined as $\alpha = h/d$.

Over the water area $\quad\quad\quad A = d^2(\theta - \sin\theta)/8 \quad\quad\quad (9-16\text{a})$

Wetted perimeter $\quad\quad\quad \chi = \dfrac{\theta d}{2} \quad\quad\quad (9-16\text{b})$

Hydraulic radius $\quad\quad\quad R = \dfrac{d}{4}\left(1 - \dfrac{\sin\theta}{\theta}\right) \quad\quad\quad (9-16\text{c})$

Velocity $\quad\quad\quad v = 0.397\dfrac{\sqrt{i}}{n}\left(d\,\dfrac{\theta - \sin\theta}{\theta}\right)^{2/3} \quad\quad\quad (9-16\text{d})$

Flow rate $\quad\quad\quad q_v = 0.0496\,\dfrac{d^{8/3} i^{1/2}}{n}\,\dfrac{(\theta - \sin\theta)^{5/3}}{\theta^{2/3}} \quad\quad\quad (9-16\text{e})$

Filling degree $\quad\quad\quad \alpha = \dfrac{h}{d} = \sin^2(\theta/4) \quad\quad\quad (9-16\text{f})$

Surface width $\qquad B = d \cdot \sin(\theta/2) \qquad (9-16\text{g})$

In the case of other quantities are known, $q_v$, $d$, $i$ can be solve directly as an unknown quantity. But if $\theta$ is an unknown quantity, we need to adopt an iterative solution.

Rewrite equation (9 – 16e) to the following form:

$$\left(\frac{q_v n}{0.049\ 6 d^{8/3} i^{1/2}}\right)^{5/3} = \frac{\theta_{j+1} - \sin\theta_j}{\theta_j^{2/5}}$$

The iterative equation is:

$$\theta_{j+1} = k\theta_j^{0.4} + \sin\theta_j \qquad (9-17)$$

### 9.3.2 Characteristics of non-pressure pipe uniform flow

In addition to the aforementioned common uniform flow characteristics, there still are characteristics as follow.

(1) Flow rate and velocity are the full flow before the water reaches the maximum value. Now introduce the flow rate $q_0$ and velocity $v_0$ at full flow and compare them with the flow rate $q$ and velocity $v$ at full flow. Different filling levels correspond to a flow rate and a velocity, and a dimensionless combination is used to represent the relationship between filling level and flow rate and velocity, as shown in Fig. 9 – 5.

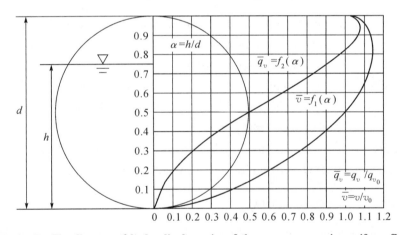

**Fig.9 – 5** The diagram of hydraulic dynamics of the non-pressure pipe uniform flow

① When the fullness is $\alpha = \dfrac{h}{d} \approx 0.95$, the flow rate is maximum which is $\dfrac{q}{q_0} \approx 1.08$. That is, $q$ reaches its maximum value, and the flow rate through the pipe is 1.08 times the full flow;

② When the fullness is $\alpha = \dfrac{h}{d} \approx 0.81$, the average velocity is maximum which is $\dfrac{v}{v_0} \approx 1.16$. That is, $v$ reaches its maximum value, and the flow rate in the pipe is 1.16 times the full flow.

Analysis shows that the maximum flow rate and maximum velocity of a pressure-free circular pipe do not occur when the pipe is full. This is because when the upper part of the circular section is filled with water, after exceeding a certain water depth, its wetted perimeter grows faster than the flow cross-sectional area, and the hydraulic radius begins to decrease, resulting in a decrease in flow rate and velocity.

(2) When the fullness of open channel flow close to 1, uniform flow is not easy to keep stability. Once outside fluctuations is interfered, it is easy to form a alternating flow of pressurized stream and no pressure stream.

Therefore, the sewers of hydraulic calculation need to comply with the relevant provisions in the engineering, and it is not allowed to exceed the maximum design fullness.

## 9.4 Concepts of non-uniform flow in open channel

### 9.4.1 Non-uniform flow in open channel

Non-uniform flow in open channel is a flow in which the velocity and depth of water through the open channel vary along the process. The flow in natural and man-made open channels is mostly non-uniform. Because the natural open channel does not exist prismatic channel, even artificial open channel, its section shape, size, roughness, bottom slope may also change along the process, or in the channel in the construction of hydraulic structures (culverts, gates, bridges), so that the flow of water in the open channel occurs non-uniform flow.

Compared with the uniform flow in open channel, the characteristics of the non-uniform flow in open channel are as follows:

① The average cross-sectional velocity, depth of water change along the process;

② Stream lines are not mutually parallel straight lines, the same stream line on the size and direction of the velocity is different. That is, the total head line $J$, water surface line $J_p$ and bottom slope $i$ are not equal, $J \neq J_p \neq i$.

To distinguish between uniform flow and non-uniform flow, the parameter corresponding to uniform flow will be indicated by subscript "0" later.

### 9.4.2 Unit energy of cross section

The flow state of the open channel flow can be analyzed and judged from the perspective of energy. As shown in Fig. 9-6, let the open channel non-uniform gradually varied flow, datum 0-0 by unit gravity of the liquid's mechanical energy is

$$E = z + \frac{p}{\rho g} + \frac{\alpha v^2}{2g} \qquad (9-18)$$

If the datum of the section is raised $z_1$ so that it passes through the lowest point of the section, the mechanical energy of the liquid subjected to unit gravity with respect to the new datum $0_1-0_1$ is

$$E_s = E - z_1 = h + \frac{\alpha v^2}{2g} \qquad (9-19)$$

or
$$E_s = h + \frac{\alpha q_v^2}{2gA^2} \quad (9-20)$$

where $E$ is called section unit energy, or section specific energy, is relative to the section through the lowest point of the datum, the fluid subject to unit gravity has the mechanical energy.

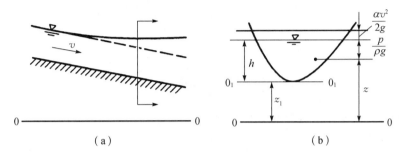

**Fig. 9-6  Unit energy of cross section**

The section unit energy $E_s$ and the mechanical energy $E$ of the fluid subjected to unit gravity are two different energy concepts. The mechanical energy $E$ of the fluid subjected to unit gravity is the mechanical energy of all the water relative to the same datum along the process, and its value decreases along the process. And section unit energy $E_s$ is calculated according to the lowest point of the respective section through the datum of mechanical energy, its value along the process may increase, may decrease, and only in the uniform flow, its value along the process is the same.

The water depth of non-uniform flow in open channel is variable, and a certain amount of flow rate can pass through a flow cross section in different water depth and have different section unit energy. For prismatic channels, when the flow rate is certain, the section unit energy $E_s$ varies only with the change of water depth $h$, i.e.

$$E_s = h + \frac{\alpha q_v^2}{2gA^2} = f(h)$$

It can be seen from the above equation: under the conditions of the section shape, size and flow rate are certain, when $h \to 0$, $A \to 0$, $v \to \infty$, then $E_s \to \infty$, and if taking the water depth $h$ as the vertical coordinates, section unit energy $E_s$ as the horizontal coordinates, the horizontal coordinate axis is the asymptote to the function curve $E_s = f(h)$; when $h \to \infty$, $A \to \infty$, $v \to 0$, then $E_s \approx h \to \infty$, the straight line through the origin of the coordinates at an angle of 45° with the horizontal axis is the asymptote to the curve. There is a minimum value of section unit energy, $E_{smin}$, which corresponds to the critical water depth $h_c$.

### 9.4.3  Critical depth

1. Meaning and universal equation

Critical depth refers to the water depth, in which the value of section unit energy is minimum when the section form and flow rate are certain. That is, $E_s = E_{smin}$, when the corresponding water depth $h = h_c$, and the hydraulic elements corresponding to $h_c$ are prensented by subscript "c". So that $\dfrac{dE_s}{dh} = 0$, and taking into account, that $\dfrac{dA}{dh} \approx B$, there are

$$\frac{dE_s}{dh} = \frac{d}{dh}\left(h + \frac{\alpha q_v^2}{2gA^2}\right) = 1 - \frac{\alpha q_v^2}{gA^3}\frac{dA}{dh} = 1 - \frac{\alpha q_v^2}{gA^3}B = 0$$

then

$$\frac{\alpha q_v^2}{g} = \frac{A_c^3}{B_c} \tag{9-21}$$

The value of $h_c$ can be obtained from the above equation when the channel flow rate, cross-section shape and size are given. From the above equation, it can be seen that the critical water depth is only related to the cross-section shape, dimensions and flow rate, but not to the channel bottom slope $i$ and wall roughness coefficient $n$.

### 2. Calculation of critical depth in rectangular cross section

For a rectangular cross section open channel with water surface width $B_c = b$ and single width flow rate $q_l = \dfrac{q_v}{b}$, substituting them into equation (9-21), yields

$$\frac{\alpha q_v^2}{g} = \frac{A_c^3}{B_c} = \frac{b_c^3 h_c^3}{b_c}$$

then

$$h_c = \sqrt[3]{\frac{\alpha q_l^2}{b_c}} \tag{9-22}$$

### 3. Calculation of critical depth in trapezoidal cross section

For trapezoidal cross sections, $A_c = h_c(b + mh_c)$, $B_c = b + 2mh_c$ substituting them into equation (9-21), yieldes

$$\frac{[h_c(b + mh_c)]^3}{b + 2mh_c} = \frac{\alpha q_v^2}{g}$$

This equation is a high order implicit equation which is not easy to solve and can be written in iterative form

$$h_{c(j+1)} = \left[\frac{\alpha q_v^2}{g}\frac{b + 2mh_{cj}}{b + mh_{cj}}\right]^{1/3} \quad (j = 0, 1, 2, \cdots) \tag{9-23}$$

### 4. Calculation of critical depth in circular cross section

Substituting equation (9-16a) into equation (9-21) leads to an iterative equation for solving $\theta$:

$$\theta_{j+1} = \left[8\left(\frac{\alpha q_v^2}{gd^5}\sin\frac{\theta_j}{2}\right)^{1/3} + \sin\theta_j\right]^{0.7} \cdot \theta_j^{0.3} \quad (j = 0, 1, 2, \cdots) \tag{9-24}$$

Critical water depth is

$$h_c = d \cdot \sin^2\left(\frac{\theta}{4}\right)$$

The weighted exponent of the accelerated convergence of the iterative process in equation (9-24) is 0.7, indicating that it converges quickly.

### 9.4.4 Critical slope

In a prismatic channel, when the section shape, size and flow rate are fixed, if the normal water depth $h_0$ of the water flow is exactly equal to the critical water depth $h_c$, the corresponding channel bottom slope is called the critical bottom slope, denoted by $i_c$. It can be solved by combining basic equation of open channel uniform flow, $q_v = A_c C_c \sqrt{R_c i_c}$, and critical water depth relationship equation $\dfrac{\alpha q_v^2}{g} = \dfrac{A_c^3}{B_c}$, that is

$$i_c = \frac{q_v^2}{A_c^2 C_c^2 R_c} = \frac{g}{\alpha C_c^2} \cdot \frac{\chi_c}{B_c} \tag{9-25}$$

Critical bottom slope $i_c$ is not the actual existence of the channel bottom slope, it is the hypothetical bottom slope of a hypothetical uniform flow ($h_0 = h_c$) introduced only in order to facilitate the analysis and calculation of non-uniform flow. If the actual bottom slope of the open channel is less than the critical slope under a certain flow rate, that is, $i < i_c (h_0 > h_c)$, the channel bottom slope is known as a slow slope; if $i > i_c (h_0 < h_c)$, the channel bottom slope is known as a sharp or steep slope; if $i = i_c (h_0 = h_c)$, the channel bottom slope is known as the critical slope. It is worth noting that the above on the bottom of the channel bottom slope of slow, sharp said, is corresponding to a certain flow rate. For a certain channel, the bottom slope is certain, but when the flow rate increases or changes in time, the corresponding $h_c$ (or $i_c$) may change, so that the names of slow slope or sharp slope of the channel should be changed.

### 9.4.5 Slow flow, rapid flow, critical flow and criterion

The velocity of flow in an open channel at a critical depth of water is called the critical velocity and is denoted by $v_c$. Such a state of open channel flow is called critical flow. When the velocity of the open channel is less than the critical velocity, the flow is called slow flow, which occurs mostly in the plains and off-shore rivers. Greater than the critical velocity, known as rapid flow, occurring in mountainous rivers and steep slopes. Commonly used discriminatory methods are:

#### 1. The method of critical depth

Comparing the water depth in the open channel with the critical water depth $h_c$ can determine the flow state of open channel water flow. When the actual depth in open channel flow is $h > h_c$, $v < v_c$, it is slow flow. When the actual depth in open channel flow is $h = h_c$,

$v=v_c$, it is critical flow. When the actual depth in open channel flow is $h<h_c$, $v>v_c$, it is rapid flow.

### 2. The method of earthquake wave velocity

The stone is thrown into the calm lake water, produce tiny interference waves, interference waves to the stone landing point as the center, with a certain velocity $c$ to propagate in all directions, the lake surface forms a series of concentric circles of waveforms, this propagation of the microwave velocity in static water is called the relative wave velocity, expressed in terms of $c$:

$$c=\sqrt{gh}$$

It is possible to determine which flow regime a stream belongs to by comparing the magnitude of the cross-sectional average velocity $v$ and the relative wave velocity $c$. When $v<c$, the disturbing wave can propagate upstream, and the current is a slow flow; when $v=c$, the disturbing wave can not propagate upstream exactly, and the current is a critical flow; when $v>c$, the disturbing wave can not propagate upstream, and the current is a rapid flow.

### 3. The method of Froude number

Define the ratio of the open channel velocity $v$ to the relative wave velocity $c$ as the Froude number, denoted by $Fr$, i.e.

$$Fr=\frac{v}{c}=\frac{v}{\sqrt{gh}}$$

It can be used as a discriminant of the flow pattern of water: $Fr<1$, the water is slow flow; $Fr=1$, the water is critical flow; $Fr>1$, the water is rapid flow.

The Froude number is an important discriminant in fluid dynamics, and to deepen the understanding of its physical significance, it can be rewritten in the form of

$$Fr=\frac{v}{\sqrt{gh}}=\sqrt{\frac{\frac{2v^2}{2g}}{h}} \qquad (9-26)$$

The above methods are applicable to the flow pattern discrimination of both uniform and non-uniform flow. For the open channel uniform flow, it can also be discriminated based on the bottom slope; when the bottom slope $i<i_c$, it is a slow flow; when $i=i_c$, it is a critical flow; when the bottom slope $i>i_c$, it is a rapid flow.

All the various discrimination methods mentioned above are equivalent, and one of them can be used according to the engineering conditions.

**Example 9-3** A long and straight rectangular cross-section channel has a bottom width $b=1$ m, a roughness coefficient $n=0.014$, a bottom slope $i=0.0004$, and a normal water depth of uniform flow in the channel $h_0=0.6$ m. Try to determine the flow state of the water flow.

**Solution:** The average velocity at the section is

$$v = C\sqrt{Ri}$$

where
$$R = \frac{bh_0}{b+2h_0} = 0.273 \text{ m}$$

$$C = \frac{1}{n}R^{1/6} = 57.5 \text{ m}^{1/2}/\text{s}$$

So we get
$$v = 0.6 \text{ m/s}$$

(1) Use wave speed method to determine

$$c = \sqrt{g\bar{h}} = \sqrt{gh_0} = 2.43 \text{ m/s} > v$$

The flow is slow flow.

(2) Use Froude number to determine

$$Fr = \frac{v}{\sqrt{gh_0}} = 0.25 < 1$$

The flow is slow flow.

(3) Use critical water depth to judge

$$q_v = vh_0 = 0.36 \text{ m}^2/\text{s}$$

$$h_c = \sqrt[3]{\frac{\alpha q_v^2}{g}} = 0.24 \text{ m} < h_0$$

The flow is slow flow.

(4) Use critical bottom slope to judge

Since the flow is uniform, the critical bottom slope can also be used to identify the water flow state and calculate the corresponding quantity.

$$B_c = b = 1 \text{ m}$$

$$\chi_c = b + 2h_c = 1.48 \text{ m}$$

$$R_c = \frac{bh_c}{\chi_c} = 0.16 \text{ m}$$

$$C_c = \frac{1}{n}R_c^{1/6} = 52.7 \text{ m}^{1/2}/\text{s}$$

$$i_c = \frac{g}{\alpha C_c^2}\frac{\chi_c}{B_c} = 0.005\,2 > i = 0.000\,4$$

This channel has a gentle slope and the uniform flow is slow flow.

## 9.4.6 Hydraulic jump and energy dissipation

The water flow in the open channel is transitioned from rapid to slow flow, the free surface of the water flow will suddenly jump up and form a spinning roll on the surface. This rapid change flow phenomenon is called hydraulic jump or water leap. It can be formed not only under the spillway, under the spillway gate, under the water drop, but also in the flat slope channel under the gate when the outflow is formed.

Hydraulic jump area on the part of the rapids are rushing to the slow flow of the surface caused by the whirlpool, tossing and rolling, full of air, known as the "surface spinning". Below the spinning roll is the main stream of the section forward expansion. Due to the formation of surface tumbling at the bottom of the mainstream, water turbulence, fluid mass collision with each other, mixing strong. Rotating roll and the main stream between the quality of continuous exchange, resulting in the hydraulic jump section has a greater energy loss, up to 60% to 70% of the energy of the section rapids before the leap. Therefore, commonly used hydraulic jump to eliminate the huge energy of high-speed water flow downstream of the relief building, that is, the hydraulic jump is commonly used in the downstream of the relief building energy dissipation, is an effective means of energy dissipation.

### 1. The basic equations of the hydraulic jump

Taking a complete hydraulic jump on a flat slope as an example, we establish the hydraulic jump equation. During the derivation process, the following assumptions are made based on the actual conditions of the hydraulic jump occurrence:

(1) The length of hydraulic jump is shorter, channel bottom friction can be negligible.

(2) Flow cross sections of before and after jump are gradually varied sections, which subject to hydrostatic pressure distribution.

(3) Momentum correction factors of before and after the jump are $\beta_1 = \beta_2 = \beta$.

Taking the hydrodynamic space between the hydraulic jump sections of before and after jump as the control body, list the momentum equation for the total flow in the flow direction as follows:

$$\rho v(\beta_2 v_2 - \beta_1 v_1) = \rho g h_{c1} A_1 - \rho g h_{c2} A_2 \tag{9-27}$$

where $h_{c1}$, $h_{c2}$ are the water depths at the center of the hydraulic jump sections of before and after jump, respectively; $\rho g h_{c1} A_1$, $\rho g h_{c2} A_2$ are the dynamic water pressure acting on the hydraulic jump sections of before and after jump.

By the continuity equations, we can get the average velocity of the flow cross section of before and after jump:

$$v_1 = \frac{q_v}{A_1}, \quad v_2 = \frac{q_v}{A_2}$$

Then
$$\frac{q_v^2 \beta}{g A_1} + h_{c1} A_1 = \frac{q_v^2 \beta}{g A_2} + h_{c2} A_2 \tag{9-28}$$

Equation (9-28) is the basic equation of hydraulic jump in flat slope prismatic channel. It shows that in the hydraulic jump area in unit time, the sum of the momentum into the section before the jump and the total dynamic pressure of the section is equal to the sum of the momentum out of the section after the jump and the total dynamic pressure of the section. In the equation, $A$ and $h_c$ are functions of water depth, and the rest of the quantities are constants, so it can be written as follows:

$$J(h) = \frac{q_v^2 \beta}{gA} + h_c A \tag{9-29}$$

The basic equations of the hydraulic jump can also be written as follows:

$$J(h') = J(h'') \tag{9-30}$$

The above basic equation for hydraulic jump shows that for a certain flow rate $q_v$, there are two water depths with the same hydraulic jump function $J(h)$, and this pair of water depths is the conjugate water depth.

### 2. The basic calculation of hydraulic jump in rectangular open channel

(1) The calculation of conjugate depths

In rectangular open channels, the depth of before or after the hydraulic jump can be directly solved by a hydraulic jump equation. Assuming $\alpha = \beta = 1.0$ in rectangular open channels, and substituting $A_1 = bh'$, $A_2 = bh''$, $h_{c1} = \frac{h'}{2}$, $h_{c2} = \frac{h''}{2}$, $q_l = \frac{q_v}{b}$, $h_c^3 = \frac{\alpha q^2}{g}$ into the equation (9-28), eliminating $b$, we get

$$\frac{q_l^2}{gh'} + \frac{h'^2}{2} = \frac{q_l^2}{gh''} + \frac{h''^2}{2} \tag{9-31}$$

After organizing, the quadratic equation is obtained:

$$h'h''(h' + h'') = \frac{2q_l^2}{g}$$

Taking the water depth $h'$ before jump and the water depth $h''$ after the jump as unknown quantities, respectively, we solve for

$$h' = \frac{h''}{2}\left(\sqrt{1 + 8\frac{q_l^2}{gh''^3}} - 1\right) = \frac{h''}{2}\left[\sqrt{1 + 8\left(\frac{h_c}{h''}\right)^3} - 1\right] \tag{9-32}$$

or

$$h'' = \frac{h'}{2}\left(\sqrt{1 + 8\frac{q_l^2}{gh'^3}} - 1\right) = \frac{h'}{2}\left[\sqrt{1 + 8\left(\frac{h_c}{h'}\right)^3} - 1\right] \tag{9-33}$$

Due to

$$Fr_1^2 = \frac{v_1^2}{gh'} = \frac{q_l^2}{gh'^3}, \quad Fr_2^2 = \frac{v_2^2}{gh''} = \frac{q_l^2}{gh''^3}$$

Then

$$h' = \frac{h''}{2}(\sqrt{1 + 8Fr_2^2} - 1) \tag{9-34}$$

$$h'' = \frac{h'}{2}(\sqrt{1 + 8Fr_1^2} - 1) \tag{9-35}$$

Ratio of conjugate depth is
$$\eta = \frac{h'}{2}(\sqrt{1+8Fr_1^2}-1) \tag{9-36}$$

$\eta$ is the ratio of conjugate depths. $\eta$ increases with $Fr_1$.

(2) Calculation of hydraulic jump length

The hydraulic jump length is one of the main bases for the design of energy dissipation of water-discharging buildings. Due to the complexity of the hydraulic jump phenomenon, the current theoretical research is still immature, the determination of the hydraulic jump length is still based on experimental research. Now introduces the empirical equation used to calculate the hydraulic jump length of flat bottom slope rectangular channel.

① Expressed in terms of water depth after the jump, for cases where $4.5 < Fr_1 < 10$, $l_j = 6.1h''$;

② Expressed in terms of hydraulic jump height, $l_j = 6.9(h''-h')$;

③ Expressed as Froude number, $l_j = 9.4(Fr_1-1)h'$.

(3) Calculation of hydraulic jump energy consuming

It is found that the energy loss caused by the hydraulic jump is mainly concentrated in the hydraulic jump section, and only a very small part occurs in the section of before jump. For the flat-bottomed rectangular channel, the total energy equations for the hydraulic jump sections of before and after jump can be obtained as follows:

$$\Delta E_j = \left(h' + \frac{\alpha_1 v_1^2}{2g}\right) - \left(h'' + \frac{\alpha_2 v_2^2}{2g}\right)$$

Substituting $v_1 = \frac{q_l}{h'}$, $v_2 = \frac{q_l}{h''}$ and $h'h''(h'+h'') = \frac{2q_l^2}{g}$ into the above equation and taking $a_1 = a_2 = 1.0$, we have

$$\Delta E_j = \frac{(h''-h')^3}{4h'h''} \tag{9-37}$$

Equation (9-37) explaned that for given flow rate, the water depth ratio of after jump and before jump is greater, the energy consumption values is greater.

## 9.4.7 Hydraulic drop

In the slow-flow state of the open channel water flow, due to the bottom of the channel suddenly become steep slopes or downstream of the channel cross-sectional shape of the sudden expansion of the water surface caused by the landing, the slow flow transforms into rapid flow. This local hydraulic phenomenon of the transition from the slow flow to rapid flow is known as hydraulic drop, or water fall. Understanding the phenomenon of hydraulic drop is important for analyzing and calculating the water surface curve of the steady non-uniform flow in the open channel. For example, the slow slope channel connected to a sharp slope channel, water flow through the connection section of the depth of water can be considered to be the critical depth of water, this section is called the control section, the depth of water is called the control depth of water. In the water surface curve analysis and calculation, the control depth of water can be used as a known depth of water, which provids a known condition for analysis and calculation.

**Example 9-4** A rectangular cross-section channel downstream of a drainage building, with a single width flow rate $q_l = 1.5$ m²/s. A hydraulic jump occurs, and the water depth before the jump is $h' = 0.8$ m. Try to find: (1) water depth $h''$ after the jump; (2) hydraulic jump length $l_j$; (3) hydraulic jump energy dissipation rate $\Delta E_j / E_1$.

**Solution:**

(1) Find the water depth after the jump $h''$

$$Fr_1^2 = \frac{q^2}{gh'^3} = \frac{15^2}{9.8 \times 0.8^3} = 44.84$$

$$h'' = \frac{h'}{2}(\sqrt{1+8Fr_1^2} - 1) = \frac{0.8}{2} \times (\sqrt{1+8 \times 44.84} - 1) = 7.19 \text{ m}$$

(2) Find the hydraulic jump length $l_j$

$$l_j = 6.1 h'' = 6.1 \times 7.19 = 43.86 \text{ m}$$

$$l_j = 6.9(h'' - h') = 6.9 \times 6.39 = 44.09 \text{ m}$$

$$l_j = 9.4(Fr_1 - 1)h' = 42.83 \text{ m}$$

(3) Find the water jump energy dissipation rate

$$\Delta E_j = \frac{(h'' - h')^3}{4h'h''} = \frac{(7.19 - 0.8)^3}{4 \times 0.8 \times 7.19} = 11.34 \text{ m}$$

$$\frac{\Delta E_j}{E_1} = \frac{\Delta E_j}{h' + \frac{q^2}{2gh'^2}} = \frac{11.34}{0.8 + \frac{15^2}{2 \times 9.8 \times 0.8^2}} = 61\%$$

## 9.5 Gradually-varied flow water surface curve analysis in open channel

The water depth $h$ of non-uniform gradually-varied flow in open channel changes with the process $s$. The free water surface line is called water surface curve $h = f(s)$, which is a curve not parallel to the bottom of the channel. The main task of analysising the steady non-uniform gradually-varied flow surface curve of open channels is to determine the water surface curve along the process of the trend and the scope of change according to the channel conditions of the groove, the incoming flow conditions and the situation of hydraulic buildings, etc., and qualitatively draw the water surface curve.

## 9.5.1 Differential equation of non-uniform gradually-varied flow in prismatic open channel

Fig. 9-7 shows a non-uniform gradually-varied flow section of a open channel, a tiny flow section $ds$ is taken along the direction of water flow, and the elevation of the channel bottom in its upstream 1-1 section relative to the 0-0 datum is $z$, the depth of water is $h$, and the average velocity of the section is $v$; the corresponding parameters of the downstream 2-2 section are $z+dz$, $h+dh$, $v+dv$. List the energy equations of the 1-1 and 2-2 sections:

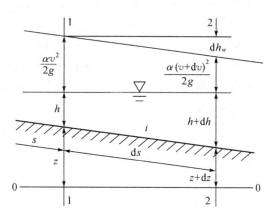

Fig. 9-7 Open-channel gradually-varied flow

$$(z+h)+\frac{\alpha v^2}{2g}=(z+dz)+(h+dh)+\frac{\alpha(v+dv)^2}{2g}+dh_w \qquad (9-38)$$

With the type of $(v+dv)^2$ and leave $(dv)^2$ out, the equation can be simplified to

$$dz+dh+d\left(\frac{\alpha v^2}{2g}\right)+dh_w=0$$

After being divideed by $ds$, then

$$\frac{dz}{ds}+\frac{dh}{ds}+\frac{d}{ds}\left(\frac{\alpha v^2}{2g}\right)+\frac{ah_w}{ds}=0 \qquad (9-39)$$

The first term of the equation

$$\frac{dz}{ds}=-\frac{z_1-z_2}{ds}=-i$$

The third term of the equation

$$\frac{d}{ds}\left(\frac{\alpha v^2}{2g}\right)=\frac{d}{ds}\left(\frac{\alpha q_v^2}{2gA^2}\right)=-\frac{\alpha q_v^2 dA}{gA^3 ds}$$

Whereas prismatic channel overflow cross-sectional area varies only with water depth, that is, $A=f(h)$, then

$$\frac{dA}{ds}=\frac{\partial A}{\partial h}\frac{dh}{ds}=B\frac{dh}{ds}$$

In the fourth term of the equation, the local head loss in the gradually-varied flow section is very small and negligible, there are $dh_w=dh_f$, get

$$\frac{dh_w}{ds}=\frac{dh_f}{ds}=J$$

Based on the equation (9-39),

$$\frac{dh}{ds} - \frac{\alpha q_v^2}{gA^3}\left(B\frac{dh}{ds}\right) = i - J$$

Transformed,

$$\frac{dh}{ds}\left(1 - \frac{\alpha q_v^2 B}{gA^3}\right) = i - J$$

The gradually-varied flow cross section changes slowly along the process and the head loss can be calculated as uniform flow, i.e.

$$J = \frac{q_v^2}{A^2 C^2 R} = \frac{q_v^2}{K^2} \tag{9-40}$$

Finally,

$$\frac{dh}{ds} = \frac{i-J}{1-\dfrac{\alpha q_v^2 B}{gA^3}} = \frac{i-J}{1-F_r^2} = \frac{i-\dfrac{q_v^2}{K}}{1-F_r^2} \tag{9-41}$$

The above equation is the equation for steady non-uniform gradually-varied flow in prismatic open channel.

### 9.5.2 The flow phenomenon and space partition of open-channel steady gradually-varied flow

#### 1. The flow phenomenon of non-uniform flow in open channel

The differential equations for open channel non-uniform flow contain the interrelationships of $h$, $h_0$, $h_c$ and $i$. Because of the different channel bottom slopes, there are different combinations of the above three water depth values, thus forming a variety of changes in the surface curve of the non-uniform flow in the open channel: $dh/ds > 0$, $dh/ds = 0$, $dh/ds < 0$, $dh/ds \to i$, and $dh/ds \to \pm\infty$ and so on.

(1) $\dfrac{dh}{ds} > 0$, the depth of open channel non-uniform flow increasing with the flow, and the water surface profile is called back water curve, it's similar to the flow between the cross sections 1-1 and 2-2, as shown in Fig. 9-8.

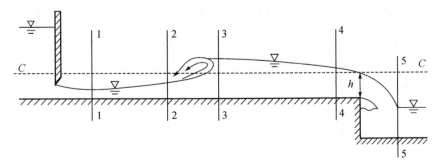

Fig. 9-8  The flow phenomenon of non-uniform flow in open channel

(2) $\dfrac{dh}{ds} < 0$, the depth of open channel non-uniform flow decreasing with the flow, and the water surface profile is called dropdown water curve, it's similar to the flow between the cross sections 3–3 and 4–4, as shown in Fig.9–8.

(3) $\dfrac{dh}{ds} \to +\infty$, the water surface profile is incontinuity and called hydraulic jump. It's similar to the flow between the cross sections 4–4 and 5–5, as shown in Fig.9–8.

(4) $\dfrac{dh}{ds} \to -\infty$, the water surface profile is incontinuity and called hydraulic drop. It's similar to the flow between the cross sections 4–4 and 5–5, as shown in Fig.9–8.

(5) $\dfrac{dh}{ds} = 0$, the depth of open channel non-uniform flow is constant along the process and belong to uniform flow.

(6) $\dfrac{dh}{ds} = i$, so $\dfrac{dh}{ds} = -\dfrac{dz}{ds}$, and the water surface profile is a horizontal line, as shown in Fig. 9–9.

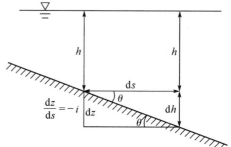

Fig. 9–9  Water level line of open channel flow

### 2. The flow space partition

In order to facilitate the analysis of the change of the water surface curve along the process, generally draw two straight lines parallel to the channel bottom on the analysis map of the water surface curve. One is from the bottom of the $h_0$, for the normal depth line $N-N$; and the other is from the bottom of the $h_c$, for the critical depth line $C-C$. The channel water flow is divided into three different parts by these two auxiliary lines ($N-N$ and $C-C$). These three parts are referred to as Part 1, Part 2 and Part 3, and each part is characterized as follows:

Part 1: $h > h_0$ and $h > h_c$;

Part 2: $h_0 > h > h_c$ (mild slope) or $h_0 < h < h_c$ (steep slope);

Part 3: $h < h_0$ and $h < h_c$.

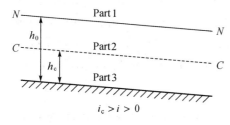

Fig. 9–10  The open channel partition

## 9.5.3  The twelve water surface profile of open channel steady gradually-varied flow

Open channel steady gradually-varied flow is actually located in different areas and different bottom slopes, its water surface curve has different forms, which will be analyzed separately below.

### 1. Mild slope channel ($i_c > i > 0$)

The normal water depth $h_0$ in the mild slope channel is greater than the critical water depth $h_c$, and the flow space is divided into three parts, Parts 1, 2, and 3, by the two auxiliary lines $N-N$ and $C-C$, and the water surface curves appearing in each zone are different, as shown in Fig. 9-11a.

(1) The depth of water is situated on Part 1 ($h > h_0 > h_c$)

In equation (9-41), numerator $h > h_0$, $J < i$, $i - J > 0$, denominator $h > h_c$, $Fr < 1$, $1 - Fr^2 > 0$, so $dh/ds > 0$, the water depth increases along the process and the water surface line is a congestion curve, called the $M_1$-type congestion curve.

As shown in Fig. 9-11a, upstream $h \to h_0$, $J \to i$, $i - J \to 0$; $h > h_c$, $Fr < 1$,

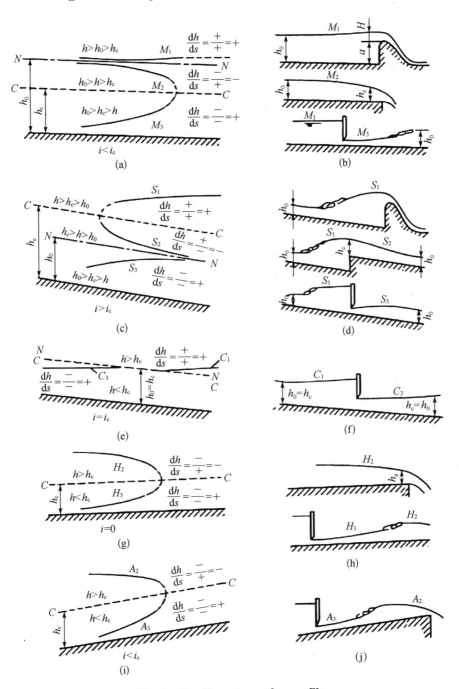

**Fig. 9-11 The water surface profile**

$1-Fr^2 > 0$, so $dh/ds \to 0$, the water depth remains constant along the process, and the water surface line is asymptotic to the $N-N$ line. Downstream $h \to \infty$, $J \to 0$, $i-J \to i$; $h \to \infty$, $Fr \to 0$, $1-Fr^2 \to 1$ so $dh/ds \to i$, the increase in water depth per unit distance is equal to the decrease in elevation at the bottom of the channel, and the water surface line is horizontal.

Constructing a water retaining building on a mild slope channel to raise the control water depth $h$, above the normal water depth $h_0$, for that flow rate, will result in an $M_1$-type water surface curve upstream of the water retaining building, as shown in Fig. 9-11b.

(2) The depth of water is situated on Part 2 ($h_0 > h > h_c$)

In equation (9-41), the numerator $h < h_0$, $J > i$, $i-J < 0$; the denominator $h > h_c$, $Fr < 1$, $1-Fr^2 > 0$, so $dh/ds < 0$, the water depth decreases along the process, and the water surface line is a precipitation curve, called the $M_2$-type precipitation curve.

As shown in Fig. 9-11a, upstream $h \to h_0$, similar to the analysis of the $M_1$-type water surface line, $dh/ds \to 0$, the water depth is constant along the process, and the water surface line is asymptotic to the $N-N$ line. Downstream $h \to h_c < h_0$, $J > i$, $i-J < 0$; $h \to h_c$, $Fr \to 1$, $1-Fr^2 \to 0$, so $dh/ds \to -\infty$, the water surface line is orthogonal to the $C-C$ line, where the water depth decreases sharply, and it is no longer an gradully-varied flow, but a hydraulic drop situation.

The end of the mild slope channel is a drop can, the channel is a $M_2$-type water surface line, and the drop can section passes the critical water depth, forming a hydraulic drop, as shown in Fig. 9-11b.

(3) The depth of water is situated on Part 3 ($h_0 > h_c > h$)

In equation (9-41), the numerator $h < h_0$, $J > i$, $i-J < 0$; the denominator $h < h_c$, $Fr > 1$, $1-Fr^2 < 0$, so $dh/ds > 0$, the water depth increases along the process, the water surface line for the congestion curve, known as the $M_3$-type congestion curve.

As shown in Fig. 9-11a, the upstream $h \to h_c$, $Fr \to 1$, $1-Fr^2 \to 0$, water depth is controlled by the outflow conditions, and the downstream, so $dh/ds \to \infty$, a hydraulic jump occurs, forming a concave congestion curve.

In the mild slope channel on the gate partially open, the water depth behind the gate is less than the critical water depth, due to resistance, the velocity along the process decreases, the water depth increases, $M_3$-type water surface curve is formed, as shown in Fig. 9-11b.

## 2. Steep slope channel ($i > i_c > 0$)

The normal water depth $h_0$ in the steep slope channel is less than the critical water depth $h_c$, and the flow space is divided into three parts, Parts 1, 2 and 3, by the $N-N$ line and the $C-C$ line, and the water surface curves appearing in the parts are different from those of the steep slope, as shown in Fig. 9-11c.

The water surface curve in Part 1 is an $S_1$-type congestion curve, where a hydraulic jump occurs upstream and the downstream water surface curve levels off. Constructing a water retaining structure in a steep slope channel creates an $S_1$-type water surface curve upstream, as shown in Fig. 9-11d.

The water surface curve in Part 2 is a $S_2$-type precipitation curve, where a hydraulic jump occurs upstream and the downstream water surface line is asymptotic to the $N-N$ line. When water flows from a steep slope channel into another section of steep slope channel with an elevated channel bottom, a $S_1$-type water surface curve is formed in the upstream channel, a $S_2$-type water surface curve is formed in the downstream steep-slope channel, and a hydraulic drop is formed by passing through the critical depth of water at the top of the variable-slope section of the riser, as shown in Fig. 9-11d.

The Part 3 water surface curve is a $S_3$-type congestion curve, with the upstream water depth controlled by the outflow conditions and the downstream water surface line asymptotic to the $N-N$ line. With the construction of a retaining structure in a steep-slope channel, the constricted water depth of the downstream flow is less than the normal water depth, and an $S_3$-type water surface curve is formed downstream, as shown in Fig. 9-11d.

### 3. Critical slope channel ($i = i_c$)

In a critical slope channel, the normal water depth $h_0$ is equal to the critical water depth $h_c$. The $N-N$ line coincides with the $C-C$ line, and the flow interval is divided into two parts, Parts 1 and 3, with no Part 2. The water surface curves are all congestion curves, referred to as $C_1$-type congestion curves and $C_3$-type congestion curves, respectively, and are all nearly horizontal as they approach the $N-N$ ($C-C$) line, as shown in Fig. 9-11e.

In a critical slope channel, $C_1$- and $C_3$-type water surface curves will be formed above and below the spillway gates, as shown in Fig. 9-11f.

### 4. Horizontal slope channel ($i = 0$)

In a horizontal slope channel, uniform flow cannot occur. Only the critical water depth line, $C-C$, divides the flow into Parts 2 and 3, with no Part 1. Part 2 is a $H_2$-type drop curve, and Part 3 is a $H_3$-type congestion curve, as shown in Fig. 9-11g.

$H_2$-type water surface curve will be formed upstream of the drop can at the end of the horizontal slope channel, and $H_3$-type water surface curve will be formed downstream of the gate in the horizontal slope channel where the opening height of the spillway gate is less than the critical water depth, as shown in Fig. 9-11h.

### 5. Adverse slope channel ($i < 0$)

In the adverse slope channel, only the critical water depth line exists as well, dividing the flow into Parts 2 and 3. The water surface lines are the $A_2$-type drop curve and the $A_3$-type congestion curve, respectively, as shown in Fig. 9-11i.

$A_2$-type water surface curves will be formed upstream of the drop canals at the end of the adverse slope channel, and $A_3$-type water surface curves will be formed downstream of the spillway gates in the adverse slope channel when the opening height of the gates is less than the critical water depth, as shown in Fig. 9-11j.

## 9.5.4 The steps of drawing the water surface profile

(1) Plot $N-N$ and $C-C$ lines (no $N-N$ lines for mild slope and adverse slope channels) to partition the flow space, with only one type of water surface curve

corresponding to each area.

(2) Select the control section. Control section should be selected in the water depth is known, and the location of the section to determine, and then control section as a starting point for analysis and calculation, to determine the type of water surface curve, and with reference to the shape of its depth increase, depth reduction and boundary conditions, to depict.

(3) If the water surface curve is interrupted, there is a discontinuity and produce a drop or hydraulic jump, to be specifically analyzed, in general, the water flow to the drop can form a drop phenomenon, the water flow from the rapid to the slow flow, the phenomenon of hydraulic jump occurs.

## 9.6 The calculation of open-channel gradually-varied flow water surface curve

After qualitatively analyzing the change in the water surface profile, it must be quantitatively calculated. For prismatic channels, the general mathematical method of approximate integration, called the numerical integration method, is usually used. Another type of method is the segment sum method, the method is the whole process of the open channel is divided into a number of segments, the differential equation is replaced by a finite difference equation, the segments of the direct sum. Segmented summation method, prismatic, non-prismatic channel are applicable, respectively, are described below.

### 9.6.1 The numerical integration method

Separating the variables for the basic differential equation (9 - 41), yields

$$ds = \frac{1 - \alpha q_v^2 \dfrac{B}{g} A^3}{i - J} dh \qquad (9-42)$$

Integrating the upper expression, yields

$$l = \int_{h_1}^{h_2} \frac{1 - \alpha q_v^2 \dfrac{B}{g} A^3}{i - J} dh \qquad (9-43)$$

For a prismatic channel with constant roughness coefficient $n$ and channel bottom slope $i$, the integrand of equation (9 - 43) is only a function of the water depth $h$, there are

$$F(h) = \frac{1 - \alpha q_v^2 \dfrac{B}{g} A^3}{i - J}$$

$$l = \int_{h_1}^{h_2} \frac{1 - \alpha q_v^2 \dfrac{B}{g} A^3}{i - J} dh = \int_{h_1}^{h_2} F(h) dh \qquad (9-44)$$

According to the trapezoidal integral method,

$$l \approx \sum \Delta l = \sum_{n=1}^{m} \frac{F(h_n) + F(h_{n+1})}{2}(h_{n+1} - h_n) \quad (9-45)$$

Based on the above equation, we can draw the $F(h)-h$ curve, as showed in Fig. 9-11. The area $\Delta A$ under the curve between two water depths, $h_n$ and $h_{n+1}$, is the lenth $\Delta l$ of the two water depth sections. We can divide the $(h_1, h_2)$ into $m$ minizones and calculate a series of $\Delta h_i$ and $\Delta l_i$, on the basis of a known water depth, then draw the water surface curve quantitatively.

The numerical integration method is only suited to the prismatic channel.

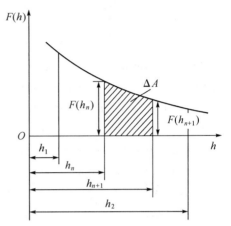

Fig. 9-12 The numerical integration method

### 9.6.2 The subsection summation method

According to the basic differential equation (9-42),

$$\frac{dh}{ds} = \frac{i - J}{1 - \left(\alpha q_v^2 \dfrac{B}{g} A^3\right)}$$

Considering the differential equation of specific section energy

$$\frac{dE_s}{dh} = 1 - \frac{\alpha q_v^2 B}{g A^3}$$

Acquirability,

$$\frac{dh}{ds} = \frac{i - J}{\dfrac{dE_s}{dh}}$$

So,

$$ds = \frac{dE_s}{i - J}$$

Rewrite the above equation to finite difference equation,

$$\Delta l = \Delta s = \frac{\Delta E_s}{i - \overline{J}} = \frac{E_{sd} - E_{su}}{i - \overline{J}} \quad (9-46)$$

Then the total process length is

$$l = \sum \Delta l = \sum \Delta s = \sum \frac{\Delta E_s}{i - \overline{J}} = \sum \frac{E_{sd} - E_{su}}{i - \overline{J}} \quad (9-47)$$

where $\Delta E_s$ difference value of specific section energy between the both ends of cross section;

$E_{sd}$, $E_{su}$ is the specific section energy between the both ends of cross section $\Delta s$; $J$ is the average value of the hydraulic slope, $\overline{J} = \dfrac{1}{2}(J_u + J_d)$ or $\overline{J} = \dfrac{\overline{v}^2}{\overline{C}^2 \overline{R}}$, where $\overline{v} = \dfrac{v_u + v_d}{2}$, $\overline{C} = \dfrac{c_u + c_d}{2}$, $\overline{R} = \dfrac{R_u + R_d}{2}$.

The equation (9 – 47) is the basic water surface curve equation based on subsection summation method. Based on the equation (9 – 47), we can gradually calculate the values of the depth of cross section $h$ and the spacing distance $\Delta l$ of the open channel nonuniform flow, and then we can draw the water surface profile along the total process length $l$. In order to get accurate the result, the total process is divided into several minizone $\Delta l$.

The subsection summation method is suited to the steady non-uniform gradually-varied flow of both prismatic channel and non-prismatic channel.

**Example 9 – 5** A long rectangular drainage channel, with bottom width $b = 2$ m, roughness coefficient $n = 0.025$, bottom slope $i = 0.0002$, drainage flow rate $q_v = 2.0$ m³/s, and the end of the channel discharges into the river (Fig.9 – 13). Try to draw a water curve.

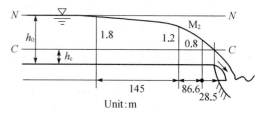

Fig.9 – 13  Water surface line drawing

**Solution:**

(1) Distinguish the nature of the channel bottom slope and the type of water surface line

The normal water depth is $h_0 = 2.26$ m calculated by equation (9 – 6); the critical water depth is $h_c = 0.467$ m calculated by equation (9 – 24). Calculate the values according to $h_0$ and $h_c$, and mark the $N-N$ line and $C-C$ line in Fig.9 – 13. $h_0 > h_c$ is a gentle slope channel, the water depth at the end (falling sill) is $h_c$, the water flow in the channel flows in the gentle slope channel 2 area, and the water surface line is an $M_2$ type precipitation curve.

(2) Calculation of water surface line

There is slow flow in the channel, and the water depth hc at the end is the control water depth, which is calculated upstream.

Take $h_2 = h_c = 0.467$ m, $A_2 = bh_2 = 0.943$ m², $v_2 = \dfrac{q_v}{A_2} = 2.14$ m/s, $\dfrac{v_2^2}{2g} = 0.234$ m, $E_2 = h_2 + \dfrac{v_2^2}{2g} = 0.7$ m, $R_2 = \dfrac{A_2}{\chi_2} = 0.32$ m, $C_2 = \dfrac{1}{n} R_2^{1/6} = 33.07$ m$^{1/2}$/s

Let $h_1 = 0.8$ m, $A_1 = bh_1 = 1.6$ m², $v_1 = \dfrac{q_v}{A_1} = 1.25$ m/s, $\dfrac{v_1^2}{2g} = 0.08$ m, $E_1 = h_1 + \dfrac{v_1^2}{2g} = 0.88$ m, $R_1 = \dfrac{A_1}{\chi_1} = 0.44$ m, $C_1 = \dfrac{1}{n} R_1^{1/6} = 34.94$ m$^{1/2}$/s

Average $\overline{v} = \dfrac{v_1 + v_2}{2} = 1.695$ m/s, $\overline{R} = \dfrac{R_1 + R_2}{2} = 0.38$ m, $\overline{C} = \dfrac{C_1 + C_2}{2} = 34$ m$^{1/2}$/s, $\overline{J} = \dfrac{\overline{v}^2}{\overline{C}^2 \overline{R}} = 0.0065$, $\Delta l_{1-2} = \dfrac{\Delta l}{i - \overline{J}} = \dfrac{-0.18}{-0.0063} = 28.57$ m

Continue to press $h = 1.2$ m、$1.8$ m、$2.1$ m, and repeat the above steps to calculate the length of each section. The calculation results of each section are shown in the table below. Based on the calculated values, the internal water surface line of the drainage channel can be drawn.

**Table 9 − 1  Water surface curve calculation table**

| Section | $h/\text{m}$ | $A/\text{m}^2$ | $v/(\text{m}\cdot\text{s}^{-1})$ | $\bar{v}/(\text{m}\cdot\text{s}^{-1})$ | $\dfrac{v^2}{2g}/\text{m}$ | $E/\text{m}$ | $\Delta E/\text{m}$ |
|---|---|---|---|---|---|---|---|
| 1 | 0.476 | 0.934 | 2.140 |       | 0.234 | 0.70  |        |
| 2 | 0.800 | 1.600 | 1.250 | 1.695 | 0.080 | 0.880 | −0.180 |
| 3 | 1.200 | 2.400 | 0.833 | 1.640 | 0.035 | 1.235 | −0.355 |
| 4 | 1.800 | 3.600 | 0.556 | 0.694 | 0.016 | 1.816 | 0.581  |
| 5 | 2.100 | 4.200 | 0.476 | 0.516 | 0.012 | 2.112 | −0.296 |

| Section | $R/\text{m}$ | $\bar{R}/\text{m}$ | $C/(\text{m}^{1/2}\cdot\text{s}^{-1})$ | $\bar{C}/(\text{m}^{1/2}\cdot\text{s}^{-1})$ | $\bar{J}$ | $i-\bar{J}$ | $\Delta l/\text{m}$ | $\Sigma\Delta l/\text{m}$ |
|---|---|---|---|---|---|---|---|---|
| 1 | 0.320 |       | 33.07 |        |        |         |       |          |
| 2 | 0.440 | 0.380 | 33.94 | 33.505 | 0.006 5 | −0.006 3 | 28.57 | 28.57    |
| 3 | 0.545 | 0.492 | 36.15 | 35.045 | 0.004 3 | −0.004 1 | 86.59 | 115.16   |
| 4 | 0.643 | 0.594 | 37.16 | 36.655 | 0.000 6 | −0.000 4 | 1 452 | 1 567.16 |
| 5 | 0.677 | 0.660 | 37.48 | 37.320 | 0.000 3 | −0.000 1 | 3 288 | 4 855    |

# Exercise

1. What are the hydraulic characteristics of uniform flow in an open channel? Under what conditions is it possible to produce open channel uniform flow?

2. The cross-sectional shape and size of the two open channels are the same, but the bottom slope and roughness are different. When the flow rate passing through is equal, is the critical water depth of the two open channels equal?

3. A certain trapezoidal long straight prism drainage channel has been built, with a length of $l = 1.0$ km, a bottom width of $b = 3$ m, a slope coefficient of $m = 2.5$, a bottom drop of 0.5 m, and a design flow rate of $q_v = 9.0$ m³/s. Trial check whether the channel can meet the design flow rate requirements when the actual water depth $h = 1.5$ m (roughness coefficient $n$ is 0.025).

4. The flow rate of a trapezoidal cross-section loam channel is $q_v = 10.5$ m³/s, the bottom width $b = 8.9$ m, the slope coefficient $m = 1.5$, the normal water depth $h_0 = 1.25$ m, and the roughness coefficient $n = 0.025$. Find the bottom slope $i$ and judge the channel. Is the bottom slope steep or gentle? Is the flow pattern rapid or slow?

5. A rectangular cross-section flat-slope channel has a bottom width $b = 2.0$ m and a flow rate $q_v = 10$ m³/s. When a hydraulic jump occurs in the channel, the water depth before the jump is $h' = 0.65$ m. Find the water depth $h''$ after the jump and the hydraulic jump length $l_j$.

# Chapter 10　Weir Flow

Weir is main water diversion and drainage building in water engineering used to control the flow rate and water level in rivers. It's also a kind of water equipment, widely used in various laboratories, water supply engineering and related industrial enterprises. The main purpose of the research is to explore the relationship between the flow rate and the basic characteristic quantities of the weir, which provides scientific basis for the design of the weir and the calculation of the flow capacity.

This chapter mainly introduces the basic knowledge of the hydraulic characteristics, basic equations, application characteristics, and hydraulic calculation methods of various weir flows, as well as the design and application of bridge openings. In learning, it is required to master the hydraulic characteristics and basic equations of weir flow, the forms and classifications of sharp-crested weirs, ogee weirs, and broad-crested weirs; and the flow rate calculation and influencing factors, and to understand the design method of small bridge aperture.

## 10.1　Definition and Classification of Weir Flow

### 10.1.1　Weir and weir flow

#### 1. Definition

In the subcritical flow in open channel set hydraulic structure of the weir walls or on both sides of the beam narrow side wall to form a barrier, water flows through the discharge at the top of the barrier, overflow surface is not subject to constraints on the open water known for the weir flow (shown in Fig.10 – 1). The barrier called for overflow weir, referred to as weir.

#### 2. The basic characteristic quantities of weir flow

The main purpose of the research is to explore the relationship between the flow rate $q_v$ and the basic characteristic quantities of weir flow. The basic characteristic quantities of weir flow including (Fig.10 – 1):

(1) Correlation head $H$, $H_0$, $H_d$

$H$ is the top of the weir, and the upstream water level of the $L = (3 \sim 5)H$ (not the top surface of the weir crest) is the difference between the upstream water level and the weir crest.

$H_0$ is the head of the weir, which considers the effect of the velocity of the weir when

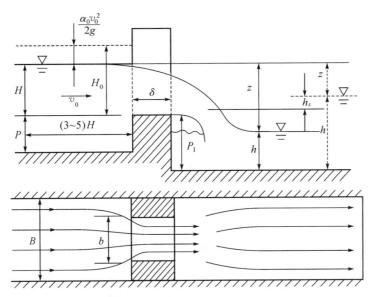

Fig.10 – 1  Weir flow and the basic equation

the velocity is near, $H_0 = H + \dfrac{\alpha_0 v_0^2}{2g}$.

$H_d$ is designed to design the head of the weir. When the actual water head $H_0$ and $H_d$ are equal, the weir flow is designed.

(2) Correlation width $b$, $B$

$b$ is the width of weir, the width of the weir's top.

$B$ is the width of channel, the width of the upper reaches of the overflow weir.

(3) Weir height $P$, $P_1$

The upstream weir height $P$ and the downstream weir $P_1$, respectively, for the top of the weir and the upper and lower reaches of the river bed elevation difference.

(4) Weir crest width $\delta$

The width of the weir crest is also known as the wall width, the width of the weir crest, the thickness of the upper and lower reaches.

(5) The upstream and downstream water level difference $z$

The upstream and downstream water level difference is the difference between (3~5) times of $H$ and the normal water level of the lower reaches of the weir wall.

(6) The downstream depth $h$ and the downstream water level higher than the height of the weir crest $h_s$ (When the water level is high, as the dotted line shown in Fig. 10 – 1), velocity of flow before weir $v_0$ etc.

## 10.1.2  Classification of weir and weir flow

According to the relationship between the thickness $\delta$ of the weir wall and the water head $H$, the weir can be divided into three types of sharp-crested weir, ogee weir and broad-crested weir.

(1) Sharp-crested weir, $\dfrac{\delta}{H} < 0.67$ (shown in Fig. 10 – 2a). The weir flow is formed by

the "water tongue", the water tongue edge to bend down after the first. It is about $0.67H$ from the upstream wall at the top of the weir.

When the width of the weir's top $\delta < 0.67H$, there is no interference to the water tongue, and the thickness of the weir is not affected the flow of water. Such a weir is called a sharp-crested weir.

(2) Ogee weir, $0.67 \leqslant \dfrac{\delta}{H} < 2.5$ (shown in Fig.10 – 2b). The thickness of the weir has a certain effect on the flow, but there is a continuous drop of the water surface. Such a weir is called a ogee weir.

(3) Broad-crested weir, $2.5 \leqslant \dfrac{\delta}{H} < 10$ (shown in Fig.10 – 2c). Weir flow at the entrance and exit of the weir to form a two drop phenomenon. Such a weir is called a broad-crested weir.

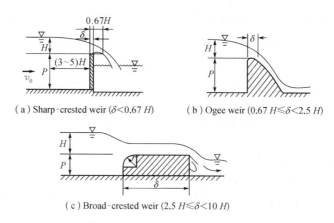

(a) Sharp-crested weir ($\delta < 0.67 H$)  
(b) Ogee weir ($0.67 H \leqslant \delta < 2.5 H$)  
(c) Broad-crested weir ($2.5 H \leqslant \delta < 10 H$)  

Fig.10 – 2 Classification of weir

When $\dfrac{\delta}{H} > 10$, water head loss along the way can not be ignored, the flow does not belong to the weir flow.

There are other methods, such as according to whether the weir width $b$ and the channel width $B$ is equal or not, the weir can be divided into side shrinkage weir ($b < B$) and non side shrinkage weir ($b = B$). According to whether the downstream water level will affect the weir's ability to flow, the weir can be divided into free weir flow (free flow) and submerged weir flow (submerged flow) (shown in Fig.10 – 3).

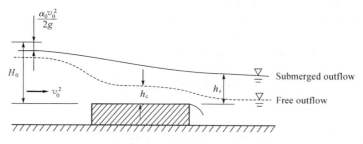

Fig.10 – 3 Forms of outflow

## 10.2 Basic equation of weir flow

### 10.2.1 Basic equation of free flow at the non side shrinkage weir

The basic equation of the weir flow is derived from the case study on the non side shrinkage weir free flow of sharp-crested weir.

As shown in Fig. 10-4, reference plane 0-0 is over weir crest, section 1-1 of gradually varied flow sets at (3~5) times of $H$ distant from upstream weir surface, cross section 2-2 of the flow sets at the intersection of reference plane and the centerline of the water tongue. Bernoulli equation:

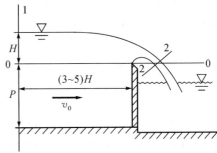

Fig. 10-4 Weir flow

$$H + \frac{P_a}{\lambda g} + \frac{\alpha_0 v_0^2}{2g} = \frac{P_2}{\lambda g} + \frac{\alpha_2 v_2^2}{2g} + \zeta \frac{v_2^2}{2g}$$

Because the water tongue is in contact with the atmosphere, $P_2 = P_a$, order

$$H_0 = H + \frac{\alpha_0 v_0^2}{2g}, \; \alpha_2 = \alpha, \; v_2 = v$$

so

$$v = \frac{1}{\sqrt{\alpha + \zeta}} \sqrt{2gH_0} = \varphi \sqrt{2gH_0}$$

$$q_v = Av = \varphi b e \sqrt{2gH_0}$$

where $\varphi = \dfrac{1}{\sqrt{\alpha + \zeta}}$ is velocity factor, $B = b$ is overflow width, $e$ is the thickness of water tongue.

If order $e = kH_0$,

$$q_v = \varphi k e \sqrt{2g} H_0^{3/2} = mb \sqrt{2g} H_0^{3/2} \qquad (10-1)$$

where $m$ is called weir flow factor, $m = \varphi k$. Upper equation is the basic equation of the non side shrinkage weir free flow.

If the influence of approach velocity is considered included in the flow factor, upper equation can be written:

$$q_v = m_0 b \sqrt{2g} H_0^{3/2} \qquad (10-2)$$

where $m_0$ is weir flow rate factor considering the influence of approach velocity.

### 10.2.2 Weir flow equation of submerged flow and the side shrinkage flow

When the downstream water level exceeds a certain height of the weir, the submerged overflow will occur, when the flow rate is constant, the upstream water level is elevated by the backwater of downstream water level, and the discharge capacity of the weir began to decrease. The right side of the equation (10-1) needs to be multiplied by the coefficient of the flood $\sigma(\sigma<1)$.

When the width of ogee weir is less than the width of the channel, affecting by the sluice pier and the side pier, the side wall is shrinking after the flow of water into the weir. So the effective flow over the flow section is reduced, head loss increase, weir flow capacity decreased. The right side of the equation (10-1) needs to be multiplied by the coefficient of side shrinkage $\varepsilon$ ($\varepsilon<1$). The coefficient of side shrinkage relates to the shape of sluice pier and the side pier.

Considering the influence of submerged flow and the side shrinkage, upper equation (10-1) can be written:

$$q_v = \sigma \varepsilon m b \sqrt{2g} H_0^{3/2} \tag{10-3}$$

Upper equation is suitable for various types of weir and various influencing factors of the weir.

All equations above is derived from sharp-crested weir, but the factor of the equation is based on the different weir type, different influence factors and influence degree derived from the experiment. Different weir, different flow conditions have different values. So all equations above are suitable for various types of weir and various influencing factors of the weir.

## 10.3 Sharp-crested weir

### 10.3.1 Sharp-crested rectangular weir

As shown in Fig.10-5, when weir shape is rectangular, it is called rectangular weir and commonly used as flow measuring equipment. The flow rate is usually calculated according to the measured water head. If the influence of approach velocity is considered included in the flow rate factor, weir width is often equal to the width of the weir, without side effects. So flow rate calculation equation is the equation (10-2).

$$q_v = m_0 b \sqrt{2g} H^{3/2}$$

$$m_0 = \left(0.405 + \frac{0.0027}{H}\right)\left[1 + 0.55\left(\frac{H}{H+P}\right)^2\right] \tag{10-4}$$

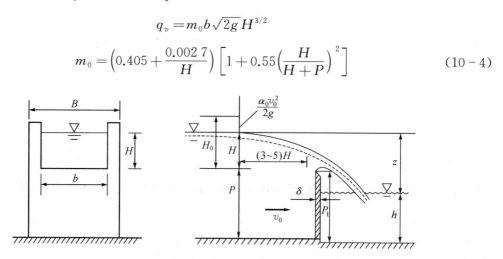

Fig.10-5 Thin-walled rectangular weir

where $H$ is weir head, $P$ is upstream weir height, $b$ is weir width, meter gauge. The equation suitable scope: $H = 0.1 \sim 1.24$ m, $b = 0.2 \sim 2.0$ m, $P = 0.2 \sim 1.13$ m.

If there is a side shrinkage, the flow factor $m_0$ is included in the side effects. It should be calculated according to the following empirical equation:

$$m_0 = \left(0.405 + \frac{0.0027}{H} - 0.03\frac{B-b}{B}\right)\left[1 + 0.55\left(\frac{H}{H+P}\right)^2\left(\frac{b}{B}\right)\right]^2 \quad (10-5)$$

where $B$ is channel width. $B$, $H$, $P$ are in m gauge.

### 10.3.2 Sharp-crested triangular weir

When weir shape is triangular, it is called triangular weir. As shown in Fig.10-6, it is commonly used as small flow measuring equipment in laboratory ($q_v < 0.1$ m³/s). When weir shape is at right angle ($\theta = 90°$), it is called right angled triangular weir. Thompson experiential equation is used as flow rate calculation equation.

$$q_v = 1.4 H^{5/2} \quad (10-6)$$

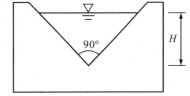

**Fig.10-6 Sharp-crested triangular weir**

where $q_v$ is in m³/s; $H$ is in m, equation suitable scope:

$$0.05 \text{ m} < H < 0.25 \text{ m}, \ P \geqslant 2H, \ B \geqslant (3 \sim 4)H$$

The experimental results show that the submerged condition of the sharp-crested weir is as followed:

(1) Downstream water level is over the top of the weir.

(2) Submerged hydraulic jumps happen in downstream.

When sharp-crested weir is used as flow measuring equipment, it should pay attention to:

(1) Because of the large fluctuation of the water jump, it is not suitable to work under the submerged condition.

(2) The lower part of the water tongue should be kept in contact with the atmosphere, otherwise it will affect the accuracy of the measurement.

## 10.4 Ogee weir

According to the profile of ogee weir, the utility weir can be divided into the curve shape (Fig. 10-2b) and the fold line shape (Fig.10-7). Only the curve shape of ogee weir mentioned here.

### 10.4.1 Hydraulic characteristic of weir

Fig.10-8 shows that sectional shape of the

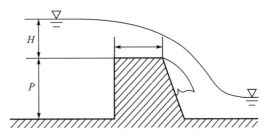

**Fig.10-7 Fold line shape ogee weir**

curved weir consists of the following components.
(1) Upstream straight section $AB$;
(2) Crest curve segments $BOC$;
(3) Downstream slope segment $CD$ and ogee section $DE$.

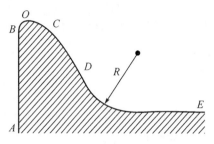

Fig.10-8  Curved sectional weir

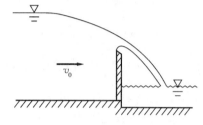

Fig.10-9  Curve of sublingual sharp-crested weir edge

Ogee weir is actually designed by the curve of sharp-crested weir edge curve compared with Fig.10-9. If it is slightly higher fat, crest of water form a non-vacuum. If it is a little slightly thin, cross-sectional of weir form a vacuum. The experiment shows that surface pressure of weir decreases, the flow increases capacity. Vacuum weir can increase discharge. However, the vacuum is too large which leading to the weir vibration. So weir vacuum in the actual project should be avoided.

The role of the head increases with the flow coefficient from the experiment. Pressure of weir decreases with the role of head. A weir on the same terms, when the role of the head $H_0$ is greater than the design $H_d$, it could create a vacuum weir even under non-vacuum design head.

There are a variety of mature curved-shaped practical weirs, among which WES profile weir given by American Waterway Experiment Station is the most widely used in the engineering field at present.

### 10.4.2  Flow calculation equation

The equation of weir flow rate remains the equation (10-3). But the flow coefficient $m$ take different values.

$$m = -0.024 \frac{H_0}{H_d} + 0.185\,03 \sqrt{\frac{H_0}{H_d}} + 0.340\,6 \qquad (10-7)$$

Conditions are that cross-sectional surface perpendicular to the upstream weir. And $\frac{P}{H_d} \geqslant 1.33$, $\frac{H_0}{H_d} < 1.5$. When $\frac{H_0}{H_d} \leqslant 1.0$, dam pressure is non-negative. When $\frac{H_0}{H_d} > 1.0$, dam pressure generates negative. When $\frac{H_0}{H_d} = 1.33$, negative pressure can be achieved $0.5H_d$.

When $\frac{H_0}{H_d} = 1.0$, $m = m_d = 0.502$.

Conditions of submerged ogee weir and sharp-crested weir are the same.

(1) Downstream water level is over the top of the weir.

(2) Submerged hydraulic jumps happen in downstream.

## 10.5 Broad-crested weir

### 10.5.1 Hydraulic characteristics of broad-crested weir

Broad-crested weir can be divided into broad-crested weir with sill and broad-crested weir without sill, such as flow cross small bridge, short culvert over the water and so on. Characteristic of stream flow freely out over time has typical features as shown in Fig. 10 - 10. Not far from the water in the weir import occurs first landing, crest forming a shrink depth. Then the formation of flow lines approximately parallel to the crest of the gradient flow. In engineering, if hydraulic characteristics in line with broad crested wier, and meet the conditions $2.5 < \dfrac{\delta}{H} < 10$, it can be considered as broad-crested weir.

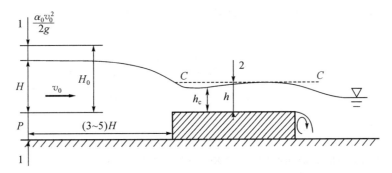

Fig. 10 - 10   Broad-crested weir

### 10.5.2 Flow rate calculation equation

Flow rate calculation equation of broad-crested weir remains the equation (10 - 3).

**1. Flow coefficient of broad-crested weir**

Establish the energy equation by taking sections 1 - 1 and 2 - 2 in Fig. 10 - 10. When the head loss is ignored, and $\alpha = 1.0$,

$$v = \sqrt{2g(H_0 - h)}$$

$$q_l = h\sqrt{2g(H_0 - h)}$$

Because excluding the loss, flow through the weir is maximum at this time.

$$\frac{\partial q_l}{\partial h} = \frac{\partial(h\sqrt{2g(H_0 - h)})}{\partial h} = 0 \qquad (10 - 8)$$

$$H_0 = \frac{3}{2}h \qquad (10 - 9)$$

It can be obtained from equation (10 - 8) and equation (10 - 9):

$$h = \sqrt[3]{\frac{q_l^2}{g}} = h_c \quad (10-10)$$

Equation shows that, excluding resistance, crest depth is critical depth.

$$m = 0.385$$

Excluding flow resistance, $m = 0.385$ is the maximum flow coefficient of broad-crested weir. Generally, value of $m$ is between 0.32 and 0.385.

When $0 < \frac{P}{H} \leqslant 3$ in rectangular broad-crested weir imports,

$$m = 0.32 + 0.01 \times \frac{3 - \frac{P}{H}}{0.46 + 0.75\frac{P}{H}} \quad (10-11)$$

When $\frac{P}{H} > 3$, $m = 0.32$.

And when $\frac{r}{H} \geqslant 0.2$ in crest arc imports,

$$m = 0.36 + 0.01 \times \frac{3 - \frac{P}{H}}{1.2 + 1.5\frac{P}{H}} \quad (10-12)$$

When $\frac{P}{H} > 3$, $m = 0.36$.

### 2. Lateral shrinkage coefficient of broad-crested weir ($\varepsilon$)

$\varepsilon$ is calculated by using empirical equation of broad-crested weir hole as follows:

$$\varepsilon = 1 - \frac{a}{\sqrt[3]{0.2 + \frac{P}{H}}} \sqrt[4]{\frac{b}{B}} \left(1 - \frac{b}{B}\right) \quad (10-13)$$

where $a$ is a mound-like factor, $a = 0.19$ in rectangular pier, $a = 0.1$ in round pier.

### 3. Submerged factor of broad-crested weir ($\sigma$)

$\sigma$ is calculated by using empirical equation as follows:

$$\sigma = -96.01\left(\frac{h_s}{H_0}\right)^3 + 235.32\left(\frac{h_s}{H_0}\right)^2 - 193.23\left(\frac{h_s}{H_0}\right) + 54.14 \quad \left(0.8 \leqslant \frac{h_s}{H_0} \leqslant 0.98\right) \quad (10-14)$$

Submerged criterion of broad-crested weir is:

$$\frac{h_s}{H_0} \geqslant 0.8 \quad (10-15)$$

**Example 1** A rectangular cross-section channel was constructed with a single hole broad-crested weir for water diversion irrigation, as shown in Fig. 10-11. Known: flow velocity $v_0 = 1.48$ m/s in the canal in front of the weir, channel width $B = 3$ m, weir crest width $b = 2$ m, weir height $h_1 = h_2 = 1$ m, weir crest head $H = 2$ m, weir crest inlet is a right angle inlet, pier head shape is rectangular, and downstream water depth $h = 2$ m. Try to calculate the flow rate through the weir.

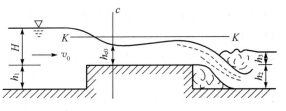

Fig.10-11 Single hole wide top weir

**Solution:** (1) Distinguish the flow form.

$$h_s = h - h_2 = 2 - 1 = 1 \text{ m} > 0$$

$$H_0 = H + \frac{a_0 v_0^2}{2g} = 2 + \frac{1.48^2}{2 \times 9.8} = 2.11 \text{ m}$$

Thus $\dfrac{h_s}{H_0} < 0.8$, Not meeting the submerged outflow conditions, it is free outflow, $\sigma_s = 1$.

(2) Calculate flow factor.

Due to $b < B$, there is contraction phenomenon. The weir mouth is a right angled inlet, $\dfrac{h_1}{H} = 0.5 < 3$, which can be obtained from the equation

$$m = 0.32 + 0.01 \frac{3 - \dfrac{h_1}{H}}{0.46 + 0.75 \dfrac{h_1}{H}} = 0.35$$

(3) Calculate shrinkage coefficient.

The weir is a single hole broad-crested weir, so it can be concluded that

$$\varepsilon = 1 - \frac{a}{\sqrt[3]{0.2 + \dfrac{h_1}{H}}} \sqrt[4]{\frac{b}{B}} \left(1 - \frac{b}{B}\right)$$

$$= 1 - \frac{0.19}{\sqrt[3]{0.2 + 0.5}} \times \sqrt[4]{\frac{2}{3}} \times \left(1 - \frac{2}{3}\right) = 0.936$$

(4) Calculate flow.

$$q_v = \sigma \varepsilon m b \sqrt{2g} H_0^{3/2} = 1 \times 0.936 \times 0.35 \times 2 \times \sqrt{2 \times 9.8} \times 2.11^{1.5} = 8.88 \text{ m}^3/\text{s}$$

## 10.6 Hydraulic computation of flow cross small bridge

### 10.6.1 Hydraulic calculation of the bridge hole overcurrent

1. Submerged standard

Bridge over the stream can be divided into free outflow and submerged flow. By the experiment that, when the bridge downstream water depth is $h < 1.3h_c$, the flow is rapids. Downstream water level does not affect the flow of water across the bridge, the bridge over the stream is freestyle. And when the bridge downstream water depth is $h \geqslant 1.3h_c$, the flow is subcritical flow (as shown in Fig.10 – 11). Downstream water level affect the flow of water across the bridge, the bridge over the stream is submerged (as shown in Fig.10 – 12).

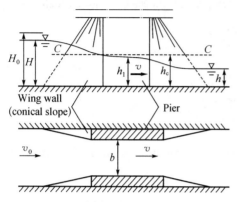

Fig.10 – 11  Overcurrent of freestyle bridge hole       Fig.10 – 12  Overcurrent of submerged bridge hole

### 2. Hydraulic calculation

(1) The depth of bridge opening $h_1$

In engineering design

$$h_1 = \begin{cases} h & \text{(submerged)} \\ \psi h_c & \text{(freestyle)} \end{cases} \quad (10-16)$$

where $\psi$ is shrinkage coefficient by experience. $\psi = 0.75 \sim 0.80$ in non-smooth import, and $\psi = 0.8 \sim 0.85$ in smooth import. Some design methods think that $\psi = 1.0$.

(2) Hydraulic calculation

Hydraulic calculation is deducted by the Bernoulli equation and the continuity equation.

$$v = \varphi \sqrt{2g(H_0 - h_1)} \quad (10-17)$$

$$q_v = \varepsilon b h_1 \varphi \sqrt{2g(H_0 - h_1)} \quad (10-18)$$

## 10.6.2 The principles of hydraulic calculation

In order to design a safe and economical bridge, hydraulic calculation should meet the following requirements.

(1) Aperture reach hydrology calculation to determine the design flow.

(2) Flow rate of bridge opening can not exceed the allowable flow rate.

(3) The water depth of bridge is no more than the water depth of allow.

## 10.6.3 Hydraulic calculation method

### 1. Calculate the critical depth of bridge opening ($h_c$)

Critical depth is:

$$h_c = \sqrt[3]{\frac{\alpha q_v^2}{g(\varepsilon b)^2}} \qquad (10-19)$$

Considering $h_1 = \psi h_c$

$$q_v = \varepsilon b h_1 v_{max} = \varepsilon b \psi h_c v_{max}$$

We can get the relationship between allowable velocity and critical depth from the equation (10-19).

$$h_c = \frac{\alpha \psi^2 v_{max}^2}{g} \qquad (10-20)$$

## 2. Calculate the bridge opening aperture ($b$)

According to $q_v = \varepsilon b h_1 v_{max}$, pore size of bridge is:

$$b = \frac{q_v}{\varepsilon h_1 v_{max}} \qquad (10-21)$$

In engineering, bridge opening aperture is generally used standard aperture $B$ ($B > b$). Standard aperture is often used in practical engineering, and there are various standard apertures for railways, highways bridges, including 4 m, 5 m, 6 m, 8 m, 10 m, 12 m, 16 m, 20 m, etc.

## 3. Check the water depth of backwater in front of the bridge ($H$)

The water depth of backwater in front of the bridge can be checked by the following equation:

$$H \approx H_0 = h_1 + \frac{q_v^2}{2g\varphi^2(\varepsilon B h_1)^2} \leqslant H' \qquad (10-22)$$

where $H$ is the allowable depth of backwater; $h_1$ is the water depth under the bridge aperture. If it is the free outflow, $h_1 = \psi h_c$; If it is the submerged flow, $h_1 = h$.

**Example 10-2** According to hydrological calculations, the design flow rate of the small bridge $q_v = 30 \text{ m}^3/\text{s}$. According to the flow level relationship curve of the downstream river section, the downstream water depth $h = 1.0$ m at this flow rate is calculated. The allowable depth of backwater in front of the bridge is $H' = 2$ m, and the allowable flow velocity under the bridge is $v' = 3.5$ m/s. Based on the imported form of small bridge, the coefficients of each item were found to be: $\varphi = 0.9$, $\varepsilon = 0.85$, $\psi = 0.8$. Design the aperture of this small bridge.

**Solution:** (1) Calculate the critical water depth.

$$h_c = \frac{\alpha \psi v'^2}{g} = \frac{1.0 \times 0.8^2 \times 3.5^2}{9.8} = 0.8 \text{ m}$$

Therefore, the crossing of the small bridge is a free outflow.

(2) Calculate the aperture of the small bridge.

$$b = \frac{q_v}{\varepsilon v' \psi h_c} = \frac{30}{0.85 \times 0.8 \times 0.8 \times 3.5} = 15.8 \text{ m}$$

Take standard aperture $B = 16$ m.

(3) Recalculate the critical water depth.

$$h_c = \sqrt[3]{\frac{\alpha q_v^2}{g(\varepsilon b)^2}} = \sqrt[3]{\frac{1 \times 30^2}{9.8 \times (0.85 \times 16)^2}} = 0.79 \text{ m}$$

$$1.3 h_c = 1.3 \times 0.792 = 1.03 \text{ m} > h$$

Therefore, the flow over the small bridge is still free flowing. The actual flow velocity of the bridge opening is

$$v = \frac{q_v}{\varepsilon B \psi h_c} = \frac{30}{0.85 \times 16 \times 0.8 \times 0.792} = 3.48 \text{ m/s}$$

$v < v'$, no erosion will occur.

(4) Check the depth of backwater in front of the bridge.

$$H \approx H_0 = \frac{v^2}{2g\varphi^2} + \psi h_c = \frac{3.48^2}{2 \times 9.8 \times 0.9^2} + 0.8 \times 0.792 = 1.396 \text{ m} \leqslant H'$$

$H < H'$, meeting design requirements.

# Exercises

1. Install a rectangular thin-walled weir at the end of a rectangular cross-section water tank, with a width of $B = 2.0$ m, a width of $b = 1.2$ m, and a height of $P = P' = 0.5$ m. Calculate the flow rate of non-submerged weir flow at a water head of $H = 0.25$ m.

2. There is a wide crest weir with a thickness of the crest $\delta = 16$ m, with a weir crest head of $H = 2$ m. If the wpsteam and downstream water level and weir height remain unchanged. When $\delta$ reduced to 8 m and 4 m respectively, does the weir still belong to a wide crest weir?

3. Known design flow rate of $q_v = 15 \text{ m}^3/\text{s}$ and an allowable velocity of $v' = 3.5$ m/s. The downstream water depth of the bridge is $h = 1.3$ m. Taking $\varepsilon = 0.9$, $\psi = 1.0$, $\varphi = 0.9$, allowable height of backwater $H' = 2.2$ m, Try to design small bridge aperture $B$.

4. A right angled inlet with no side shrinkage and a wide top weir, with a width of $b = 4.0$ m and a height of $P_1 = P_2 = 0.6$ m, with a water head $H = 1.2$ m in front of the weir. What are the flow rates $q_v$ that pass through the downstream water depths $h = 0.8$ m and 1.7 m, respectively?

5. Constructing a curved practical overflow weir in a rectangular river channel, with downstream weir height $h_1 = 6$ m, overflow width $B = 60$ m, weir flow rate $q_v = 480 \text{ m}^3/\text{s}$, weir flow factor $m = 0.45$, flow velocity factor $\varphi = 0.95$. Find the water depth $h_c$ of the downstream contraction section of the weir.

# 第1章 引　　言

## 1.1 概论

### 1.1.1 流体力学的任务

流体力学是力学的一个独立分支，是研究流体（液体、气体和等离子体）作用在其上的力学规律及其应用的一门科学。

流体力学包括两个分支。一是关于静止的流体，包括研究流体处于静止或相对平衡的条件，这个分支称为流体静力学或静水力学。二是关于流体流动规律——运动中流体（液体和气体）的自然科学，这个分支称为流体动力学。

流体力学按研究内容又可分为理论流体力学和应用流体力学（也称为工程流体力学）。理论流体力学侧重于理论研究和数学推理，而应用流体力学则侧重于如何解决实际工程问题。本教材以应用为主，属于后者。

### 1.1.2 流体力学的发展历史

**1. 流体力学发展的主要阶段**

流体力学起源于西西里岛的希腊学者阿基米德所著的《论浮体》一文。他首次总结了静止液体的力学性能（静止液体的机械性能），奠定了流体静力学的基础。

1687年，牛顿在其名著《自然哲学的数学原理》一书中讨论了流体阻力和波浪运动等内容，标志着流体力学进入了主要发展时期，并经历了三个主要发展阶段。

第一阶段：伯努利提出的液体运动的能量估计和欧拉提出的液体运动分析方法为液体运动规律的研究奠定了坚实的基础。由此，属于数学的经典"流体力学"形成了。

第二阶段：纳维和斯托克斯在经典"流体力学"的基础上，提出了著名的粘性流体的基本运动方程，为流体力学的长远发展奠定了理论基础。然而，由于其数学求解的复杂性和流体模型的局限性，无法完美解决工程问题。因此，形成了用实验方法总结概括经验公式的"实验流体力学"。由于缺乏理论基础，最初的"实验流体力学"的发展受到限制。

第三阶段：自19世纪末以来，经典流体力学和实验流体力学的内容不断更新。同时，理论分析方法与实验分析方法相结合，产生了理论与实践并重的现代流体力学。20世纪60年代以后，由于计算机科学的发展和普及，流体力学得到了越来越广泛的应用。

**2. 流体力学领域的重大事件**

1738年，瑞士数学家、物理学家伯努利在其名著《流体力学》中提出了伯努利方程。

1757年，欧拉提出了一组重要的非粘性流动方程，即现在的欧拉方程。

1781年，拉格朗日首次提出了"流函数"的概念。

1822 年,法国工程师、物理学家纳维,1845 年,英国数学家、物理学家斯托克斯提出了著名的纳维-斯托克斯方程。

1876 年,英国物理学家雷诺发现了流体的两种流动状态:层流和湍流。

1886 年,德国物理学家亥姆霍兹提出了亥姆霍兹定理,用于无粘性流体中的涡流动力学。

19 世纪后期,相似性理论被提出。

1904 年,德国工程师普朗特在其开创性的论文《非常小摩擦下的流体流动》中描述了边界层。

20 世纪 60 年代以来,计算流体力学得到了迅速发展,流体力学的内涵也得到了不断的丰富与完善。

中国的水利行业历史悠久,人们通过长期的实践逐渐加深了对流体运动规律的认识。例如:

早在 4 000 年前,"大禹治水"的故事表明,当时的人们就已经知道治水要顺应水性。

秦朝在公元前 256—前 210 年,修建了三大水利工程(都江堰、郑国渠和灵渠);隋朝的大运河建于公元 587—610 年,这表明当时人们对明渠的水流和堰流的认识达到了相当高的水平。郑和凭借对洋流和气流的认识及利用七下西洋,创造了人类航海史上的奇迹。

清朝雍正年间,何梦瑶在《算迪》一书中提出,水的流量等于过水断面面积乘以断面平均流速的计算方法。

### 1.1.3 流体力学研究方法

**1. 理论方法**

理论方法就是:通过对流体的物理性质和流动特性进行科学抽象,提出合理的理论模型,并根据机械运动的普遍规律,建立控制流体运动的封闭方程。这样,特定的流动问题可以转化为数学问题,从而根据相应的边界条件和初始条件进行求解。理论方法论的关键在于提出理想模型,通过数学方法得出理论结果,从而揭示流体运动规律。然而,由于数学上的困难,很难精确地解决许多实际的流动问题。

在理论方法中,流体力学中引用的主要定律有:

(1) 质量守恒定律:$\dfrac{dm}{dt}=0$

(2) 动量定律:$\sum F = \dfrac{d(mu)}{dt}$

(3) 牛顿第二运动定律:$F = ma$

(4) 机械能转换和守恒定律:动能+压能+位能+能量损失=常数

因为纯理论方法存在数学困难,在计算机尚未发展的时代,一套理论与实验相结合的分析求解方法得到了逐渐完善。这套分析方法简单实用,直到计算机高度发达的今天,它仍然适用。

**2. 实验方法**

应用流体力学是一门理论与实践相结合的基础学科。许多实用的公式和因数都是从实验中推导出来的。许多可以通过现代理论分析和数值计算解决的工程问题最终也需要通过实验加以检验和修正。

(1) 实验研究的主要方法有:a. 原型测量(直接观察实际工程或自然流量分布);b. 统实

验(在实验室系统研究人工流动现象);c. 模型研究(模拟实际工程条件,通过预览或重复流动现象进行研究)。

(2) 实验研究的基本理论是相似理论和量纲分析。

**3. 数值方法**

数值方法是通过各种离散化方法建立各种数值模型,然后通过数值计算和数值实验在时间和空间上收集大量数据,最终得到定量描述流场的数值解。在过去的二三十年里,这种方法学得到了极大的发展,形成了一门新学科——计算流体力学。

**4. 简化流动分析**

由于流体运动状态的复杂性,考虑流体运动中涉及的所有因素是非常困难且没有必要的,因此,有必要对各种运动状态进行分类和简化。例如,流动可分为一维流和多维流、恒定流和非恒定流、均匀流和非均匀流、急变流和渐变流、层流和湍流等。

## 1.1.4 土建工程中的流体力学

**1. 课程性质和目标**

(1) 性质:本课程是土木工程和水利专业的一门必修专业基础课。研究对象为水、气体及可压缩流体。研究内容包括流体平衡、流体机械运动规律及其工程应用。

(2) 目的:本课程为学生提供流体运动的基本概念、理论、计算方法和实验技能,有助于培养学生分析问题的能力以及在其未来的学习和工作中所需的创新能力。

(3) 地位:本课程以流体力学原理为基础,开设了水文学、水力学、土力学、工程地质学、土木工程、水工建筑物和建筑设备等课程,帮助学生深入理解土木工程、水利与大气以及水环境等学科间的关系。

(4) 其他:本课程是建筑工程专业人员的必修课。

**2. 流体力学在土木工程中的应用**

流体是人类生活和生产中经常会遇到的一种物质形式。许多科学技术部门或专业都涉及流体力学。水利工程、土木工程、交通运输、机械制造、石油开采、化学工业和生物工程等领域的流体问题的解决均依赖于流体力学相关知识。因此,流体力学课程中的内容已广泛应用在生活中的各个领域。

(1) 应用于建筑工程,如地基降水、路基排水、地下水渗流、水下和地下建筑物的受力分析、围堰施工、海洋平台的浮力和抗外部干扰能力等。

(2) 应用于市政工程,如桥涵孔径设计、给排水、管网计算、泵站和水塔设计、隧道通风等。给水排水系统设计和运行控制的理论基础是流体力学理论。

(3) 应用于河道泄流能力、堤坝受力和渗流、闸坝防洪泄流能力等防洪工程。

(4) 应用于建筑环境和设备工程如供热、通风和泵站设计等。

(5) 应用于水利方面。水利比其他课程更依赖于流体力学。对于水利领域的专业人员来说,工程水力学也是一门必修课。

**3. 本课程的基本要求**

(1) 完全理解基本理论

① 理解、掌握流体力学领域的基本概念。

② 掌握总流量的分析方法,了解量纲分析与实验相结合的方法以及求解简单平面势流的方法。

③ 掌握流体运动的能量转换和水头损失规律。
（2）具备对一般流动问题进行分析和计算的能力，包括：
① 计算水力负荷。
② 计算隧道、渠和堰的过流能力以及井的渗流。
③ 分析和计算水头损失。
（3）掌握水位、压力、流速和流量的常规测量方法，知道如何观察流动现象、分析实验数据和编写实验报告。
（4）掌握流体力学的基本知识，即基本概念、基本方程和基本应用。

## 1.2 流体与建模

### 1.2.1 物质的三种状态

物质在自然界中通常以三种状态存在：气体、液体和固体。液体和气体统称为流体。从力学分析方面来看，流体和固体对外力抵抗能力的不同是两者之间的主要区别。固体的特征是形状和体积恒定，这是因为其分子之间的平均距离很小而内聚力很大使其既能承受压力，又能承受拉力和抵抗拉伸变形。但流体由于分子间距离大、内聚力小，因而只能承受压力，几乎不能承受张力，不能抵抗拉伸变形。在任何微小的切应力下，流体都容易变形或流动。

虽然液体和气体都是流体，但两者有一定的区别。与气体分子相比，液体分子之间的距离小而内聚力大。因此，液体可以保持相对恒定的体积并形成自由表面。同时，它们之间的另一个本质区别在于气体的可压缩性更大。

### 1.2.2 流体质点与连续介质模型

从分子物理学的观点来看，和其他物质一样，流体也是由大量作无规则运动的分子组成的，分子之间存在空隙。从技术上讲，由于分子之间存在空隙，导致流体是不连续的，因而描述流体的物理量（质量密度 $\rho$、速度 $v$、压力 $p$）在空间中的分布也不连续。流体分子运动的随机性也会导致任一空间点的物理量在时间上的不连续性。

在标准条件下，大约有 $3.3 \times 10^{22}$ 个分子存在于 $1 \text{ cm}^3$ 的流体中，分子之间的距离为 $3 \times 10^{-8} \text{ cm}$。流体力学的任务是研究整个流体的宏观性质和宏观力学运动规律，而不是研究分子的微观运动。所研究的流体力学宏观问题中，在流动空间和时间中使用的所有特征尺度和特征时间都远大于分子距离和碰撞时间。因此，引入了流体质点和连续介质模型两个概念以便于研究流体运动规律。

**1. 流体质点**

与所考虑的整个流体部分占据的体积大小相比，流体质点所占的体积很小。另一方面，流体质点分子数量又必须很大，具有一定质量。这种流体质点也称为流体微团。

**2. 连续介质模型**

流体连续介质模型是假定流体是由充满整个空间的流体粒子组成的连续体，流体的物理性质和物理量是连续的。因此，物理量可以看作是时空连续函数。数学分析中的连续函数理论可以用于分析流体运动，这一假设适用于大多数流体。然而，当气体很稀薄时，流体应被视为不连续体,此时连续介质模型不适用。

## 1.3 流体的主要物理性质

流体的重要物理性质包括惯性、压缩性、膨胀性、粘性和表面张力等。

### 1.3.1 惯性

惯性是任何物理物体对其保持原有运动状态的任何变化(包括其速度、方向或静止状态的变化)的阻力。物体保持匀速直线运动是一种趋势。质量是物体惯性的定量量度,物体的质量越大,惯性就越大。质量密度是指流体单位体积的质量,通常以 $\rho$ 表示,单位为 $kg/m^3$。对于均质流体,质量密度定义为质量除以体积,如下式:

$$\rho = \frac{m}{V} \tag{1-1a}$$

式中,$m$ 为质量,$V$ 为体积。

如果流体不均匀,则其密度在流体的不同区域之间变化。在这种情况下,任何给定位置周围的密度都是通过计算该位置周围小体积($\Delta V$)的密度来确定的。在无限小体积的限制下,非均匀流体在某一点的密度变为

$$\rho = \lim_{\Delta V \to 0} \frac{\Delta m}{\Delta V} = \frac{dm}{dV} \tag{1-1b}$$

式中,$dV$ 为位置 $r$ 处的基本体积。

物体的密度随温度和压强的变化而变化,这种变化对于液体来说通常很小。液体的密度一般可看作常数,例如,水的密度为 $1\,000\ kg/m^3$,汞的密度为 $13\,600\ kg/m^3$。

气体的密度随温度和压强的变化较大。在 $0\ ℃$ 和一个标准大气压下,空气的密度为 $1.29\ kg/m^3$。

### 1.3.2 压缩性

**1. 液体的压缩性**

在流体力学中,液体的压缩性是液体的相对体积变化对其所受压力(或平均应力)变化的度量。液体的压缩性可以用体积压缩率来量化,压缩率可用 $\kappa$ 表示,单位为 $Pa^{-1}$。液体的压缩率可由下式确定:

$$\kappa = -\frac{1}{V}\frac{dV}{dp} \tag{1-2}$$

式中,$V$ 为体积,$p$ 为压力。

随着压力的增加,液体体积变小,质量不变($dm=0$),导致密度增加。相对增量之间的关系可以通过 $dm = d(\rho V) = \rho dV + V d\rho = 0$ 来确定,即:

$$-\frac{dV}{V} = \frac{d\rho}{\rho}$$

代入式(1-2)，变为

$$\kappa = \frac{1}{\rho}\frac{\mathrm{d}\rho}{\mathrm{d}p} \tag{1-3}$$

$\kappa$ 值越小，压缩液体的难度就越大。

在工程中，体积模量 $K$ 用来表示液体的压缩性。体积模量是压缩率（$\kappa$）的倒数，即

$$K = \frac{1}{\kappa} = -V\frac{\mathrm{d}p}{\mathrm{d}V} = \rho\frac{\mathrm{d}p}{\mathrm{d}\rho} \tag{1-4}$$

体积模量的单位为 $Pa(N/m^2)$，与压强的单位相同。$K$ 值越大，压缩液体的难度就越大。当 $K$ 趋于无穷大时，液体不可能被压缩。

$K$ 和 $\kappa$ 值随液体种类不同而不同。例如，汞的压缩性约为水的 8%，而硝酸的压缩性约为水的 6 倍。对于同一种液体，$\kappa$ 和 $K$ 的值随温度和压强而变化，然而，这种变化很小。

水的体积模量在一定温度和中等压强下变化不大，其体积模量可近似表示为

$$\frac{V_2 - V_1}{V_1} = -\frac{p_2 - p_1}{K} \quad 或 \quad \frac{\Delta V}{V_1} = -\frac{\Delta p}{K} \tag{1-5}$$

在工程设计领域，水的体积模量约为 $2.1 \times 10^9$ Pa，表明当压强改变一个大气压时，体积的相对变化值 $\left(\frac{\Delta V}{V_1}\right)$ 约为 $\frac{1}{20\,000}$。因此，当 $\Delta p$ 不大时，水的压缩性可以忽略，水的密度可以视为常数。然而，在讨论管道水流的水击问题时，水的压缩性是不容忽视的，必须考虑。

液体的膨胀系数用 $\alpha_V$ 来表示，它表示当液体温度变化为 1 K 而压强不变时液体体积的相对变化率，可表示为

$$\alpha_V = \frac{1}{V}\frac{\mathrm{d}V}{\mathrm{d}T} = -\frac{1}{\rho}\frac{\mathrm{d}\rho}{\mathrm{d}T} \tag{1-6}$$

式中，$\mathrm{d}T$ 表示温度变化量；$\alpha_V$ 的单位为 $K^{-1}$。

液体的膨胀系数随压强和温度的变化而变化。在 1 标准大气压和 20 ℃ 条件下，$\alpha_V$ 为 $2.1 \times 10^{-4}\,K^{-1}$，一般情况下可以忽略。然而，当工作温度变化较大时，必须考虑水的膨胀。

**2. 气体的压缩性**

气体密度由状态方程决定。理想气体是一种假想气体，由许多随机移动的点粒子组成，这些粒子只有在弹性碰撞时才会相互作用。其状态方程可描述如下：

$$\frac{p}{\rho} = RT \tag{1-7}$$

式中，$p$ 为气体的绝对压强；$R$ 为理想气体常数，其值为 287 N·m/(kg·K)；$T$ 表示热力学温度。

实际上，理想气体是不存在的。然而，在标准温度和压强等正常条件下，大多数真实气体的性质与理想气体类似。许多气体，如氮、氧、氢、惰性气体，以及一些较重的气体，如二氧化碳，可以在合理的范围内被视为理想气体加以研究。

**例 1-1** 在 20 ℃ 时，水的体积为 2.5 m³，当温度上升到 80 ℃ 时，其体积增加多少？

**解：** 在 20 ℃ 时水的密度 $\rho_1 = 998.23\,kg/m^3$，而在 80 ℃ 时水的密度 $\rho_2 = 971.83\,kg/m^3$。由

于质量守恒,当温度升高时,水的体积随着密度的降低而增加。可以得到

$$-\frac{\mathrm{d}V}{V}=\frac{\mathrm{d}\rho}{\rho}$$

将数据代入上述方程式,得

$$\Delta V=-\frac{\Delta\rho}{\rho}V=-\frac{\rho_2-\rho_1}{\rho_1}V_1=-\frac{971.83-998.23}{998.23}\times 2.5=0.066\ 1\ \mathrm{m}^3$$

那么

$$\frac{\Delta V}{V_1}=\frac{0.066\ 1}{2.5}\times 100\%=2.64\%$$

水的体积增加了 2.64%。

**例 1-2** 若要使水的体积减少 0.1% 和 1%,则压强应分别增加多少?水的体积模量 $K$ 为 2 000 MPa。

**解**:式(1-5)可变形为

$$\Delta p=-K\frac{\Delta V}{V_1}$$

当 $\frac{\Delta V}{V_1}=-0.1\%$ 时,$\Delta p=-2\ 000\times(-0.1\%)=2.0\ \mathrm{MPa}$

当 $\frac{\Delta V}{V_1}=-1\%$ 时,$\Delta p=-2\ 000\times(-1\%)=20\ \mathrm{MPa}$

所以,压强分别增加 2.0 MPa 和 20 MPa,以使水的体积相应减少 0.1% 和 1%。

**例 1-3** 输水管的长度和直径分别为 200 m 和 0.4 m。进行水压试验时,当管内压强达到 $5.39\times 10^6$ Pa 后停止加压,1 h 后管内压强降至 $4.90\times 10^6$ Pa。在不考虑管道变形的情况下,每秒从管道裂缝中漏出的水量是多少?水的压缩率 $\kappa$ 为 $4.83\times 10^{-10}$ Pa$^{-1}$。

**解**:管道破裂漏水后,管道内压强下降,导致水体膨胀。膨胀的水体积为

$$\mathrm{d}V=-\kappa V\mathrm{d}p$$
$$=-4.83\times 10^{-10}\times\left[\frac{\pi}{4}\times 0.4^2\times 200\right]\times(4.90-5.39)\times 10^6\ \mathrm{Pa}$$
$$=5.95\times 10^{-3}\ \mathrm{m}^3$$

1 h 内从管道裂缝中漏出的水量即水的膨胀量为 $5.95\times 10^{-3}$ m³。每秒从管道裂缝中漏出的水的体积为

$$q_v=\frac{5.95\times 10^{-3}}{3\ 600}=1.65\times 10^{-6}\ \mathrm{m}^3/\mathrm{s}$$

### 1.3.3 粘度

**1. 粘度与流体的内摩擦力**

流体,顾名思义,是具有流动性的。即使是非常小的切应力也会使静止的流体变形,而流体却可以抵抗运动状态下的切应力的变形。流体的粘度是其抵抗切应力或拉伸应力逐渐变形

的能力的度量。

现用平板实验来说明流体的粘度。考虑两块平行板之间的流体,并在恒定切应力 $F$ 的作用下附着在它们之上。两块平行板之间的距离为 $Y$,假设这些板非常大,因此无需考虑靠近它们边缘的情况。当下板保持不动时,上板在外力 $F$ 的作用下做以匀速 $U$ 作平行于下板运动(见图1-1)。如果上板的速度足够小,流体粒子将与其平行移动,它们的速度将从底部的零线性变化到顶部的 $U$。如果板间距离 $Y$ 不大或速度 $U$ 不高,则流体在垂直于板的方向上的速度分布是线性的。那么,可以得到如下方程:

$$u(y) = \frac{U}{Y} y \tag{1-8}$$

则

$$\frac{\mathrm{d}u}{\mathrm{d}y} = \frac{U}{Y} \tag{1-9}$$

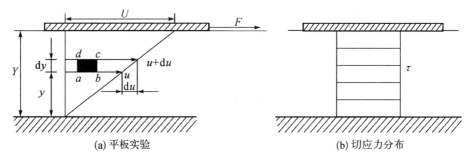

(a) 平板实验  (b) 切应力分布

图1-1 流动流体的内摩擦

可以认为流动流体是由一系列薄层组成的。每一层流体的移动速度都比它下面的一层要快,它们之间的摩擦会产生抵抗它们相对运动的力,即内摩擦,这是粘度的表征。快速移动的流体层在与其运动方向相同的方向上对慢速移动的流体层施加内摩擦力,内摩擦力使其加速运动;另一方面,较慢移动的流体层在与其运动方向相反的方向上对快速移动的流体层施加内摩擦力,内摩擦力减慢了运动。因此,可以改变流体的流动模式。由于粘度的存在,为了克服流体运动过程中的摩擦力,做了功并消耗了机械能。因此,流体粘度是流体机械能损失的原因。

**2. 牛顿内摩擦定律**

1686年,艾萨克·牛顿提出并由后人通过实验证明的定律是:对于大多数流动流体,切应力与接触面 $A$ 和流速梯度 $\frac{\mathrm{d}u}{\mathrm{d}y}$ 成正比。因此,我们可以写成

$$F \propto A \frac{\mathrm{d}u}{\mathrm{d}y}$$

通过引入比例因子 $\mu$,然后可以得到下式:

$$\tau = \frac{F}{A} = \mu \frac{\mathrm{d}u}{\mathrm{d}y} \tag{1-10}$$

根据公式(1-10),当流速分布呈线性时,粘性切应力 $\tau$ 为常数,如图1-1b所示。

流速梯度$\dfrac{\mathrm{d}u}{\mathrm{d}y}$是垂直于流动层方向的流速变化率。为了解释流速梯度的物理含义，在两块板之间加入一个矩形流体质点($abcd$)，矩形的厚度为$\mathrm{d}y$。在经过一段时间($\mathrm{d}t$)之后，该质点发生了位移，从$abcd$移动到$a'b'c'd'$，如图1-2所示。由于两个运动层之间存在速度差$\mathrm{d}u$，流体质点除了发生平移运动外，还具有角变形$\mathrm{d}\theta$。但是$\mathrm{d}t$与$\mathrm{d}\theta$成正比，$\mathrm{d}t$很小，所以$\mathrm{d}\theta$也很小。因此，我们可以写成

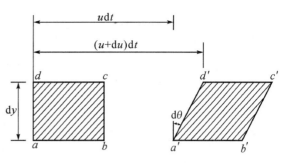

图1-2 流体质点的剪切变形率

$$\mathrm{d}\theta \approx \tan(\mathrm{d}\theta) = \dfrac{\mathrm{d}u\,\mathrm{d}t}{\mathrm{d}y}$$

$$\dfrac{\mathrm{d}u}{\mathrm{d}y} = \dfrac{\mathrm{d}\theta}{\mathrm{d}t} \tag{1-11}$$

根据式(1-11)，$\dfrac{\mathrm{d}u}{\mathrm{d}y}$为流体质点的局部剪切速度。由于切应力导致角变形的发生，所以角变形率也称为剪切应变速率。

因此，牛顿内摩擦定律可以改写为

$$\tau = \mu\dfrac{\mathrm{d}u}{\mathrm{d}y} = \mu\dfrac{\mathrm{d}\theta}{\mathrm{d}t} \tag{1-12}$$

表明流体的切应力与切应力速率成正比，这也是一个流体不同于固体的非常重要的特征。固体的切应力的大小是与角变形成正比，但是并不受变形速率的影响。

比例因子$\mu$被称为动力粘度(简称粘度)，它是用来衡量流体粘滞性大小的，其单位是Pa·s，即N·s/m²。$\tau$是表示单位面积的内部摩擦力，即粘性切应力，$\tau$的单位是Pa。流体的粘度因流体的种类而异。在相同条件下，液体的粘度随温度和压力的变化而变化，并且大于气体的粘度。液体的粘度与温度成反比，而气体则刚好相反，是正比的关系。在液体中，分子在排斥力允许的情况下尽可能紧密地堆积在一起，每个分子都在其他几个分子的吸引力范围内。负责可变形性的分子之间的点位交换受到来自相邻分子的吸引力的阻碍，表明分子间的吸引力(内聚力)很大，内聚力是产生粘度的一个十分重要的原因。对于气体，分子之间的平均距离很大(大约10个有效分子直径)，内聚力很小。气体的粘度(其中动量传递基本上是其唯一来源)随着温度的升高而增加，因为升高温度会增加分子的热速度，因此有利于动量交换。虽然气体和液体的粘度都受温度的影响，但影响是不同的。温度对液体粘度的影响要大于对气体粘度的影响。例如，当温度从15℃持续升高到50℃时，水的粘度降低了约一半，但是空气的粘度增加了约9%。压力对流体的粘度几乎没有影响，对于水和空气等常见流体，压力的影响可以忽略不计。

运动粘度也可以用来表示流体粘度。运动粘度(也称为"动量扩散率")是动态粘度$\mu$与流体密度$\rho$的比值。它通常由希腊字母$\nu$ $\left(\nu = \dfrac{\mu}{\rho}\right)$ 表示，单位为m²/s。运动粘度也受温度变化的影响。水的运动粘度$\nu$可以通过如下经验方程式计算：

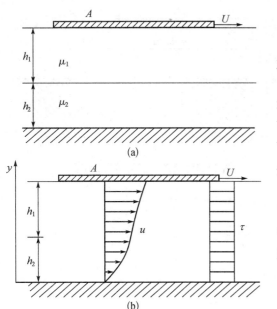

图 1-3 平行板间液体的流速和切应力分布

$$\nu = \frac{0.01775 \times 10^{-4}}{1 + 0.0337t + 0.000221 t^2} \quad (1-13)$$

式中，$t$ 为摄氏温度，以℃计。

**例 1-4** 尝试绘制液体流速和切应力的分布曲线。如图 1-3a 所示，假设两个水平板的上下层液体的流速分布是线性的。已知条件有上、下层液体的粘度分别为 $\mu_1$ 和 $\mu_2$，液体层的高度分别为 $h_1$ 和 $h_2$，水平板的运动速度为 $U$。

**解**：设液层界面的流速为 $u$。

因为液体的流速分布是线性的，因此，根据牛顿内摩擦定律，上层和下层的切应力可以描述如下：

上层：$\tau_1 = \mu_1 \dfrac{U-u}{h_1}$

下层：$\tau_2 = \mu_2 \dfrac{u-0}{h_2}$

在液层界面，已知：$\tau_1 = \tau_2 = \tau$.

那么

$$\mu_1 \frac{U-u}{h_1} = \mu_2 \frac{u-0}{h_2}$$

解得

$$u = \frac{\mu_1 h_2 U}{\mu_2 h_1 + \mu_1 h_2}$$

因此，上、下层流速分布如下：

上层：$u_1 = u + \dfrac{U-u}{h_1}(y-h_2)$

下层：$u_2 = \dfrac{u}{h_2} y$

由于各层流体的流速分布是线性的，只要定性绘制界面上的流速，就可以得到液体流速分布曲线，切应力分布均匀，如图 1-3b 所示。

**例 1-5** 一块质量为 5 kg，底面积和高度分别为 $0.4 \times 0.45 \text{ m}^2$、0.01 m 的木板，沿着涂有润滑剂的斜面匀速向下运动，如图 1-4 所示。已知该板的移动速度 $u$ 为 1 m/s。润滑剂的厚度 $\delta$ 为 0.1 mm。若木板带动油层运动的速度分布是线性的。计算润滑剂的粘度。

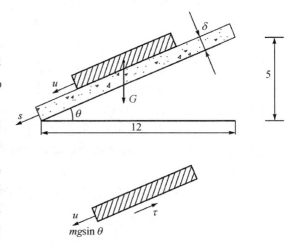

图 1-4 木板的受力分析

**解：** 对木板进行受力分析。

运动方向的重力分力为 $mg\sin\theta$，切应力可以表示为

$$F = -\tau A = -\mu \left(\frac{\mathrm{d}u}{\mathrm{d}y}\right) A = -\mu \left(\frac{u}{\delta}\right) A$$

根据牛顿运动定律，我们可以写成

$$\sum F_s = m\, a_s = 0$$

那么可以推导出下面的等式：

$$mg\sin\theta - \mu \frac{u}{\delta} A = 0$$

且

$$\theta = \arctan\frac{5}{12} = 22.62°$$

那么 $\mu$ 可以计算如下：

$$\mu = \frac{mg\sin\theta \cdot \delta}{Au} = \frac{5 \times 9.8 \times \sin 22.62° \times 0.000\,1}{0.4 \times 0.45 \times 1} = 0.010\,5 \text{ Pa·s}$$

**例 1-6** 由转轴驱动的圆盘（直径为 0.1 m）在固定板上旋转。盘与木板之间夹有一层厚度 $\delta$ 为 1.5 mm 的油膜。当圆盘以 50 r/min 的速度旋转时，测得的扭矩 $M$ 为 $2.94 \times 10^{-4}$ N·m。且油膜中的速度分布在垂直于圆盘的方向上是线性的。推算出圆盘边缘处油的粘度和切应力 $\tau_0$。

**解：** 对于微元，流速为

$$u = \omega r = \frac{n}{60} \cdot 2\pi r = \frac{\pi n}{30} r$$

则切应力为

$$\mathrm{d}F = \tau \mathrm{d}A = \mu \mathrm{d}A \frac{u}{\delta} = \mu \cdot 2\pi r \mathrm{d}r \frac{1}{\delta} \cdot \frac{\pi n}{30} r$$

施加在该盘的微元上的粘性摩擦力矩为

$$\mathrm{d}M = r\mathrm{d}F = \frac{\mu \pi^2 r^3 n \mathrm{d}r}{15\delta}$$

那么总摩擦力矩为

$$M = \int_0^R \mathrm{d}M = \mu \pi^2 \frac{n}{15\delta} \int_0^R r^3 \mathrm{d}r = \frac{\mu \pi^2 n R^4}{60\delta}$$

$$\mu = \frac{6\delta M}{\pi^2 n R^4} = \frac{60 \times 0.001\,5 \times 2.94 \times 10^{-4}}{3.14^2 \times 50 \times 0.05^4} = 85.8 \times 10^{-4} \text{ Pa·s}$$

并且 $\tau = \mu \dfrac{1}{\delta} \cdot \dfrac{\pi n}{30} r$，$\tau$ 分布是线性的，如图 1-5 所示。

因此，当 $r = 0.05$ m 时，$\tau_0 = 1.50$ Pa。

### 1.3.4 表面张力

在液-气界面上，液体分子之间的吸引力大于空气中的分子力，从而产生表面张力。由于不平衡的力，表面处于张力之下，因此也称之为表面张力。表面张力可使水龙头出口处半球形水滴悬浮而不滴落。当管子插入液体中时，如果管子足够窄，且液体与管壁的粘附力足够强，表面张力会以一种称为毛细现象将液体吸上管子。

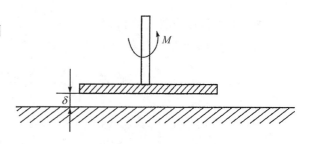

图 1-5 转盘粘度计

表面张力，通常用符号 $\sigma$ 表示，其单位为 N/m。表面张力 $\sigma$ 随流体类型和温度而变化。20 ℃ 时，水和汞在空气中的表面张力分别为 0.074 N/m 和 0.51 N/m。

表面张力的数值不是很大，因此在实际工程中其影响可以忽略不计。但是，对于毛细上升、水滴和气泡形成、液体射流的分散和小水力模型的实验，表面张力很重要。

## 1.4 流体的分类

### 1.4.1 牛顿流体和非牛顿流体

牛顿内摩擦定律只适用于一般流体，不适用于某些特殊流体。符合牛顿内摩擦定律的流体称为"牛顿流体"，反之则称为"非牛顿流体"。在连续介质力学中，牛顿流体是这样一种流体，它在每一点流动时产生的粘性应力与局部应变率（即其变形随时间的变化率）成线性比例。虽然没有真正的流体完全符合定义，但许多常见的液体和气体，如水和空气，可以认定为牛顿流体，以便在普通条件下进行实际计算。

图 1-6 描述了流体的切应力和切应变率之间关系的曲线，所有这些曲线都称为流变曲线，曲线的斜率是流体的粘度 $\mu$。牛顿流体的粘度 $\mu$ 在恒温恒压下保持不变，因此，流体切应力与剪切应变率的关系曲线具有固定斜率。

在非牛顿流体中，宾厄姆体是这些物质的一个典型例子，在达到一定切应力之前，宾厄姆体不会表现出任何剪切变形（没有流动，因此没有速度）。泥浆、血浆和牙膏都是宾厄姆体。另一类非牛顿流体是假塑性流体，如锦纶、橡胶、纸浆、血液、牛奶和水泥浆，其特征是粘度随切应变率的增加而降低。此外，膨胀流体也是一种非牛顿流体，它（包括面团和厚淀粉糊）的粘度随着切应变率的增加而增加。本书只讨论牛顿流体。

### 1.4.2 理想流体

粘度，即抵抗切应力逐渐变形的能力，是所有真实流体的特性。考虑流体的粘度使流体问题复杂化。为了简化流体问题，因此引入了"理想流体"的概念。理想流体是具有零粘度状态

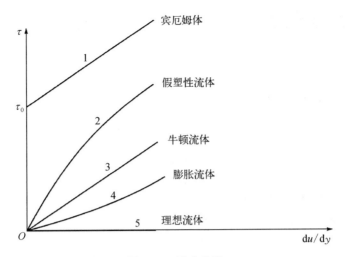

图 1-6 流变曲线

的假设流体。而且理想流体是理想化的流体，并不存在。然而，流场或流区中的流体，粘度影响较小，可被认为是理想流体。基于理想流体得出的结论可以进行修正，使其适用于实际流体。

### 1.4.3 可压缩流体和不可压缩流体

根据流体在受压条件下体积减小的特性，流体可分为可压缩流体和不可压缩流体。流体的密度随压力的变化是非常小的。密度恒定的流体可以认为是不可压缩流体，反之则是可压缩流体。从理论上来说，完全不可压缩的流体是不存在的，但在通常情况下，液体的密度可以被认为是一个常数，那么液体可以被认为是不可压缩的流体。当压力变化相对较小时，气体可以被认为是不可压缩的流体。当空气或水蒸气高速流过长管道时，压降和密度变化很大，在这种情况下，气体必须被视为可压缩流体。

## 1.5 作用在流体上的力

在对流体进行受力分析时，从流体中取出的隔离体是被任何封闭表面包围的流体的一部分。作用在流体隔离体上的力可分为表面力和质量力。

### 1.5.1 表面力

表面力是相邻流体或其他物体直接作用于流体隔离体表面的接触力。换句话说，表示为 $F_s$ 的表面力是作用在材料体中的内部或外部表面上的力。表面力的大小与作用于运动流体的面积成正比。表面力可以分解为两个垂直分量：压力和应力。压力通常垂直作用在一个区域上，而应力则平行地作用在一个区域上。如图 1-7 所示，将隔离体作为运动流

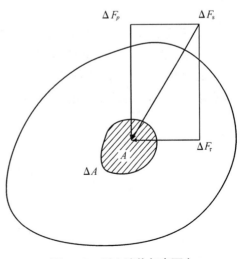

图 1-7 孤立流体与表面力

体中的研究对象,在隔离体表面的 $A$ 点处取微面积 $\Delta A$。那么压力为

$$p = \lim_{\Delta A \to 0} \frac{\Delta F_p}{\Delta A}$$

式中,$\Delta F_p$ 为作用在微面积 $\Delta A$ 上的压力。

切应力为

$$\tau = \lim_{\Delta A \to 0} \frac{\Delta F_\tau}{\Delta A}$$

式中,$\Delta F_\tau$ 为作用在面元 $\Delta A$ 上的切应力。

压力和切应力的单位都是 $\text{Pa}(\text{N}/\text{m}^2)$。

### 1.5.2 质量力

质量力是作用于隔离体中每个流体质点的力。质量力的大小与质量成正比。

单位质量力是作用在单位质量流体上的质量力:

$$\boldsymbol{f} = \frac{\boldsymbol{F}}{m} = f_x \boldsymbol{i} + f_y \boldsymbol{j} + f_z \boldsymbol{k} = \frac{F_x}{m}\boldsymbol{i} + \frac{F_y}{m}\boldsymbol{j} + \frac{F_z}{m}\boldsymbol{k}$$

式中,$f_x$、$f_y$ 和 $f_z$ 分别为在 $x$、$y$ 和 $z$ 坐标方向上作用在单位质量流体上的质量力。单位质量力的单位与加速度的单位相同。最常见的质量力是重力和惯性力。

对于重力作用下的静流,若 $z$ 轴垂直于上部,则流体在各方向的单位质量力分量为

$$f_x = f_y = 0, \ f_z = -g$$

## 习 题

1. 粘性流体在静止时有切应力吗?理想流体在运动时有切应力吗?静止流体有没有粘性?

2. 如果煤油的密度为 $808 \text{ kg/m}^3$,那么体积为 $2 \times 10^{-3} \text{ m}^3$ 的煤油的质量和重量是多少?

3. 手枪加压时,压力为 $0.1 \text{ MPa}$ 时缸体内液体体积为 $10^{-3} \text{ m}^3$,压力为 $10 \text{ MPa}$ 时为 $9.94 \times 10^{-4} \text{ m}^3$。计算液体的体积模量。

4. 如图 1-8 所示,两个水平板之间的距离为 $0.5 \text{ mm}$,两个水平板之间有液体被截留。上板在水平方向以 $0.25 \text{ m/s}$ 的速度移动,压力为 $2 \text{ Pa}$。计算液体的粘度。

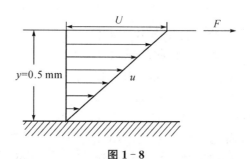

图 1-8

5. 计算轴与轴套之间的流体粘度(见图1-9)。

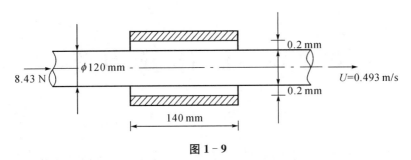

图1-9

# 第 2 章 流体静力学

流体静力学是研究静止不可压缩流体的流体力学分支。它包括研究流体处于稳定平衡状态的静止条件,不同于流体动力学,研究运动中的流体。

停滞是相对的,表明处于相对静止或相对平衡状态的流体粒子之间没有相对运动。在静止的流体中,所有的摩擦应力都消失了。流体静力学主要研究与静止流体有关的力学规律。流体静力学阐述了静压强的特性和分布规律、欧拉方程、作用在平面或曲面上的静总压的计算方法、潜体和浮体的稳定性等。

本章学习要求:掌握绝对压强、相对压强、真空度、等压面、测压头、压强棱镜、压强中心、稳心半径、偏心距等基本概念;了解静止流体中应力的特征和静压强分布的原理;了解潜体和浮体的稳定性;了解液体相对平衡的分析方法;掌握等压面、压强分布图的确定方法及压强棱镜的绘制方法;掌握流体静力学中的基本方程,掌握如何使用这些方程并理解它们的物理意义;学习欧拉微分平衡方程;掌握如何计算作用在平面和曲面上的总静压强(解析法和图解法)以及如何运用流体静力学的基本原理来解决工程问题。

## 2.1 静止流体中的应力特性

### 2.1.1 静止流体中应力的两个特性

与固体和流动流体相比,静止流体任意点的应力具有两个重要特征:

(1) 静止流体中的表面应力是压应力,即压强,静压强的方向与作用表面的法线方向相同。

在静止流体中,相邻流体层之间没有相对运动。因此,静止的流体中没有切应力。一个静止的流体被 N—N′面分成两部分。阴影部分被认为是一个分离体。对于分离体表面上的任何点 A,切应力 τ 为零,角度 α 为 90°,如图 2-1 所示。压强 $p$ 与作用力垂直。

由于流体的流动性,它们不能承受拉伸应力。因此,静压强的方向只能指向其受压面。结果表明,静止流体的表面应力只能是压应力,其方向与作用面的内法线方向相同。

(2) 如果静止流体中的一个点被认为是一个无穷小的立方体,那么它遵循平衡原理,即这个流体单位每一侧的压强必须相等。因此,静止流体上的压强是各

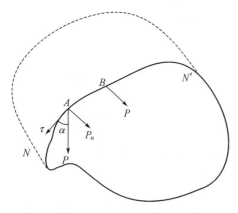

图 2-1 静止流体中任意点的应力特征一

向同性的；即它在所有方向上的作用力大小相等。

这一特点可以证明如下：

从处于平衡状态的流体中提取一个包含点 $O$ 的无穷小四面体作为研究对象，点 $O$ 被指定为原点，如图 2-2 所示。$n$ 是斜面 $ABC$ 的外法线方向。然后对其受力平衡进行分析。

首先，确定表面力。$p_x$、$p_y$、$p_z$ 和 $p_n$ 分别代表三个不同坐标面和斜面上的静压强。因为这个四面体的每个侧面的面积都是无穷小的，所以表面上所有点的压强都是相同的。因此，$x$、$y$、$z$ 和 $n$ 方向上的表面力可以表示为 $p_x \cdot \dfrac{\mathrm{d}y\mathrm{d}z}{2}$、$p_y \cdot \dfrac{\mathrm{d}z\mathrm{d}x}{2}$、$p_z \cdot \dfrac{\mathrm{d}x\mathrm{d}y}{2}$ 和 $p_n \cdot \mathrm{d}A_n$（$\mathrm{d}A_n$ 是斜面 $ABC$ 的面积）。

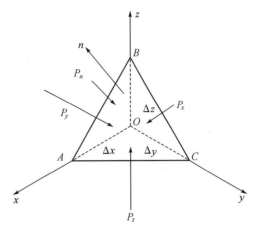

图 2-2　静止流体中任意点的应力特征二

除了施加在四面体 $OABC$ 上的表面力外，四面体还承受质量力。流体密度为 $\rho$，$f_x$、$f_y$、$f_z$ 表示单位质量力在相应坐标轴方向上的分量。在三个坐标轴上，四面体上的质量力分量分别为 $f_x \cdot \dfrac{\rho \mathrm{d}x\mathrm{d}y\mathrm{d}z}{6}$、$f_y \cdot \dfrac{\rho \mathrm{d}x\mathrm{d}y\mathrm{d}z}{6}$ 和 $f_z \cdot \dfrac{\rho \mathrm{d}x\mathrm{d}y\mathrm{d}z}{6}$。

在平衡条件下，$\sum F = 0$，表示其各向异性分量在 $x$ 轴上的投影之和也为零。因此，在 $x$ 轴的方向上，可以表示为：

$$\sum F_x = p_x \cdot \frac{\mathrm{d}y\mathrm{d}z}{2} - p_n \mathrm{d}A_n \cos(n, x) + f_x \cdot \rho \frac{\mathrm{d}x\mathrm{d}y\mathrm{d}z}{6} = 0$$

式中，$\cos(n, x)$ 表示角平面外法线与正 $x$ 轴之间夹角的余弦，$\mathrm{d}A_n \cos(n, x)$ 表示 $\mathrm{d}A_n$ 在垂直于 $x$ 轴的 $Oyz$ 平面上的投影。因此：

$$\mathrm{d}A_n \cos(n, x) = \frac{\mathrm{d}y\mathrm{d}z}{2}$$

将其代入上述方程式，可得到以下方程式：

$$p_x - p_n + \frac{1}{3}\mathrm{d}x \cdot \rho f_x = 0$$

四面体的边长趋于零，上述方程中的第三项也趋于零。因此，上述方程式可以简化为：

$$p_x = p_n$$

同理可得：

$$p_y = p_n, \ p_z = p_n$$

因此，可以得出：

$$p_x = p_y = p_z = p_n$$

斜面是随机选择的，表明 $n$ 方向也是任意的。因此，可以得出结论，静止时流体上的压强是各向同性的，即它在所有方向上的作用力大小相等。不同点的静压强不同，是位置的连续函数，即 $p = p(x, y, z)$。

### 2.1.2 推论——运动流体和理想流体中的压强

对于运动中的真实流体，由于流体层之间的相对运动，粘性导致切应力的产生。同一点不同方向的法向应力不相等。流体的动态压强可以通过平均三个相互垂直的压缩应力来确定，即：

$$p = \frac{1}{3}(p_x + p_y + p_z)$$

然而，如果运动的流体是理想流体，就没有切应力。因此，理想流体中只存在压应力，且：

$$p_x = p_y = p_z = p$$

## 2.2 微分平衡方程

### 2.2.1 流体微分平衡方程——欧拉方程

为了分析流体平衡定律，将处于平衡状态的流体中的无穷小立方体作为研究对象，立方体的每一侧分别平行于相应的笛卡儿坐标轴（见图 2-3）。立方体中心点 $M$ 处的压强设定为 $p = p(x, y, z)$。根据连续函数的特点，当自变量（坐标位置）稍有变化时，函数的大小（压强）也会随之变化。函数的大小可以用泰勒级数计算，并忽略高阶微量。之后，中心点处左右表面的压强可以分别确定为

$$\left(p - \frac{\partial p}{\partial x}\frac{\mathrm{d}x}{2}\right) \text{ 和 } \left(p + \frac{\partial p}{\partial x}\frac{\mathrm{d}x}{2}\right)$$

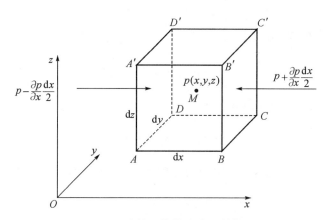

图 2-3 欧拉平衡微分方程的推导

因为立方体的每个面都是无穷小的，所以可以近似地表示，施加在表面上每个点上的压强与施加在表面中心点上的压强相同。表面 $AA'D'D$ 和 $BB'C'C$ 上的表面力可推断如下：

$$\left(p - \frac{\partial p}{\partial x}\frac{\mathrm{d}x}{2}\right)\mathrm{d}y\mathrm{d}z$$

$$\left(p + \frac{\partial p}{\partial x}\frac{\mathrm{d}x}{2}\right)\mathrm{d}y\mathrm{d}z$$

立方体的质量为 $\rho\mathrm{d}x\mathrm{d}y\mathrm{d}z$，$f_x$、$f_y$、$f_z$ 表示单位质量力在相应坐标轴方向上的分量。质量力在 $x$ 轴方向上的分量为 $f_x\rho\mathrm{d}x\mathrm{d}y\mathrm{d}z$。

在平衡条件下，在 $x$ 轴方向上有 $\sum F_x = 0$，即：

$$\left(p - \frac{1}{2}\frac{\partial p}{\partial x}\mathrm{d}x\right)\mathrm{d}y\mathrm{d}z - \left(p + \frac{1}{2}\frac{\partial p}{\partial x}\mathrm{d}x\right)\mathrm{d}y\mathrm{d}z + f_x\rho\mathrm{d}x\mathrm{d}y\mathrm{d}z = 0$$

整理得

$$f_x - \frac{1}{\rho}\frac{\partial p}{\partial x} = 0 \qquad (2-1\mathrm{a})$$

同理，可以在 $y$ 轴和 $z$ 轴方向上获得以下方程式：

$$f_y - \frac{1}{\rho}\frac{\partial p}{\partial y} = 0 \qquad (2-1\mathrm{b})$$

$$f_z - \frac{1}{\rho}\frac{\partial p}{\partial z} = 0 \qquad (2-1\mathrm{c})$$

方程(2-1)是流体的微分平衡方程，也称为欧拉方程，因为这些方程是由瑞士学者欧拉在 1755 年首次推导出来的。这些方程表明，对于处于平衡状态的流体，单位质量流体所受到的表面力 $\left(\frac{1}{\rho}\frac{\partial p}{\partial x}, \frac{1}{\rho}\frac{\partial p}{\partial y}, \frac{1}{\rho}\frac{\partial p}{\partial z}\right)$ 与单位质量力分量 $(f_x, f_y, f_z)$ 彼此对应相等，压强沿轴向的变化率 $\left(\frac{\partial p}{\partial x}, \frac{\partial p}{\partial y}, \frac{\partial p}{\partial z}\right)$ 等于作用于单位体积的轴向质量力分量 $(\rho f_x, \rho f_y, \rho f_z)$。

### 2.2.2 流体微分平衡方程的积分形式

如果将等式(2-1)分别乘以 $\mathrm{d}x$、$\mathrm{d}y$ 和 $\mathrm{d}z$，然后相加，可以得出：

$$f_x\mathrm{d}x + f_y\mathrm{d}y + f_z\mathrm{d}z = \frac{1}{\rho}\left(\frac{\partial p}{\partial x}\mathrm{d}x + \frac{\partial p}{\partial y}\mathrm{d}y + \frac{\partial p}{\partial z}\mathrm{d}z\right)$$

静压强 $p$ 是坐标的函数，即 $p = p(x, y, z)$，其总微分为

$$\mathrm{d}p = \frac{\partial p}{\partial x}\mathrm{d}x + \frac{\partial p}{\partial y}\mathrm{d}y + \frac{\partial p}{\partial z}\mathrm{d}z$$

联立上述两个方程，得下式：

$$\mathrm{d}p = \rho(f_x\mathrm{d}x + f_y\mathrm{d}y + f_z\mathrm{d}z) \qquad (2-2)$$

式(2-2)称为流体微分平衡方程的积分形式。当施加在流体上的质量力已知时，可以计算流体中的压强分布。

等式(2-2)的物理含义可描述如下：

对于处于平衡状态的流体，$\rho$ 是一个常数。令 $\frac{p}{\rho} = \omega$。因为 $p = p(x, y, z)$，所以 $\omega = \omega(x, y, z)$。根据等式(2-2)，可得

$$d\left(\frac{p}{\rho}\right) = f_x dx + f_y dy + f_z dz = \frac{\partial \omega}{\partial x}dx + \frac{\partial \omega}{\partial y}dy + \frac{\partial \omega}{\partial z}dz$$

因此,可以得出如下结论:

$$f_x = \frac{\partial \omega}{\partial x}, \ f_y = \frac{\partial \omega}{\partial y}, \ f_z = \frac{\partial \omega}{\partial z}$$

如果坐标函数 $\omega(x,y,z)$ 的偏导数与相应坐标轴上力场中的力分量相等,则该函数称为力函数或势函数。这种力称为有势力。因此,等式(2-2)中的质量力是势能力。换句话说,密度恒定的流体只有在有势质量力的作用下才能保持平衡。

**推论** 流体在重力或惯性力的作用下可以达到平衡,说明重力和惯性力是有势力。

### 2.2.3 等压面

**1. 等压面概念**

等压面是由压强相等的点组成的面。例如,液体和气体之间的界面以及平衡时两种液体之间的界面都是等压面。

在等压面上,$p=$常数,$dp=0$。因此:

$$f_x dx + f_y dy + f_z dz = 0 \quad (2-3)$$

**2. 等压面特性**

等压面上任意点的质量力与等压面始终正交,可以证明如下:

在等压面上有一流体粒子 $M$,如图 2-4 所示。该流体粒子的单位质量力为: $\boldsymbol{f} = f_x \boldsymbol{i} + f_y \boldsymbol{j} + f_z \boldsymbol{k}$,该流体粒子的线性矢量为: $d\boldsymbol{s} = dx\boldsymbol{i} + dy\boldsymbol{j} + dz\boldsymbol{k}$。根据等式(2-3),可以得出:

$$\boldsymbol{f} \cdot d\boldsymbol{s} = f_x dx + f_y dy + f_z dz = 0$$

上式表明 $\boldsymbol{f}$ 与 $d\boldsymbol{s}$ 正交,$\theta = 90°$,其中 $d\boldsymbol{s}$ 的方向在等压面上是任意的。因此,可以得出结论,施加在静止

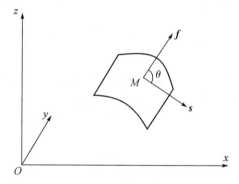

图 2-4 等压面特征

流体中任何一点上的质量力必须垂直于通过该点的等压面。这是等压面的一个重要特征。根据这一特性,如果质量力的方向已知,则可以确定等压面的形状;如果等压面的形状已知,则可以确定质量力的方向。

## 2.3 重力作用下的流体静压强分布

质量力包括重力和惯性力。在不同质量力的作用下,静压强的分布规律不同。在实际工程或日常生活中,流体平衡意味着流体相对于地球是静止的,或者施加在流体上的质量力只是重力。所以本节主要讨论重力作用下的静压强分布。

### 2.3.1 流体静力学基本方程

**1. 静力学基本方程的三个表达式**

将质量力仅包括重力的静止流体作为研究对象。在直角坐标系中，$z$ 轴垂直向上，如图 2-5 所示。液体表面上的压强为 $p_0$，任意点 $A$ 处的压强为 $p$，单位质量力为：$f_x = f_y = 0$，$f_z = -g$。因此，流体微分平衡方程的积分形式[式(2-2)]可改写为

$$dp = -\rho g \, dz$$

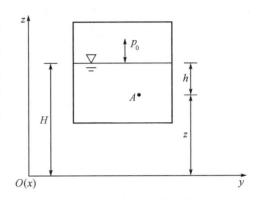

**图 2-5 静止液体中某点的压强**

将该方程积分后，可得出下式：

$$p = -\rho g z + C' \qquad (2-4)$$

在自由表面上有：$z = H$，$p = p_0$，于是，

$$C' = p_0 + \rho g H$$

且 $h = H - z$。通过将上述两个方程代入式(2-4)，式(2-4)可以写成

$$p = p_0 + \rho g h \qquad (2-5)$$

每项除以 $\rho g$ 后，等式(2-4)变为

$$z + \frac{p}{\rho g} = C \qquad (2-6)$$

式中，$p$ 为静止流体中某一点的压强，$p_0$ 为液体表面压强。对于开放式大气容器中的液体，$p_0$ 为大气压强，表示为 $p_a$。$h$ 为浸没深度，$z$ 为该点高于坐标平面的高度。

式(2-5)和式(2-6)是流体静力学基本方程的两个表达式。式(2-5)表明：(1) 对于静止流体，压强随深度线性增加，同一水平面上各点的静压强相等，与容器形状无关；(2) 流体中任意点的压强 $p$ 等于液体表面压强 $p_0$ 和垂直液柱从该点到液体表面每单位面积的重量 $\rho g h$ 之和。

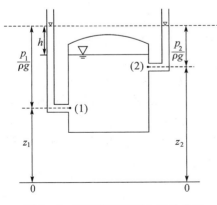

**图 2-6 高度和静压强之间的关系**

$C$ 是等式(2-6)中的常数，其大小可根据边界条件确定。如图 2-6 所示，随机选取基准面 0—0。距离基准面 0—0 的两点(1)和(2)的高度分别为 $z_1$ 和 $z_2$。这两点的静压强分别为 $p_1$ 和 $p_2$。根据式(2-6)，可推导出：

$$z_1 + \frac{p_1}{\rho g} = z_2 + \frac{p_2}{\rho g} \qquad (2-7)$$

公式(2-7)是流体静力学基本方程的第三个表达式。该方程适用于静止、连续和均质流体，其中质量力仅为重力。

**2. 流体静力学基本方程各项的基本概念和其物理意义**

等式(2-6)衍生出了几个重要的基本概念,如下所述:

**测压管** 当测点处的绝对压强大于当地大气压强时,在该测点处设置一个向上开口的透明管。这种透明管叫作测压管。

**位置水头** $z$ 为远离参照平面0—0的任意点处的位置高度。$z$ 还表示单位重量流体离基准面的位置势能。

**测压管高度** $\dfrac{p}{\rho g}$ 为测压管中液体的上升高度,也称为压强水头,表示单位重量流体在大气压强下的压力势能(压力能)。

**测压管水头** $\left(z+\dfrac{p}{\rho g}\right)$ 为单位重量流体离基准面的总势能。

式(2-6)的物理意义可概括如下:在重力作用下,在静止的连续均匀流体中,单位重量流体在任意点的总势能是相同的。换句话说,测压管水头在任何地方都是相等的。这是静止流体中能量分布的规律。

**3. 推论**

(1) 等压面判别

根据式(2-7),静止和连续流体的水平面(其中质量力仅为重力)为等压面。换句话说,为了区分水平面是否为等压面,有必要分析水平面内的流体是否满足以下四个条件:静止、连续、均匀和质量力仅为重力。如果流体不连续,或两种或两种以上流体共存,或流体处于磁场中,或流体处于相对平衡状态(参考第2.4节),则水平面不一定是等压面。例如,如图2-7所示,水平面 $B$—$B'$ 是等压面,而表面 $C$—$C'$ 不是等压面。

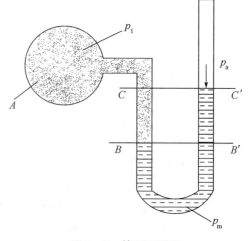

图 2-7 等压面测定

(2) 对于有限空间内气体的压强分布,由于气体密度小,高度 $z$ 有限,重力对气体压强的影响可以忽略,因此气体中任意点的压强相等($p_{气}=C_0$)。

### 2.3.2 压强度量

**1. 压强表示**

可以从不同的基准面计算压强。因此,有不同的表达方式,如:绝对压强、相对压强和真空度。

绝对压强是相对于完美真空的零参考,使用绝对刻度,所以它等于表压加大气压强。

在水利工程中,水流和建筑物表面的压强是环境空气压强。因此,环境空气压强被视为参考。以环境空气压强为基准计算的压强称为相对压强,用 $p$ 表示。相对压强也称为表压。压强表测量的压强是相对压强,因为压强表中的环境空气压强假定为零。表压相对于环境空气压强为零,因此它等于绝对压强减去大气压强。显然,绝对压强大于或等于零,而相对压强可以是正、负或零。绝对压强、相对压强和环境空气压强之间的关系可确定为

$$p = p_{abs} - p_a \tag{2-8}$$

如图 2-8 所示，测点 1 处的压强为负值。压强值可加上"真空"一词，表示绝对压强小于环境空气压强。真空度通常用 $p_v$ 表示，可根据下式计算：

$$p_v = p_a - p_{abs} \quad (2-9)$$

如图 2-8 所示，在测量点处，真空度可确定为：$p_{v2} = p_a - p_{abs2}$。

真空度可用水柱高度表示为

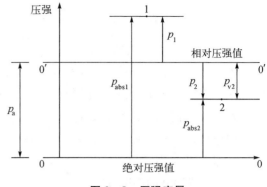

图 2-8 压强度量

$$p_v = -p = \rho g h_v \quad (2-10)$$

式中，$h_v$ 被称为真空高度。

绝对压强、相对压强和真空度之间的关系如图 2-8 所示。绝对压强参考值和相对压强参考值相差一个当地大气压 $p_a$。应注意，除非另有说明，否则本书中提到的压强是指相对压强。

**2. 压强计量单位**

(1) 国际单位制压强单位为 Pa(N/m²)。

(2) 大气压强的表示方法包括标准大气压和工程大气压。工程大气压与标准大气压略有不同。1 标准大气压(atm)等于 101.325 kPa，而 1 工程大气压(at)为 98 kPa。工程科学通常使用工程大气压代替标准大气压。不同地点的大气压强略有不同。

(3) 通常用水柱或水银柱的高度来表示液柱的高度。液柱高度的常用单位是 $mH_2O$ 或 mmHg。

**例 2-1** 如图 2-9 所示，试着在装满液体的容器中的 $A$、$B$ 和 $C$ 点标记高程头、测压管和测压头的高度。将平面 0—0 设置为参照平面。

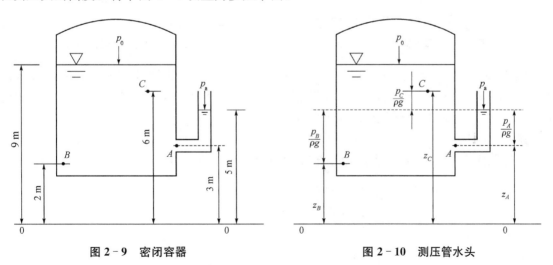

图 2-9 密闭容器　　　　图 2-10 测压管水头

**解**：高程水头、测压管高度和 $A$ 点测压管水头分别为 3 m、2 m 和 5 m(见图 2-10)。

因为 $z + \dfrac{p}{\rho g} = C$，根据 $A$ 点的测压管水头，$B$ 点测压管的高程水头和高度可分别确定为 2 m 和 3 m。

在 $C$ 点，测压管水头为 $z_C + \dfrac{p_C}{\rho g} = z_A + \dfrac{p_A}{\rho g} = 5\text{ m}$，位置水头 $z_C = 6\text{ m}$，测压管的高度 $\dfrac{p_C}{\rho g} = -1\text{ m}$，$p_C < 0$，处于真空中。

**例 2 - 2** 如图 2 - 11 所示，$h_v = 2\text{ m}$，容器 $B$ 中的液体为水。计算密封容器 $A$ 中的真空度。如果真空度固定，且密度 $\rho'$ 为 820 kg/m³ 的油代替水，则计算测压管中油柱的高度 $h_v'$。

**解**：(1) 计算密封容器 $A$ 中的真空度 $p_v$。

根据公式(2 - 10)，$p_v = \rho g h_v = 9\,800 \times 2 = 19.6 \text{ kN/m}^2$

容器 $A$ 中的真空度为 19.6 kN/m²。

(2) 计算油柱的高度 $h_v$。

$$h_v \rho g = h_v' \rho' g$$

$$h_v' = \dfrac{h_v \rho}{\rho'} = \dfrac{2 \times 1\,000}{820} = 2.44 \text{ m}$$

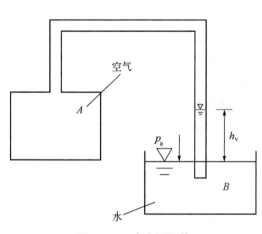

图 2 - 11 真空测压管

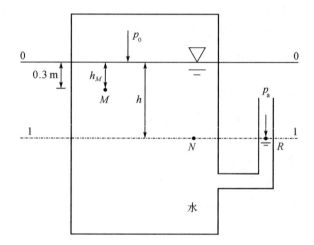

图 2 - 12 测点压强计算

**例 2 - 3** 有一个密封水箱，如图 2 - 12 所示。如果水面上的相对压强为 −44.5 kPa，计算：(1) $h$；(2) 水下 0.3 m 处 $M$ 点的相对压强、绝对压强和真空度，并分别用大气压和水柱高表示；(3) 以平面 0—0 为基准的 $M$ 点测压管水头。

**解**：(1) 计算 $h$。

在等压面 1—1 上，$p_N = p_R = p_a$，在下面的计算中使用相对压强。

$$p_0 + \rho g h = 0$$
$$-44.5 \times 10^3 + 9\,800 \times h = 0$$
$$h = 4.54 \text{ m}$$

(2) 计算 $p_M$。

相对压强：

$$p_M = p_0 + \rho g h_M = -44.5 \times 10^3 + 9\,800 \times 0.3 = -41.56 \text{ kPa}$$

或

$$p_M = \frac{-41.56}{98} \times 1 = -0.424 \text{ at } (1 \text{ at} = 98 \text{ kPa})$$

或

$$p_M = \frac{-41.56}{9.8} \times 1 = -4.24 \text{ mH}_2\text{O}$$

绝对压强：

$$p_{Mabs} = p_M + p_a = -41.56 + 98 = 56.44 \text{ kPa} = 0.576 \text{ at} = 5.76 \text{ mH}_2\text{O}$$

真空度：

$$p_v = -p = 41.56 \text{ kPa} = 0.424 \text{ at} = 4.24 \text{ mH}_2\text{O}$$

以水柱高度表示的真空度：

$$h_v = \frac{p_v}{\rho g} = \frac{41.56 \times 10^3}{9\,800} = 4.24 \text{ m}$$

(3) 以平面 0—0 为基准的 $M$ 点测压头为

$$z_M + \frac{p_M}{\rho g} = -0.3 + \frac{-41.56 \times 10^3}{9\,800} = -0.3 - 4.24 = -4.54 \text{ m}$$

**例 2-4** 如图 2-13 所示，通过水银压强计测量 $A$ 点和 $B$ 点之间的测压管水头差，试着写出它的表达式。

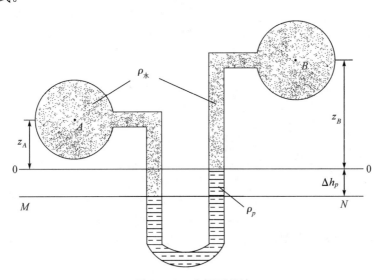

**图 2-13 水银压差计**

**解**：以汞柱上液面所在的水平面 0—0 为基准。将下液面所在的平面视为等压平面，则

$$p_M = p_A + \rho_\text{水} g z_A + \rho_\text{水} g \Delta h_p$$
$$p_N = p_B + \rho_\text{水} g z_B + \rho_p g \Delta h_p$$

因为 $p_M = p_N$，则有

$$p_A - p_B = (\rho_p g - \rho_\text{水} g)\Delta h_p + \rho_\text{水} g(z_B - z_A)$$

$$\left(z_A + \frac{p_A}{\rho_\text{水} g}\right) - \left(z_B + \frac{p_B}{\rho_\text{水} g}\right) = \left(\frac{\rho_p - \rho_\text{水}}{\rho_\text{水}}\right)\Delta h_p$$

由于汞密度与水密度之比为 13.6,因此可以得到以下表达式:

$$\left(z_A + \frac{p_A}{\rho_\text{水} g}\right) - \left(z_B + \frac{p_B}{\rho_\text{水} g}\right) = 13.6\Delta h_p$$

## 2.4 液体的相对平衡

容器中的液体以地球为基准作相对运动,液体的不同部分之间或液体与容器之间没有相对运动。因此,如果坐标系建立在容器中,液体相对于坐标系处于平衡状态。对于处于相对平衡状态的流体,质量力包括重力和惯性力。根据达朗贝尔原理,液体的相对运动可以转化为形式上的能量平衡问题,如果惯性力包含在质量力中,则可以应用式(2-2)来分析液体的相对运动。

### 2.4.1 匀加速直线运动中液体的相对平衡

如图 2-14 所示,液体随容器作匀加速直线运动。原始液体表面的中心点被指定为原点。$z$ 轴垂直向上,加速度和 $x$ 轴之间的夹角为 $\alpha$。单位质量力包括重力和加速度惯性力。任意点 $M(x, y, z)$ 处的单位质量力为

$$f_x = -a\cos\alpha, \quad f_y = 0, \quad f_z = -g - a\sin\alpha$$

将上述方程式代入式(2-2),即可得到以下方程式:

$$dp = \rho[-a\cos\alpha dx - (g + a\sin\alpha)dz]$$

积分后,可得出以下方程式:

$$p = -\rho[a\cos\alpha \cdot x + (g + a\sin\alpha)z] + C$$

在液体表面,$x = z = 0$, $p = p_0$

将其代入上述方程式,即可得到 $C = p_0$,于是有

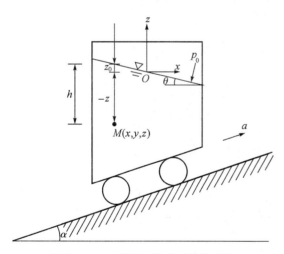

图 2-14 匀加速运动中的液体平衡

$$p = -\rho(g + a\sin\alpha)\left(\frac{a\cos\alpha}{g + a\sin\alpha}x + z\right) + p_0 \tag{2-11}$$

这是液体随容器匀加速直线运动处于相对平衡状态时压强分布规律的一般表达式。根据式(2-11),可以得到等压面方程:

$$\frac{a\cos\alpha}{g + a\sin\alpha}x + z = C_1 = 常数 \tag{2-12}$$

因此，等压面是与水平面成 $\theta$ 角的斜面。$\theta$ 可以计算为

$$\theta = \arctan\left(-\frac{a\cos\alpha}{g + a\sin\alpha}\right)$$

液体表面的垂直坐标值可计算为

$$z_0 = -\frac{a\cos\alpha}{g + a\sin\alpha}x \tag{2-13}$$

通过将式(2-13)代入式(2-11)，可以得到下式：

$$p = p_0 + \rho(g + a\sin\alpha)(z_0 - z)$$

式中，$z_0 - z$ 为质点在液面以下的浸入深度，用 $h$ 表示。因此

$$p = p_0 + \rho(g + a\sin\alpha)h \tag{2-14}$$

这表明，对于匀加速直线运动的液体，任意点的压强随水深的变化是线性的。

式(2-14)还表明，如果液体随容器自由下落，液体任意点的压强为 $p_0$，开放容器中液体任意点的压强为零。

### 2.4.2 旋转液体的相对平衡

如图 2-15 所示，在装满液体的直立圆柱形容器以匀角速度 $\omega$ 旋转的情况下，靠近容器壁的液体在开始时随容器旋转，随后所有液体以匀角速度 $\omega$ 随圆柱形容器旋转。可以达到相对平衡状态。在这种情况下，液体的自由表面从水平面转变为旋转抛物面。在旋转圆柱体中建立坐标轴，原点与旋转抛物面的顶点重合，$z$ 轴垂直向上。旋转物体中任何粒子 $A$ 的向心加速度为 $-\omega^2 r$，$x$ 轴和 $y$ 轴上的分量分别为 $-\omega^2 x$ 和 $-\omega^2 y$。重力的单位质量力为 $f_x = f_y = 0$，$f_z = -g$。惯性力的单位质量力为 $f_x = \omega^2 x$，$f_y = \omega^2 y$，$f_z = 0$。因此，总单位质量力为 $f_x = \omega^2 x$，$f_y = \omega^2 y$，$f_z = -g$。

将上述方程代入欧拉方程，即可得到以下方程：

$$dp = \rho(\omega^2 x dx + \omega^2 y dy - g dz)$$

积分后，上述方程可转化为

$$p = \rho\left(\frac{1}{2}\omega^2 x^2 + \frac{1}{2}\omega^2 y^2 - gz\right) + C$$

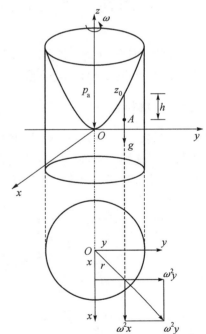

**图 2-15** 旋转液体的相对平衡

在原点 ($x = y = z = 0$)，$p = p_0$，所以 $C = p_0$。然后可得

$$p = p_0 + \rho g\left(\frac{\omega^2 r^2}{2g} - z\right) \tag{2-15}$$

这是液体在以匀速旋转的垂直容器中相对平衡时压强分布规律的一般表达式。

含自由表面的等压面簇方程为

$$\frac{\omega^2 r^2}{2g} - z = C_1 = 常数 \tag{2-16}$$

这表明等压面簇是一个具有中心轴的旋转抛物面。

在液体的自由表面上，$p = p_a = p_0$，自由表面方程可用相对压强表示：

$$z_0 = \frac{\omega^2 r^2}{2g} \tag{2-17}$$

式中，$z_0$ 为自由表面的垂直坐标。通过将公式(2-17)代入公式(2-15)，可获得以下方程式：

$$p = p_0 + \rho g(z_0 - z)$$

式中，$z_0 - z$ 为粒子 $A$ 在液体自由表面以下的深度，用 $h$ 表示。则上式变为

$$p = p_0 + \rho g h \tag{2-18}$$

式(2-18)表明，在相对平衡的旋转液体中，任意点处的压强随液体深度的变化呈线性。然而，在旋转容器中，任意位置的测压管水头不是常数。

## 2.5 作用在水平面上的液体总静压强

确定作用于水平面的液体总静压强的大小、方向和作用点是许多工程技术必须解决的力学问题。计算水平面总静压强的方法有解析法和图解法。

### 2.5.1 解析法

如图 2-16 所示，$MN$ 是倾斜平面与水平面成 $\theta$ 角的投影。水平面上的受压面面积为 $A$。大气压强施加在水面上。$MN$ 的延伸平面和液体自由表面之间的交线被设定为 $Ox$ 轴。$Oy$ 轴向下并垂直于 $Ox$ 轴。水平面所在的坐标平面围绕 $Oy$ 轴旋转 $90°$，以显示平面 $xOy$ 的几何关系。$h$ 为任意点的水深。$y$ 为从该点到 $Ox$ 轴的距离。$C$ 代表受压面的质心。

**1. 力的大小和方向**

在水深 $h$ 处，以水平条带中的无穷小面积 $\mathrm{d}A$ 为研究对象。在这个无限小的表面上，压强分布是均匀的。因此，施加在无限小表面上的压强为

$$\mathrm{d}F = p\,\mathrm{d}A = \rho g h\,\mathrm{d}A = \rho g y \sin\theta\,\mathrm{d}A$$

压强方向与 $\mathrm{d}A$ 正交，与内部法线方向相同。

因为 $MN$ 是一个表面，所以作用在任何极小区域上的压强方向是相互平行的。它们的代数和可以通过积分上述方程来计算。作

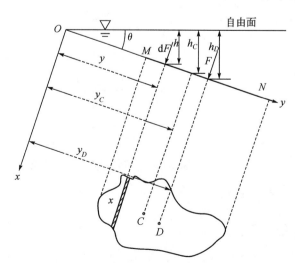

**图 2-16 解析法推导平面上的总静压强**

用在表面上的总压强为

$$F = \int dF = \int_A \rho g y \sin\theta dA = \rho g \sin\theta \int_A y dA$$

式中，$\int_A y dA$ 为受压面相对于 $Ox$ 轴的静态力矩。其大小等于受压面的面积 $A$ 乘以质心的坐标 $y_C$。

把 $\int_A y dA = y_C \cdot A$ 代入上式，且有 $h_C = y_C \sin\theta$，可得

$$F = \rho g \sin\theta \cdot y_C \cdot A = \rho g h_C A = p_C A \tag{2-19}$$

式中，$F$ 为作用在表面上的总静压强，$h_C$ 为受压面质心的淹没深度，$p_C$ 为受压面质心处的压强。

式(2-19)表示作用于任意形状表面上任意方向的总水压的大小等于受压面质心处的静压强乘以面积。换句话说，作用在任何受压面上的平均压强等于施加在其质心上的压强。请注意，总静压强的大小与倾斜角度 $\theta$ 无关，仅取决于流体的比重、表面下方质心的面积和深度。总静压强的方向与受压面的内部法线方向相同。

**2. 总静压强作用点——压力中心**

总静压强作用的点称为压力中心，表示为 $D$，如图 2-16 所示。作用点可根据合力矩定律确定，即任意轴上合力矩等于轴上各分力矩的代数和。在 $x$ 轴上，可以建立以下方程式：

$$F \cdot y_D = \int y \cdot dF = \rho g \sin\theta \int_A y^2 dA$$

式中，$\int_A y^2 dA$ 为受压面 $A$ 相对于 $Ox$ 轴的惯性矩，即 $\int_A y^2 dA = I_x$。把此方程代入上面的方程后，可得

$$F \cdot y_D = \rho g \sin\theta \cdot I_x$$

把 $F = \rho g \sin\theta \cdot y_c \cdot A$ 代入方程式后，可得

$$y_D = \frac{\rho g \sin\theta \cdot I_x}{F} = \frac{I_x}{y_C A}$$

根据惯性矩的平行移轴定律，可得

$$I_x = I_{Cx} + A y_C^2$$

$I_{Cx}$ 为受压面相对于穿过其质心并平行于 $Ox$ 轴的质心轴的惯性矩。因此，从表面总静压强作用点到 $Ox$ 轴的距离可以计算为：

$$y_D = \frac{I_x}{y_C A} = y_C + \frac{I_{Cx}}{y_C A} \tag{2-20}$$

该方程表明压力中心的位置与受压面的倾角 $\theta$ 无关。因为 $\frac{I_{Cx}}{y_C A} \geqslant 0$，压力中心始终低于质心（$y_D \geqslant y_C$），并且随着浸入深度的增加，压力中心向质心移动。只有当受压面水平时，压力中心和质心才会重叠。

在实际工程中，受压面通常是轴对称的平面，其中对称轴与 $Oy$ 轴平行。总压力 $F$ 的作用点必

须位于对称轴上,即 $x_D = x_C$。因此,只要计算 $y$ 方向上的压力中心位置,就可以确定压力中心。

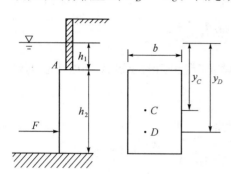

图 2-17 矩形闸门

**例 2-5** 在水中放置一扇垂直矩形门(见图 2-17)。闸门顶部到水面的距离为 $h_1 = 1$ m。闸门高度为 $h_2 = 2$ m。闸门宽度为 $b = 1.5$ m。计算闸门上的总静压强和作用点。

**解:** 根据式(2-19),总静压强为

$$F = p_C A = \rho g h_C A = \rho g \left(h_1 + \frac{h_2}{2}\right) \cdot b h_2$$

代入数据后,$F$ 可计算为

$$F = 9\,800 \times \left(1 + \frac{2}{2}\right) \times 1.5 \times 2 = 58\,800 \text{ N} = 58.8 \text{ kN}$$

根据式(2-20),压力中心的位置可通过以下公式确定:

$$y_D = y_C + \frac{I_{Cx}}{y_C A}$$

相对于矩形质心轴的惯性矩为 $I_{Cx} = \frac{1}{12} b h_2^3$。将数据代入上述方程式后,$y_D$ 可计算为

$$y_D = \left(h_1 + \frac{h_2}{2}\right) + \frac{\frac{1}{12} b h_2^3}{\left(h_1 + \frac{h_2}{2}\right) b h_2}$$

$$= \left(1 + \frac{2}{2}\right) + \frac{\frac{1}{12} \times 1.5 \times 2^3}{2 \times 1.5 \times 2} = 2.17 = h_D$$

压力中心位于水面以下 2.17 m 处,方向为水平向右。

**例 2-6** 有一个垂直半圆平面。如图 2-18 所示,正面朝水。水到达表面的直径正好位于液体的表面。计算总静压强的大小和作用点 $\left(h_C = \frac{4r}{3\pi}, I_{Cx} = \frac{9\pi^2 - 64}{72\pi} r^4\right)$。

**解:** $F = p_C A = \rho g h_C A = \rho g \times \frac{4d}{6\pi} \times \frac{1}{8} \pi d^2 = \rho g \times \frac{1}{12} d^3 = \frac{1}{12} \rho g d^3$

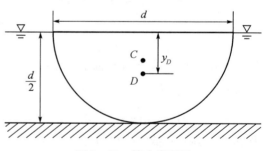

图 2-18 挡水半圆板

将 $y_C = h_C = \frac{4d}{6\pi}$,$I_{Cx} = \frac{9\pi^2 - 64}{1\,152\pi} d^4$,$A = \frac{1}{8} \pi d^2$ 代入式(2-20),然后可得

$$y_D = y_C + \frac{I_{Cx}}{y_C A} = \frac{4d}{6\pi} + \frac{(9\pi^2 - 64) d^4}{1\,152\pi \times \frac{4d}{6\pi} \times \frac{\pi d^2}{8}} = \frac{64d}{96\pi} + \frac{9\pi^2 d - 64d}{96\pi} = \frac{3}{32} \pi d$$

总静压强为 $\frac{1}{12} \rho g d^3$,压力中心位于水面以下的 $\frac{3}{32} \pi d$。

## 2.5.2 图解法

图解法更适用于计算总静压强和规范平面(例如矩形面)上的作用点。为了计算总静压强,需要绘制压强分布图。

### 1. 静压强分布图

静压强分布规律可用几何图形来描述。管线的长度表示某一点的压强大小。线端的箭头表示某一点的压强方向,即受压面内部法线的方向。该图是垂直于作用表面的线的组合,称为静压强分布图。需要注意的是,静压强分布图中的静压强是相对压强。因为建筑物周围都是大气,所以每个方向的大气压强相互抵消。众所周知,静压强与深度成正比,即静压强与深度之间呈线性关系。如果受压面是平面,压强分布图必须用直线描述,两点确定一条直线。如果受压面是曲面,则包络线和曲面之间的距离表示压强的大小,它与曲面上作用点的深度成正比。如果曲面是圆弧,则每个压强作用线必须穿过圆心。图2-19显示了不同情况下的压强分布图。

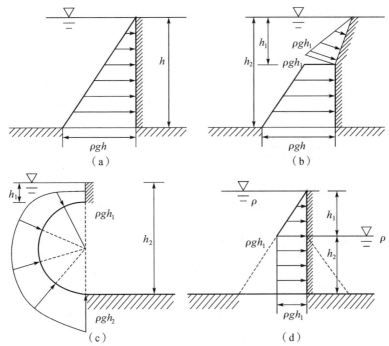

图 2-19 静压强图

### 2. 图解法

有一个垂直矩形闸门,高度为 $h$,宽度为 $b$。顶部边缘与水面齐高。水下平面为 $ABCD$,如图 2-20 所示。

根据式(2-19),

$$F = p_C A = \rho g h_C \cdot A = \rho g \times \frac{1}{2} h \cdot bh$$
$$= \frac{1}{2} \rho g h^2 \cdot b$$

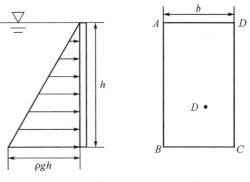

图 2-20 用图解法推导平面上的总静压强

式中，$\frac{1}{2}\rho g h^2$ 的大小等于静压强分布图 $\Omega$ 的面积，单位为 N/m。

因此，上述方程式可以改写为

$$F = \Omega \cdot b \tag{2-21}$$

式(2-21)表示总静压强的大小等于作用在平面上的压强分布图的体积。

$$h_D = h_C + \frac{I_{Cx}}{h_C A} = \frac{1}{2}h + \frac{\frac{1}{12}b h^3}{\frac{1}{2}h \cdot bh} = \frac{2}{3}h$$

式中，$\frac{2}{3}h$ 为水面以下压强分布图的质心深度。总静压强的作用线穿过压强分布图的质心并指向作用面。

**例 2-7** 使用图解法计算示例 2-5 中的总静压强和压强中心位置。

**解：** 绘制矩形闸门压强分布图：底部为受压面面积，高度为各点压强，如图 2-21b 所示。

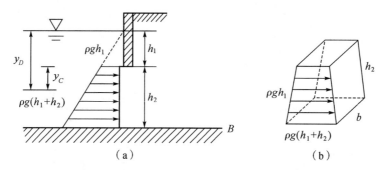

图 2-21 用图解法计算矩形闸门上的总静压强

根据图解法原理，总压的大小就是压强分布图的体积，有

$$F = \frac{1}{2}[\rho g h_1 + \rho g (h_1 + h_2)] \cdot b \cdot h_2$$

$$= \frac{1}{2} \times 9\,800 \times (1 + 1 + 2) \times 1.5 \times 2$$

$$= 58\,800 \text{ N} = 58.8 \text{ kN}$$

作用线穿过压强分布图的中心，即梯形的质心。梯形的质心方程为

$$y_C = \frac{h}{3} \cdot \frac{a + 2b}{a + b}$$

式中，$a$、$b$ 和 $h$ 分别为上部和底部的宽度以及梯形的高度（$a = \rho g h_1$，$b = \rho g (h_1 + h_2)$，$h = h_2$）。则

$$y_D = h_1 + y_C = h_1 + \frac{h_2}{3} \cdot \frac{\rho g h_1 + 2\rho g (h_1 + h_2)}{\rho g h_1 + \rho g (h_1 + h_2)}$$

$$= 1 + \frac{2}{3} \times \frac{1 + 2 \times (1 + 2)}{1 + 1 + 2} = 2.17 \text{ m} = h_D$$

**例 2-8** 一个矩形平面在水中倾斜。矩形平面的顶部在水面以下 1 m（$h = 1$ m），底部在

水面以下 3 m（$H=3$ m）。矩形的宽度为 $b=5$ m。用解析法和图解法计算作用在平面上的总静压强，并确定总静压强的作用点。

**解：**（1）解析法

沿 $y$ 轴的矩形平面长度为

$$l = \frac{H-h}{\sin 30°} = \frac{3-1}{0.5} = 4 \text{ m}$$

且

$$h_C = h + \frac{H-h}{2} = 1 + \frac{3-1}{2} = 2 \text{ m}$$

根据式(2-19)，

$$F = \rho g h_C \cdot A = 9\,800 \times 2 \times 20 = 392\,000 \text{ N} = 392 \text{ kN}$$

且

$$y_D = y_C + \frac{I_{Cx}}{y_C A}$$

式中

$$y_C = \frac{h_C}{\sin 30°} = \frac{2}{0.5} = 4 \text{ m}$$

$$I_{Cx} = \frac{1}{12} b l^3 = \frac{1}{12} \times 5 \times 4^3 = \frac{80}{3} \text{ m}^4$$

代入得

$$y_D = 4 + \frac{\frac{80}{3}}{4 \times 20} = 4.33 \text{ m}$$

$$h_D = y_D \sin 30° = 2.17 \text{ m}$$

总静压强为 392 kN。压强中心位于水面以下 2.17 m 处，压强方向为受压面的内部法线方向。

（2）图解法

因为这是一个规则的矩形平面，所以可以用图解法计算总静压强。绘制压强分布图，并将其分为矩形和三角形两部分，如图 2-22b 所示。

已知 $l = \frac{H-h}{\sin 30°} = 4$ m，梯形的面积可计算为

$$\Omega = \frac{\rho g}{2}(H+h)l = \frac{\rho g}{2} \times (3+1) \times 4 = 8 \text{ m}^2 \times \rho g$$

则

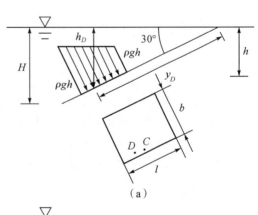

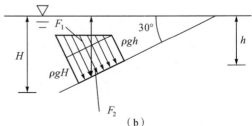

图 2-22 矩形斜面上总静压强的计算

$$F = \Omega b = 8 \times \rho g \times 5 = 40 \times 9\,800 = 392\,000 \text{ N} = 392 \text{ kN}$$

由于梯形的质心点坐标方程比较复杂，因此可以将梯形划分为简单的三角形和矩形，以计算其面积和压强中心。

对于三角形：

$$F_1 = \frac{\rho g}{2}(H-h)lb = 20 \text{ m}^3 \times \rho g$$

$$y_{D1} = \frac{2}{3}l + \frac{h}{\sin 30°} = \frac{2}{3} \times 4 + 2 = \frac{14}{3} \text{ m}$$

对于矩形：

$$F_2 = \rho g h b l = 20 \text{ m}^3 \times \rho g$$

$$y_{D2} = \frac{1}{2}l + \frac{h}{\sin 30°} = \frac{1}{2} \times 4 + 2 = 4 \text{ m}$$

根据合力矩定律：

$$F \cdot y_D = F_1 \cdot y_{D1} + F_2 \cdot y_{D2}$$

将数据代入该方程后，$y_D$ 和 $h_D$ 可计算如下：

$$y_D = \frac{1}{2}(y_{D1} + y_{D2}) = \frac{1}{2} \times \left(\frac{14}{3} + 4\right) = 4.33 \text{ m}$$

$$h_D = y_D \sin 30° = 2.17 \text{ m}$$

计算结果与解析法计算结果一致。

## 2.6 曲面上的总静压力

在工程实践中，拱坝、弧形闸门、U 形渡槽等承受水压力作用的面都是曲面。常见的曲面是双向曲面，即具有平行母线的柱面。因此，本书主要讨论施加在双向曲面上的总静水压力。

设有一个一侧承压的双向曲面 $MN$（圆柱体），母线垂直于图，曲面的面积为 $A$。建立一个坐标系，其中 $O_{xy}$ 平面和液面重合，$Oz$ 轴向下，如图 2-23a 所示。为了求解双向曲面上的总静压力 $F$，需要分别计算 $F$ 的分量 $F_x$ 和 $F_y$。

### 2.6.1 水平分力

曲面 $MN$ 在垂直平面上的投影面是 $M'N'$。投影平面的面积为 $A_x$。取液体 $M'N'NM$ 为脱离体。在脱离体上的受力平衡分析如图 2-23b 所示。在脱离体上有三个力，包括 $F'$、$F_1$ 和 $F_2$。液体作用在曲面上的反力包括水平力 $F'_x$、垂直力 $F'_z$ 和重力 $W$，因此，作用于脱离体上的水平力只有大小相等、方向相反的 $F'$ 和 $F'_x$。再根据作用力和反作用力之间的关系，可得

$$F_x = F' = p_C A_x \tag{2-22}$$

式中，$F_x$ 是液体作用在曲面上的水平分力，$p_C$ 是投影平面 $A_x$ 形心点的压强。

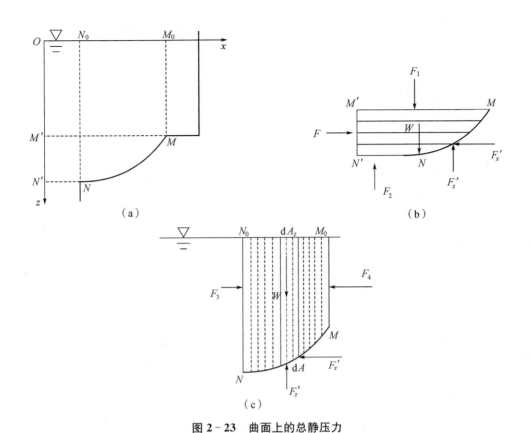

**图 2-23 曲面上的总静压力**

式(2-22)表明作用在曲面上的水平分力 $F_x$ 等于作用在该曲面的铅垂投影面上的力 $F'$。$F_x$ 的作用线也必定是 $F'$ 的作用线。

### 2.6.2 竖向分力

曲面上的竖向分力可根据被曲面 $MN$ 包围的液体(称为压力棱镜)体积、$MN$ 在自由液面的投影面 $M_0N_0$ 以及曲面边缘的竖向面来计算(图 2-23c)。该体积的流体处于静力平衡。自由面上的压强为零,竖向力只有重力 $W$ 和作用在曲面上的竖向力 $F_z$ 的反作用力 $F'_z$。因此:

$$F_z = W = \int_{A_z} \rho g h \, dA_z = \rho g V_p \qquad (2-23)$$

式中,$F_z$ 为液体作用在曲面上的竖向力;$A_z$ 为曲面在水平面上的投影面积;$\int_{A_z} h \, dA_z = V_p$ 为压力体的体积。

式(2-23)表明,作用在曲面上的静压力的竖向分力等于被 $MN$、$M_0N_0$ 和 $MN$ 边缘的竖向平面包围的液体的重量。竖向分力作用线的位置与重力 $W$ 的位置相同,必定通过压力体的中心。

### 2.6.3 曲面上的总静压力

一般来说,不规则平面上不存在单一合力。水平分力和竖向分力可能不在同一平面上。

对于二向平面,水平分力和竖向分力会在同一平面上,并可合并成力 $F$:

$$F=\sqrt{F_x^2+F_z^2} \qquad (2-24)$$

$F$ 与水平线之间的夹角为

$$\theta=\arctan\frac{F_z}{F_x} \qquad (2-25)$$

其作用线必定经过 $F_x$ 和 $F_z$ 作用线的交点。$F$ 的作用点位于曲面与 $F$ 作用线的交点。在许多实际工程中不必计算 $F$,只需计算 $F$ 的分力,包括大小、方向和作用点,就可以满足许多实际工程的需要。

### 2.6.4 压力体

在式(2-23)中,积分 $\int_{A_z} h\mathrm{d}A_z=V_p$ 表示的几何体积称为压力体。流体静压力在曲面上的竖向分力的计算实际上是压力体的计算。因此,压力体的概念非常重要。

压力体是以作用面为下表面,自由液面为上表面,用一条铅垂直线沿作用面边缘移动一圈切割而成的圆柱体。根据定义,压力体由三个曲面包围:(1)受压曲面;(2)曲面边缘的铅垂面;(3)自由液面或自由液面的延伸面。

**1. 实压力体**

当液体和压力体在曲面的同一侧时,如图 2-24a 所示,$F_z$ 方向向下,并且 $F_z$ 等于压力体的水重。这种压力体称为实压力体。

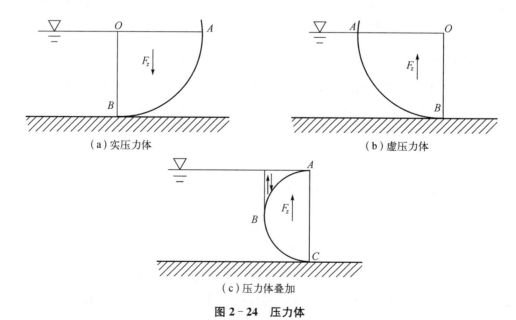

(a)实压力体　　(b)虚压力体

(c)压力体叠加

图 2-24　压力体

**2. 虚压力体**

如图 2-24b 所示,当液体和压力体位于曲面的不同侧时,$F_z$ 方向向上,并且 $F_z$ 等于压力体的水重。这种想象的压力体被称为虚压力体。

**3. 压力体叠加**

对于水平投影重叠的曲面，分别界定压力体，然后将其叠加。根据曲面$\overset{\frown}{AB}$和$\overset{\frown}{BC}$可以确定半圆柱面的压力体。叠加后得到虚压力体$ABC$，$F_z$方向向上。

### 2.6.5 推论

（1）在有限的空间内，气体压强随深度变化很小且气体重量可以忽略不计。在对分离体进行受力分析后可知，气体作用在曲面上的压力为

水平分力：
$$F_x = p \cdot A_x \tag{2-26}$$

竖向分力：
$$F_z = p \cdot A_z \tag{2-27}$$

式中，$A_x$和$A_z$分别为曲面在水平面和铅垂面上的投影面积。

（2）如果液体表面不是自由液面，而是在封闭容器中，且表面压强为$p_0$，则总压力为

水平分力：
$$F_x = (p_C + p_0) A_x \tag{2-28}$$

竖向分力：
$$F_z = \rho g V_p + p_0 A_z \tag{2-29}$$

**例 2-9** 将两个半球表面铆接成球形容器。有铆钉$n$个。球形容器装满密度为$\rho$的液体，如图2-25a所示。计算作用在每个铆钉上的压力。

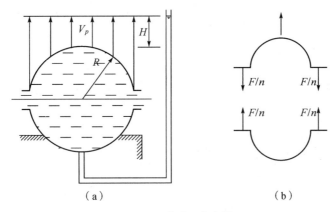

图 2-25 球形压力容器

**解：** 将球形容器的上半球体视为受压曲线。作用在曲面上的压力体如图2-25b所示。则有

$$nF = \rho g V_p = \rho g \left[ \pi R^2 (R+H) - \frac{2}{3} \pi R^3 \right]$$
$$= \rho g \left( \frac{1}{3} \pi R^3 + \pi R^2 H \right)$$

因此，

$$F = \frac{\rho g}{n} \left( \frac{1}{3} \pi R^3 + \pi R^2 H \right)$$

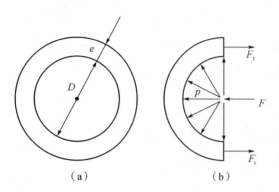

图 2-26 压力管的管壁压力

**例 2-10** 如图 2-26a 所示，用允许拉力为 $[\sigma]=150$ MPa 的钢板制造内径 $D=1$ m 的水管。该水管内的压强是 500 mH$_2$O。求水管的壁厚。（注：忽略由于管道内部不同点的高度差异所产生的压差。）

**解：** 以长度为 1 m 的管段为研究对象。管壁上任意点的压强相等。

沿着管径将管壁切开，将半管段视为一个分离体，对该分离体进行受力分析（见图 2-26b）。作用在半环内表面上的水平压力等于半环垂直投影面上的压力，即 $F=p \cdot A_z = p \cdot D \times 1$。该压力与半环壁上的拉应力保持平衡，即 $2F_t = F = pD$。假设 $F_t$ 沿管壁厚度均匀分布，则：

$$F_t \leqslant [\sigma] \cdot e \times 1$$

因此，

$$e \geqslant \frac{F_t}{[\sigma]} = \frac{\dfrac{pD}{2}}{[\sigma]} = \frac{9\,800 \times 500 \times 1}{2 \times 150 \times 10^6} = 0.016\,3 \text{ m}$$

**例 2-11** 有一个单宽（$b=1$ m）半圆柱。在浮力 $F_z$ 和水平水压力 $F_x$ 的共同作用下，是否会产生相对于轴心的转动力矩？（见图 2-27）

**解：**（1）概念分析

由于作用在半圆柱上的水压力作用线总是垂直于作用面，并通过圆心，因此水面以下的半圆柱不会对轴心产生转动力矩。

（2）计算说明

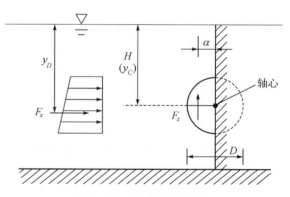

图 2-27 半圆柱受水压力

$$F_z = \rho g V_p = \rho g \times \frac{\pi}{4} D^2 \times \frac{1}{2} \times 1 = \frac{\pi}{8} D^2 \rho g$$

$$F_x = \rho g h_C A = \rho g H D \times 1 = \rho g H D$$

$$y_D = y_C + \frac{I_{Cx}}{y_C A}$$

$$y_D - y_C = \frac{I_{Cx}}{y_C A} = \frac{1 \times \dfrac{D^3}{12}}{H \times D \times 1} = \frac{D^2}{12H}$$

$$M_x = -F_x \cdot \frac{D^2}{12H} = -\frac{\rho g D^3}{12}$$

竖向力作用点到轴心的距离为

$$\alpha = \frac{2D}{3\pi}$$

于是，

$$M_z = F_z \cdot \alpha = \frac{\pi}{8} D^2 \rho g \times \frac{2D}{3\pi} = \frac{\rho g D^3}{12}$$

$$\sum M = M_x + M_z = 0$$

因此，作用在半圆柱上的水压力不会产生相对于轴心的转动力矩。

**例 2-12** 圆柱体直径为 2 m，水平放置。各部分的尺寸如图 2-28a 所示。左侧有水，右侧无水。求作用在每米圆柱体上的总静压力的水平分力 $F_x$ 和竖向分力 $F_z$。

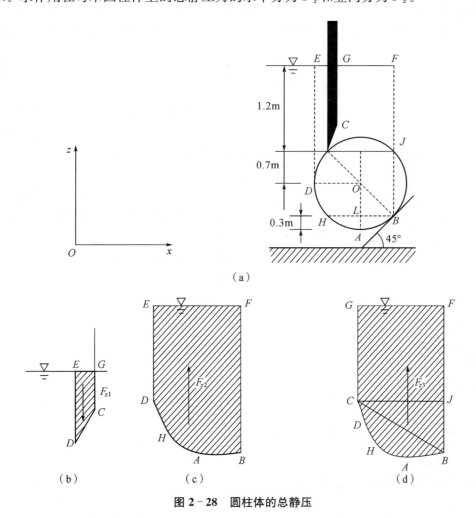

图 2-28 圆柱体的总静压

**解：** CDHAB 是圆柱体的受压面。HAB 平面两侧的水平分力相互抵消，则受压面 CDH 上的水平分力为

$$F_x = \rho g h_C A = 9\,800 \times (1.2 + 0.7) \times 1.4 \times 1 = 26\,068 \text{ N} = 26.068 \text{ kN}$$

通过绘制曲面 CDHAB 的压力体，可以计算出竖向分力 $F_z$。将曲面 CDHAB 分为 CD

和 $DHAB$ 两部分。如图 2-28b 和图 2-28c 所示,绘制了两个截面的压力体。对于 $CD$ 的压力体, $F_{z1}$ 方向向下。对于 $DHAB$ 的压力体, $F_{z2}$ 方向向上。这两者可以在一定程度上相互抵消。最终得到的压力体是如图 2-28d 所示的阴影部分。作用在每米圆柱体上的总静压力的竖向分力 $F_z$ 在数值上等于 $DHABJFGCD$ 的水重。为了方便计算, $DHABJFGCD$ 被划分为矩形、三角形和半圆形三个简单的几何图形。则:

$$F_z = (矩形\ JFGC + 三角形\ CJB + 半圆形\ CDHABC)\ 的水重$$

$$F_z = 9\ 800 \times \left(1.2 \times 1.4 + \frac{1}{2} \times 1.4 \times 1.4 + \frac{1}{2}\pi \times 1^2\right) \times 1$$

$$= 9\ 800 \times (1.68 + 0.98 + 1.57) \times 1 = 41\ 454\ \text{N} = 41.454\ \text{kN}$$

**例 2-13**  隔板上有一个矩形孔口。孔口高度 $a = 1.0$ m,宽度 $b = 3$ m,用直径 $d = 2$ m 的圆柱体堵住孔口。隔板的两侧用水装满。$h = 2$ m, $z = 0.6$ m。计算作用在圆柱体上的总静压力。

**解**: 通过绘制压力分布曲线和压力体进行计算。

隔板两侧同时施加静压力时,隔板的水平方向压强分布如图 2-19d 所示。左、右压力相互抵消后,缸内压强分布为矩形分布,如图 2-29 所示。

压力体的左侧等于面积 $BAA'B'$ 乘以圆柱体的宽度 $b$。因此,隔板两侧受压曲线的压力体之和等于圆柱体体积,为虚压力体。$F_z$ 方向向上。

绘制压力分布图和压力体后,总静压力的水平分力为

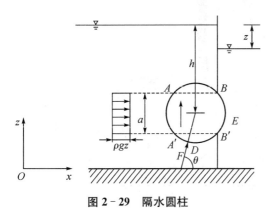

**图 2-29 隔水圆柱**

$$F_x = ab\rho gz = 1.0 \times 3 \times 1 \times 9\ 800 \times 0.6 = 17\ 640\ \text{N} = 17.64\ \text{kN}$$

式中, $F_x$ 方向向右。

总静压力的竖向分力为

$$F_z = \rho g\ V_p = \rho g \left(\frac{\pi}{4} d^2 \times b\right) = 9\ 800 \times \left(\frac{\pi}{4} \times 2^2 \times 3\right)$$

$$= 92\ 316\ \text{N} \approx 92.32\ \text{kN}$$

式中, $F_z$ 方向向上。

因此,作用在圆柱体上的总静压力为

$$F = \sqrt{F_x^2 + F_z^2} = \sqrt{17.64^2 + 92.32^2} = 93.99\ \text{kN}$$

圆柱体表面上任意点的压力均通过圆柱体的中心轴。因此,合力必须通过圆柱体的中心轴。作用线与水平面之间的夹角为

$$\theta = \arctan\frac{F_z}{F_x} = \arctan\frac{92.32}{17.64} = 79.18°$$

作用点 $D$ 在水下的深度为

$$h_D = h + \frac{d}{2}\sin\theta = 2 + \frac{2}{2}\sin 79.18° = 2.98 \text{ m}$$

## 2.7 浮力及浮潜体稳定

### 2.7.1 浮力及浸没物体的三种状态

**1. 潜体的浮力及浮心**

漂浮在水面上或浸没在水面下的物体也承受静水压力,这个压力是物体表面上每个点的静压水力之和。

如图 2-30 所示,由潜体(浸没于水中的物体)作垂直切线($AA'$,$BB'$,…),这些切线是与潜体表面相切的垂直柱体的母线。柱面和潜体表面的交线将潜体表面分成两部分:$ADB$ 和 $ACB$。作用在交线上方的潜体表面上的总静水压力的竖向分力 $F_{z1}$ 等于曲面 $ADB$ 上方的压力体的重量,方向向下。作用在交线下方的潜体表面上的总静水压的竖向分力 $F_{z2}$ 等于曲面 $ACB$ 上方的压力体的重量,方向向上。

作用于整个潜体表面的竖向分力为

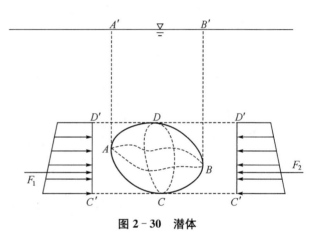

图 2-30 潜体

$$F_z = F_{z2} - F_{z1} = \rho g V_{AA'CBB'} - \rho g V_{A'ADBB'} = \rho g V_{ACBDA} \tag{2-30}$$

式中,$V_{ACBDA}$ 为被潜体排开的液体的体积。

公式(2-30)表示作用在潜体的竖向力等于潜体排开液体的重量。

同样,可将潜体分为两部分:$CAD$(左)和 $CBD$(右)。作用在物体上的水平分力 $F_x$ 是 $F_1$ 和 $F_2$ 之和。$F_1$ 和 $F_2$ 在大小上等于作用于曲面 $CAD$ 和 $CBD$ 的垂直投影面上的静水压力。这两部分在垂直平面上的投影面积相等,位置相同。因此,$F_1$ 和 $F_2$ 的大小相同,方向相反,相互抵消。结果,作用在潜体上的水平分力为零。

综上所述,作用在物体上的总静水压力只有竖向分力,它在大小上等于物体排开的液体的重量,这就是阿基米德定律。

因为 $F_z$ 倾向于将物体推到液体表面,所以将 $F_z$ 称为浮力。浮力的作用点称为浮心。浮心与所排开液体体积的形心重叠。

**2. 浸没物体的三种状态**

当浸没在液体中的物体没有其他物体支撑时,它只受到重力 $G$ 和浮力 $F_z$ 的作用。根据重力和浮力的相对大小,浸末物体有三种状态:

(1) 沉体

当 $G > F_z$ 时,物体下沉到底部。

（2）潜体
当 $G = F_z$ 时,物体处于悬浮状态。
（3）浮体
当 $G < F_z$ 时,物体浮出水面,直至物体液面以下部分排开的液体的重量等于物体的重量后保持平衡,处于漂浮状态。

### 2.7.2 潜体的平衡和稳定性

设有一个重量为 $G$ 的潜体,该潜体的重心和浮心分别位于点 $C$ 和 $D$。平衡稳定性,即当水下物体因外部扰动而倾斜时,恢复到其原始平衡状态的能力,根据重心和浮心在同一垂直线上的相对位置,可分为以下几类:

（1）随遇平衡
浮心 $D$ 与重心 $C$ 重叠(见图 2-31a)。物体在液体中的方向是任意的,即潜体在任何位置都是平衡的。

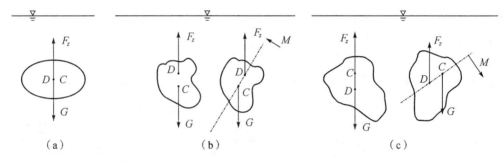

图 2-31 潜体的平衡与稳定

（2）稳定平衡
浮心 $D$ 在重心 $C$ 之上,重力 $G$ 和浮力 $F_z$ 将产生一个使潜体恢复至原始状态的转动力矩,从而使潜体恢复平衡。如果外部扰动消失,它会自动恢复平衡。也就是说,重心在浮心之下的潜体处于稳定平衡的状态。

（3）不稳定平衡
浮心 $D$ 在重心 $C$ 之下,重力 $G$ 和浮力 $F_z$ 形成的转动力矩将使潜体倾斜转动。即使外部干扰消失,潜体也会继续转动。也就是说,重心在浮心上方的潜体处于不稳定平衡的状态。

因此,为了保持潜体的平衡,浮心 $D$ 必须高于重心 $C$。

### 2.7.3 浮体的平衡和稳定性

浮体和潜体的平衡条件是相同的。平衡条件是:(1) 重力 $G$ 和浮力 $F_z$ 大小相等,方向相反;(2) 重力 $G$ 和浮力 $F_z$ 作用在同一条铅垂线上。然而,它们的稳定条件是不同的。对于浮体,即使重心 $C$ 位于浮心 $D$ 的上方,它也可以是稳定的。这是因为当物体倾斜转动时,浮力 $F_z$ 移动,穿过新形成的形心,并与重力 $G$ 大小相等,这将使潜体恢复平衡。然而,对于相对较高、细长的物体来说,很小的旋转位移就会导致浮力和重力形成使潜体倾斜的力矩。

**1. 定倾半径与偏心距**

如图 2-32a 所示,一个对称的浮体倾斜后,它的重心保持不变。如图 2-32b 所示,由于浸入水体的形状变化,浮心从 $D$ 移动到 $D'$。这里先介绍一些新概念,$H$—$H$ 线穿过浮心 $D$,重心 $C$ 称为浮动轴。浮力 $F_z$ 通过 $D$ 的作用线与浮动轴的交叉点 $M$ 称为定倾中心。从定倾中心 $M$ 到原心 $D$ 的距离是定倾半径,记为 $\rho$。重心 $C$ 和浮心 $D$ 之间的距离称为偏心距,记为 $e$。

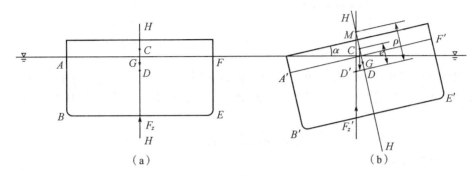

图 2-32 浮体平衡与稳定

**2. 浮体的平衡**

浮体倾斜后能否恢复到原来的平衡位置,取决于重心 $C$ 和定倾中心 $M$ 的相对位置。

从图 2-32 可以看出:如果浮体的定倾中心 $M$ 高于重心 $C$,即 $\rho > e$,倾斜后在浮体的重力 $G$ 和浮力 $F'_z$ 的共同作用下产生一个帮助浮体恢复到初始平衡位置的转动力矩,浮体处于稳定平衡状态。相反,如果浮体的定倾中心 $M$ 低于重心 $C$,即 $\rho < e$,则重力 $G$ 和浮力 $F'_z$ 的共同作用会产生一个使浮体倾向于倾斜的转动力矩。如果点 $M$ 和点 $C$ 重叠,即 $\rho = e$,则重力 $G$ 和浮力 $F_z$ 的联合作用不产生力矩。浮体处于静止平衡状态。

**3. 浮体的稳定性**

因此,浮体保持平衡的条件是重心 $C$ 必须低于定倾中心 $M$,即定倾半径 $\rho$ 大于偏心距 $e$。

对于具有固定重心的对称浮体,如果浮体的形状和重量是确定的,则可以确定偏心距 $e$。因此,浮体是否稳定取决于定倾半径的大小。

对于倾角较小的浮体($\alpha < 10°$),可得

$$\rho = \frac{I_0}{V} \tag{2-31}$$

式中,$I_0$ 为浮体浮面相对于中心纵轴的惯性矩;$V$ 为被浮体排开液体的体积。

浮体与水面相交的平面称为浮面。

式(2-31)表明,定倾半径 $\rho$ 的大小与浮面在中心纵轴上的惯性矩和被浮体排开的液体的体积有关。如果 $\rho$ 大于偏心率 $e$,则浮体是稳定的,否则就是不稳定的。定倾半径越大,浮体越稳定。

## 习 题

1. 同一容器内有两种液体($\rho_1 < \rho_2$),如图 2-33 所示。在容器的侧壁上安装了两根测压管,问图中标示的测压管内水位是否正确?

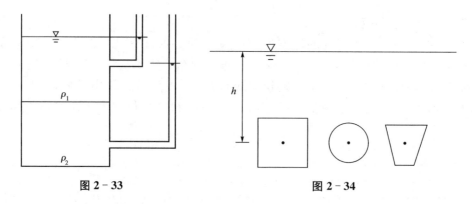

图 2-33　　　　　　　　图 2-34

2. 水中有三个平面物体，形状不同，一侧挡水，如图 2-34 所示。这三个物体的面积相同，形心处的水深相同，作用在这三个物体上的总静压力是否相等？在这三个物体中，哪个压心的位置最深？

3. 如图 2-35 所示，一个管道在静止状态下充满油（$\rho g = 8.5\,\text{kN/m}^3$）。以 $\text{mH}_2\text{O}$ 为单位计算 $A$ 点和 $B$ 点的压强。

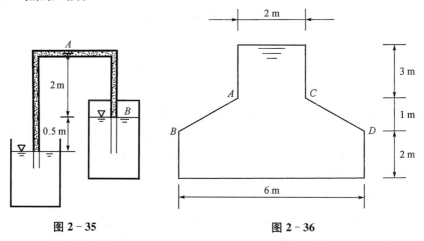

图 2-35　　　　　　　　图 2-36

4. 如图 2-36 所示，变截面圆柱体容器装满水。计算作用在锥形管段 $ABCD$ 上的竖向分力。

5. 如图 2-37 所示，求作用在弧形闸门上的（1）水平分力的大小和作用线；（2）竖向分力的大小和作用线；（3）合力的大小和方向。

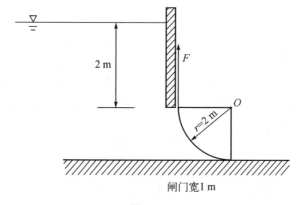

图 2-37

# 第 3 章　流体动力学基础

本章主要阐述研究流体运动的基本观点和方法。要点是流体力学分析中所用到的一些基本概念以及它们的描述方法,理解这些重要概念,包括恒定和非恒定流动、迹线、流线、流管、流量、一元流动、二元流动、三元流动、流网、势流等。推导了有旋流动和无旋流动的计算方法、连续性微分方程及伯努利方程,要求掌握速度势函数、流函数的物理含义和解法,理解流网的物理含义,理解势流叠加原理。

## 3.1 描述流体运动的方法

### 3.1.1 流体运动的两种描述方法

有两种描述流体运动的方法。

**1. 拉格朗日法**

拉格朗日法是基于对流体中每一个质点运动进行描述的方法,它是将流体中个别质点的运动分析直接扩展到整个流体质点系中,对全部质点随时间变化的位置进行标记,从而描述整个流体的总体运动情况。

由于流体质点的轨迹非常复杂,实际应用中也无需知道个别质点的运动情况,因此,拉格朗日方法在流体力学中很少使用。本书使用欧拉法。

**2. 欧拉法**

在欧拉法描述中,对流场的特性是在空间固定点或稳定区域进行监测。

欧拉描述侧重于流动空间位置或区域中的流场特性,这里涉及四个自变量,即表示位置向量的空间坐标 $x, y, z$ 和时间 $t$,如:

$$\boldsymbol{u}=\boldsymbol{u}(x, y, z, t), \text{或} \begin{cases} u_x = u_x(x, y, z, t) \\ u_y = u_y(x, y, z, t) \\ u_z = u_z(x, y, z, t) \end{cases} \tag{3-1}$$

$$p = p(x, y, z, t) \tag{3-2}$$

$$\rho = \rho(x, y, z, t) \tag{3-3}$$

### 3.1.2 欧拉加速度

**1. 欧拉加速度的组成**

在流体运动中,加速度可以分为两部分:
(1) 时变加速度(或当地加速度);
(2) 迁移加速度(或变位加速度)。

## 2. 数学描述

流速是时间的复杂函数。欧拉加速度可以从相对于时间的速度获得。

$$a = \frac{du}{dt} = \frac{\partial u}{\partial t} + \frac{\partial u}{\partial x}\frac{dx}{dt} + \frac{\partial u}{\partial y}\frac{dy}{dt} + \frac{\partial u}{\partial z}\frac{dz}{dt} = \frac{\partial u}{\partial t} + u_x\frac{\partial u}{\partial x} + u_y\frac{\partial u}{\partial y} + u_z\frac{\partial u}{\partial z} \quad (3-4)$$

所以

$$a = \frac{du}{dt} = \frac{\partial u}{\partial t} + (u \cdot \nabla)u \quad (3-5)$$

式中，$\nabla$ 是哈密顿微分算子。

$$\nabla = i\frac{\partial}{\partial x} + j\frac{\partial}{\partial y} + k\frac{\partial}{\partial z}$$

分量形式：

$$\begin{cases} a_x = \dfrac{\partial u_x}{\partial t} + u_x\dfrac{\partial u_x}{\partial x} + u_y\dfrac{\partial u_x}{\partial y} + u_z\dfrac{\partial u_x}{\partial z} \\ a_y = \dfrac{\partial u_y}{\partial t} + u_x\dfrac{\partial u_y}{\partial x} + u_y\dfrac{\partial u_y}{\partial y} + u_z\dfrac{\partial u_y}{\partial z} \\ a_z = \dfrac{\partial u_z}{\partial t} + u_x\dfrac{\partial u_z}{\partial x} + u_y\dfrac{\partial u_z}{\partial y} + u_z\dfrac{\partial u_z}{\partial z} \end{cases} \quad (3-6)$$

## 3.2 流体流动的一些基本概念

### 3.2.1 欧拉法对流体的分类

#### 1. 恒定流与非恒定流

（1）恒定流动是流场中的流体流动，其任一空间点上的运动因素（如流量、密度、压力和粘度）都不随时间变化。有

$$\frac{\partial u}{\partial t} = 0 \quad (3-7)$$

所以恒定流下的欧拉加速度为 $a = (u \cdot \nabla)u$。

（2）非恒定流是指流场中的流体流动，其空间点上至少运动要素之一随时间变化而变化。有

$$\frac{\partial u}{\partial t} \neq 0 \quad (3-8)$$

如图 3-1a 所示，当罐体液位、射流轨迹、速度和方向保持不变时，为恒定流流动。从图 3-1b 可以看出，当罐体液位、射流轨迹、速度和方向随时间变化时，为非恒定流动。

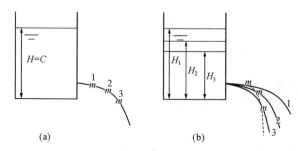

图 3-1　稳态流和非稳态流

## 2. 均匀流和非均匀流

(1) 均匀流动

流动具有相同的速度,包括沿流动方向的各个点的方向和大小。

$$(u \cdot \nabla)u = 0 \qquad (3-9)$$

(2) 非均匀流动

流动改变相同的速度,包括沿流动方向的各个点的方向或大小。

$$(u \cdot \nabla)u \neq 0 \qquad (3-10)$$

## 3. 渐变流和急变流

(1) 渐变流

渐变流:每条流线接近一条平行于流线的直线。

(2) 急变流

急变流:流线沿非均匀流急剧变化。

## 4. 一维、二维和三维流

(1) 一维流:流体运动因子是一个空间坐标的函数。所以,$u = u(s, t)$,欧拉加速度为 $a = \dfrac{\mathrm{d}u}{\mathrm{d}t} = \dfrac{\partial u}{\partial t} + u \dfrac{\partial u}{\partial s}$(见图 3-2)。

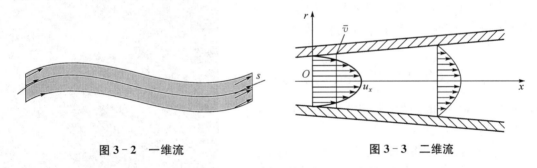

图 3-2 一维流　　　　图 3-3 二维流

(2) 二维流:流体运动因子是两个空间坐标的函数(不仅限于直角坐标,见图 3-3)。

(3) 三维流:流体运动因子是三个空间坐标的函数。例如:天然河流中的水流,其横截面形状和大小沿流向变化;水在船周围流动。

### 3.2.2 流体描述

三种类型的曲线通常用于描述流体运动——流线、迹线和脉线。这些在此处定义和描述,假设流体速度矢量 $v$ 在整个感兴趣区域的每个空间点和时间点都是已知的。当流量稳定时,流线、迹线和脉线都重合。这些曲线通常对于理解流体运动很有价值,并为跟踪种子颗粒或染料细丝的实验技术奠定了基础。

#### 1. 流线

(1) 流线是在整个流场中与流体速度瞬时相切的曲线。在非定常流动中,流线型随时间而变化。该曲线表示流线上任意点的速度矢量(图 3-4)。

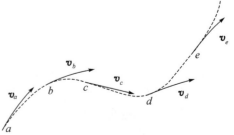

图 3-4 一瞬间的流线图

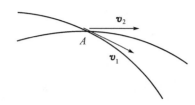

图 3-5 流线不能相交

(2) 流线特征

A. 流线：许多流体粒子，一个瞬间（迹线：一个流体粒子，一段时间）。

B. 在同一时刻，流线不能相交（图 3-5）。

C. 流线不能是折线，而是平滑的曲线。

D. 在恒定的流动中，流线和迹线重合。

(3) 流线方程

选择流线中的点 $A$，$ds$ 是微分弧长，$v$ 是点 $A$ 处的速度。在笛卡尔坐标系中，如果 $d\boldsymbol{s} = (dx, dy, dz)$ 是沿流线的弧长元且 $\boldsymbol{v} = (u, v, w)$ 是局部流体速度矢量，则 $d\boldsymbol{s}$ 和 $\boldsymbol{v}$ 的切线要求导致

$$d\boldsymbol{s} = dx\boldsymbol{i} + dy\boldsymbol{j} + dz\boldsymbol{k}, \quad \boldsymbol{v} = u\boldsymbol{i} + v\boldsymbol{j} + w\boldsymbol{k}$$

所以

$$\frac{u}{dx} = \frac{v}{dy} = \frac{w}{dz} = \frac{\boldsymbol{v}}{d\boldsymbol{s}} \tag{3-11}$$

### 2. 迹线

迹线是具有固定身份的流体粒子的轨迹。迹线是单个粒子在流场中经过一段时间后的轨迹（图 3-6）。

迹线方程为：$\frac{dx}{u} = \frac{dy}{v} = \frac{dz}{w} = dt$，$u, v, w$ 是时间 $t$ 和空间 $(x, y, z)$ 的函数。这里 $t$ 是一个自变量。

### 3. 脉线

脉线是通过连接将通过或已经通过空间中固定点的所有流体粒子而获得的曲线。在实验中，可以通过从一个小端口注入一种无源标记物（如染料或烟雾）来观察脉线，并观察它在被移动流体携带通过流场时的去向。

图 3-6 迹线

**例 3-1** 二维流场的速度为 $u = x+1$，$v = -y$，求点 $(1, 2)$ 的流线方程。

**解：**
$$\frac{dx}{u} = \frac{dy}{v}$$

所以：
$$\frac{dx}{x+1} = \frac{dy}{-y}$$

积分：$\ln(x+1) = -\ln y + \ln C$ 或 $(x+1)y = C$

点 $(1, 2)$ 的流线方程为

$$(x+1)y = 4$$

### 3.2.3 流管、流束、过流断面和流量

**1. 流管**

取流场中的任意封闭曲线(非流线)，然后通过其上的每一点画出流线，从而形成壁为流线的管状空间。该管称为流管(图3-7)。

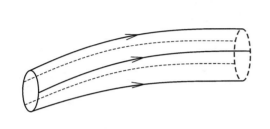

图3-7 流管和流束图

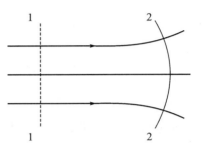

图3-8 过流断面

**2. 流束**

充满流管的流体称为流束，流束的极限是流线。

**3. 过流断面**

该断面垂直于流体流动(如管道流和渠道流)方向(如图3-8所示)。

**4. 流量**

每单位时间通过过流断面(如渠道或管道中的断面)的流体量。

体积流量($m^3/s$) $$Q = \int_A V dA \tag{3-12}$$

质量流量($kg/s$) $$Q_m = \int_A \rho V dA \tag{3-13}$$

重量流量($N/s$) $$Q_g = \int_A \rho g V dA \tag{3-14}$$

体积流量通常称为流量。因为这两个术语是同义词，所以本书可以互换地使用这两个术语。

### 3.2.4 渐变流过流断面特性

渐变流过流断面的两个重要性质：

(1) 渐变流的过流断面近似平坦。速度方向大致平行于断面中的每个点。

(2) 渐变流过流断面的流体动压近似按静压规律分布(图3-9)，即 $z + \dfrac{p}{\rho g} =$ 常数。

推导如下：

① 重力 $G = \rho g dA dn$

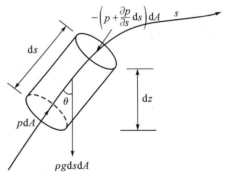

图3-9 微元体渐变断面

$$-G\cos\alpha = -\rho g dA dn \frac{z_2 - z_1}{dn} = \rho g dA(z_1 - z_2)$$

② 底部和底部上的压力分别为 $-p_2 dA$ 和 $p_1 dA$。

③ 由于流速方向垂直于断面,沿 $n$ 方向的速度为零,因此,圆柱面的切向力为零。
④ 在梯度流动条件下,仅沿 $n$ 方向的加速度为零。所以,沿 $n$ 方向没有惯性力。

根据牛顿第二定律,$\sum F_n = 0$

$$\sum F_n = p_1 \mathrm{d}A - p_2 \mathrm{d}A + \rho g \mathrm{d}A(z_1 - z_2) = 0$$

简化

$$z_1 + \frac{p_1}{\rho g} = z_2 + \frac{p_2}{\rho g}$$

$$z + \frac{p}{\rho g} = 常数$$

## 3.3 理想流体微团的运动

### 3.3.1 理想流体微团的运动类型

粒子是胶束,胶束运动被亥姆霍兹分解。流体粒子的运动可能会发生平移、旋转和变形(包括线性变形和角变形)。实际上,流体的许多运动都与运动平移、旋转和变形相结合。

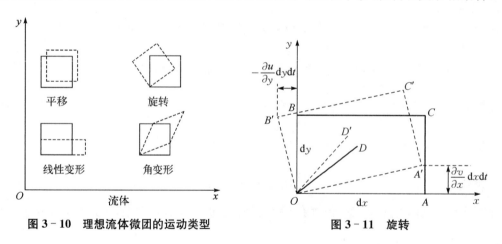

图 3-10 理想流体微团的运动类型　　图 3-11 旋转

### 3.3.2 旋转

它是反对称的,因此它的对角元素为零,并且它的非对角元素相等且符号相反。此外,它的三个独立元素可以对应一个向量。

定义 $O:(u, v)$, $A:(u_A, v_A)$(图 3-11)

$$v_A = v + \frac{\partial v}{\partial x} \mathrm{d}x$$

$\mathrm{d}t$:

$$OA \to OA', \angle AOA' \approx \frac{\partial v}{\partial x} \mathrm{d}t$$

相似地,

$$\angle BOB' \approx -\frac{\partial u}{\partial y} \mathrm{d}t$$

所以 $$\angle DOD' = \frac{\angle AOA' + \angle BOB'}{2} = \frac{1}{2}\left(\frac{\partial v}{\partial x}dt - \frac{\partial u}{\partial y}dt\right)$$

定义：
$$\omega_z = \frac{1}{2}\left(\frac{\partial v}{\partial x} - \frac{\partial u}{\partial y}\right) \tag{3-15a}$$

相似地，
$$\omega_x = \frac{1}{2}\left(\frac{\partial w}{\partial y} - \frac{\partial v}{\partial z}\right) \tag{3-15b}$$

$$\omega_y = \frac{1}{2}\left(\frac{\partial u}{\partial z} - \frac{\partial w}{\partial x}\right) \tag{3-15c}$$

### 3.3.3 有旋流和无旋流

有旋流也称为涡流。有旋流的流体粒子围绕其轴线旋转，并且与运动路径无关。旋转的三个分量($\omega_x$，$\omega_y$，$\omega_z$)至少有一个不为零。

有旋流或无旋流取决于流体粒子是否绕其轴旋转，与运动路径无关(图 3-12a)。

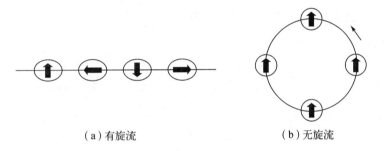

(a) 有旋流　　　　　　　　(b) 无旋流

图 3-12　有旋流和无旋流

无旋流也称为势流，其中元素会平移或变形但不会旋转，即流体粒子不会绕其任何轴旋转(图 3-12b)。如果

$$\omega_x = 0, \omega_y = 0, \omega_z = 0 \tag{3-16}$$

或者

$$\begin{cases}\dfrac{\partial w}{\partial y} - \dfrac{\partial v}{\partial z} = 0 \\ \dfrac{\partial u}{\partial z} - \dfrac{\partial w}{\partial x} = 0 \\ \dfrac{\partial v}{\partial x} - \dfrac{\partial u}{\partial y} = 0\end{cases} \tag{3-17}$$

则流动为无旋流，否则为有旋流。

## 3.4　恒定流的基本方程

恒定流的基本方程包括连续性方程、理想流体运动微分方程和粘性流体的运动微分方程。

### 3.4.1 恒定流连续性方程

**1. 恒定流基本方程的一般形式**

以一个微元六面体作为流场中的控制体,其边长为 $dx$,$dy$,$dz$。控制体是流场中的一个指定空间,其形状、位置固定,流体在其中不受影响。例如,在 $x$ 轴方向上分析该体的质量控制情况。

用泰勒级数展开取一阶微量,则有

左表面流速 $$u_M = u_x - \frac{1}{2}\frac{\partial u_x}{\partial x}dx$$

右表面流速 $$u_N = u_x + \frac{1}{2}\frac{\partial u_x}{\partial x}dx$$

$dt$ 时间内流过微面积的质量为

$$\Delta M = \rho u \Delta A dt$$

控制体在 $x$ 方向上的流进和流出质量差为

$$\Delta M_x = M_{\text{右}} - M_{\text{左}}$$
$$= \left[\rho u_x + \frac{1}{2}\frac{\partial(\rho u_x)}{\partial x}dx\right]dydzdt - \left[\rho u_x - \frac{1}{2}\frac{\partial(\rho u_x)}{\partial x}dx\right]dydzdt$$
$$= \frac{\partial(\rho u_x)}{\partial x}dxdydzdt$$

同理得

$$\Delta M_y = \frac{\partial(\rho u_y)}{\partial y}dxdydzdt$$

$$\Delta M_z = \frac{\partial(\rho u_z)}{\partial z}dxdydzdt$$

由质量守恒定律可知

$$\left[\frac{\partial(\rho u_x)}{\partial x} + \frac{\partial(\rho u_y)}{\partial y} + \frac{\partial(\rho u_z)}{\partial z}\right]dxdydzdt = -\frac{\partial \rho}{\partial t}dxdydzdt$$

故恒定流基本方程的一般形式为

$$\frac{\partial \rho}{\partial t} + \frac{\partial(\rho u_x)}{\partial x} + \frac{\partial(\rho u_y)}{\partial y} + \frac{\partial(\rho u_z)}{\partial z} = 0 \tag{3-18}$$

或

$$\frac{\partial \rho}{\partial t} + \text{div}(\rho \boldsymbol{u}) = 0 \tag{3-19}$$

连续性微分方程的推导中没有引入任何约束条件,所以方程(3-18)或(3-19)对任何类型流体都适用。

## 2. 适用不同范围的方程形式

（1）恒定流动的连续性微分方程

$$\frac{\partial(\rho u_x)}{\partial x}+\frac{\partial(\rho u_y)}{\partial y}+\frac{\partial(\rho u_z)}{\partial z}=0 \tag{3-20}$$

或

$$\mathrm{div}(\rho \boldsymbol{u})=0 \tag{3-21}$$

该方程适用范围：理想、实际、可压缩的和不可压缩的恒定流。

（2）不可压缩流体的连续性微分方程

此时，因为 $\rho$ 为常数，所以有

$$\frac{\partial u_x}{\partial x}+\frac{\partial u_y}{\partial y}+\frac{\partial u_z}{\partial z}=0 \tag{3-22}$$

或

$$\mathrm{div}(\boldsymbol{u})=0 \tag{3-23}$$

不可压缩流体连续性微分方程的物理意义是，流入单位空间的流体体积等于单位时间内的流出体积。

### 3.4.2 理想流体运动微分方程

**1. 受力分析**

如图 3-13 所示，在理想流体中取一个中心为 $(x, y, z)$ 的微元六面体。其边长分别为 $\mathrm{d}x$，$\mathrm{d}y$，$\mathrm{d}z$，其压力中心点为 $p(x, y, z)$。通过该中心的 $M$、$N$ 点的压力均作泰勒级数展开并取其一阶微量。

（1）表面力

由于理想流体没有粘度，因此在微元六面体中每一个面上的切应力皆为零。并且由于

$$p_x=p_y=p_z=p$$

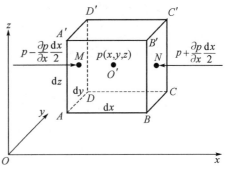

图 3-13 理想流体运动微分方程

则微元六面体

在 $x$ 方向上的左表面力 $\quad p_M A=\left(p-\dfrac{\partial p}{\partial x}\dfrac{\mathrm{d}x}{2}\right)\mathrm{d}y\mathrm{d}z$

在 $x$ 方向上的右表面力 $\quad p_N A=\left(p+\dfrac{\partial p}{\partial x}\dfrac{\mathrm{d}x}{2}\right)\mathrm{d}y\mathrm{d}z$

（2）质量力

单位质量力在各个坐标轴上的分量为 $f_x$，$f_y$，$f_z$，故 $x$ 方向上的质量力为 $f_x\rho\mathrm{d}x\mathrm{d}y\mathrm{d}z$。

**2. 运动分析**

在 $x$ 方向上应用牛顿第二定律 $\sum F_x=ma_x$，则有

$$\left(p - \frac{\partial p}{\partial x}\frac{dx}{2}\right)dydz - \left(p + \frac{\partial p}{\partial x}\frac{dx}{2}\right)dydz + f_x \rho dx dy dz = \rho dx dy dz \frac{du_x}{dt}$$

上式两边除以微元体质量 $\rho dxdydz$，得

$$f_x - \frac{1}{\rho}\frac{\partial p}{\partial x} = \frac{\partial u_x}{\partial t} + u_x\frac{\partial u_x}{\partial x} + u_y\frac{\partial u_x}{\partial y} + u_z\frac{\partial u_x}{\partial z} \quad (3-24a)$$

同理

$$f_y - \frac{1}{\rho}\frac{\partial p}{\partial y} = \frac{\partial u_y}{\partial t} + u_x\frac{\partial u_y}{\partial x} + u_y\frac{\partial u_y}{\partial y} + u_z\frac{\partial u_y}{\partial z} \quad (3-24b)$$

$$f_z - \frac{1}{\rho}\frac{\partial p}{\partial y} = \frac{\partial u_y}{\partial t} + u_x\frac{\partial u_z}{\partial x} + u_y\frac{\partial u_z}{\partial y} + u_z\frac{\partial u_z}{\partial z} \quad (3-24c)$$

上式即为理想流体的运动微分方程。

如果加速度 $\frac{du_x}{dt}$，$\frac{du_y}{dt}$，$\frac{du_z}{dt}$ 均为零，则该方程可转化为欧拉平衡微分方程。

### 3.4.3 粘性流体的运动微分方程

粘性流体也称为实际流体。粘性流体的运动微分方程的推导与上述方程的推导相同，但由于粘性的存在，其应力状态比理想流体更为复杂。这里将不作详细推导，只从动力学概念上做简要说明。

**1. 粘性流体的面积力**

粘性流体的面积力包括压应力和粘性引起的切应力。

（1）切应力由广义牛顿内摩擦定律（该定律在三元流的推广）定义。

$$\begin{cases} \tau_{xy} = \mu\left(\frac{\partial u_x}{\partial y} + \frac{\partial u_y}{\partial x}\right) = \tau_{yx} \\ \tau_{yz} = \mu\left(\frac{\partial u_y}{\partial z} + \frac{\partial u_z}{\partial y}\right) = \tau_{zy} \\ \tau_{zx} = \mu\left(\frac{\partial u_z}{\partial x} + \frac{\partial u_x}{\partial z}\right) = \tau_{xz} \end{cases} \quad (3-25)$$

式中，$\tau$ 表示切应力，其下标中第一个字母表示作用面的法线方向，第二个字母表示切应力的作用方向，如图 3-14 所示。

（2）实际流动的流体其任一点的动压强因为粘性切应力的存在，使得各向动压力大小不等，即 $p_{xx} \neq p_{yy} \neq p_{zz}$。其任意一点动压强为

$$p = \frac{1}{3}(p_{xx} + p_{yy} + p_{zz}) \quad (3-26)$$

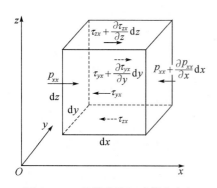

图 3-14 粘性流体运动微分方程

式中，$p$ 的下标与切应力 $\tau$ 的下标意义相同。

流体力学理论分析已证明，不可压缩流体的各向动压强有下列关系：

$$\begin{cases} p_{xx} = p - 2\mu \dfrac{\partial u_x}{\partial x} \\ p_{yy} = p - 2\mu \dfrac{\partial u_y}{\partial y} \\ p_{zz} = p - 2\mu \dfrac{\partial u_z}{\partial z} \end{cases} \tag{3-27}$$

**2. 不可压缩实际流体的运动微分方程**

另取一微元六面体作受力分析，其在 $x$ 方向上的表面力如图 3-14 所示。

由于力和流体质量相同，因此，在 $x$ 方向上，根据牛顿第二运动定律 $\sum F_x = ma_x$，可以得出

$$f_x \rho \mathrm{d}x\mathrm{d}y\mathrm{d}z + \left[ p_{xx}\mathrm{d}y\mathrm{d}z - \left( p_{xx} + \frac{\partial p_{xx}}{\partial x}\mathrm{d}x \right)\mathrm{d}y\mathrm{d}z \right] -$$
$$\left[ \tau_{yx}\mathrm{d}x\mathrm{d}z - \left( \tau_{yx} + \frac{\partial \tau_{yx}}{\partial y}\mathrm{d}y \right)\mathrm{d}x\mathrm{d}z \right] - \left[ \tau_{zx}\mathrm{d}y\mathrm{d}x - \left( \tau_{zx} + \frac{\partial \tau_{zx}}{\partial z}\mathrm{d}z \right)\mathrm{d}y\mathrm{d}x \right]$$
$$= \rho \mathrm{d}x\mathrm{d}y\mathrm{d}z \frac{\mathrm{d}u_x}{\mathrm{d}t}$$

将切应力和压应力表达方程(3-25)、(3-27)与不可压缩流体连续性微分方程联系得

$$\frac{\partial u_x}{\partial x} + \frac{\partial u_y}{\partial y} + \frac{\partial u_z}{\partial z} = 0$$

代入化简，得到不可压缩粘性流体的运动微分方程，即

$$f_x - \frac{1}{\rho} \frac{\partial p}{\partial x} + \nu \nabla^2 u_x = \frac{\mathrm{d}u_x}{\mathrm{d}t} = \frac{\partial u_x}{\partial t} + u_x \frac{\partial u_x}{\partial x} + u_y \frac{\partial u_x}{\partial y} + u_z \frac{\partial u_x}{\partial z} \tag{3-28a}$$

同理

$$f_y - \frac{1}{\rho} \frac{\partial p}{\partial y} + \nu \nabla^2 u_y = \frac{\mathrm{d}u_y}{\mathrm{d}t} = \frac{\partial u_y}{\partial t} + u_x \frac{\partial u_y}{\partial x} + u_y \frac{\partial u_y}{\partial y} + u_z \frac{\partial u_y}{\partial z} \tag{3-28b}$$

$$f_z - \frac{1}{\rho} \frac{\partial p}{\partial z} + \nu \nabla^2 u_z = \frac{\mathrm{d}u_z}{\mathrm{d}t} = \frac{\partial u_z}{\partial t} + u_x \frac{\partial u_z}{\partial x} + u_y \frac{\partial u_z}{\partial y} + u_z \frac{\partial u_z}{\partial z} \tag{3-28c}$$

或用矢量表示为

$$\boldsymbol{f} - \frac{1}{\rho}\nabla p + \nu \nabla^2 \boldsymbol{u} = \frac{\partial \boldsymbol{u}}{\partial t} + (\boldsymbol{u} \cdot \nabla)\boldsymbol{u} \tag{3-29}$$

式中，$\nabla$ 为拉普拉斯算子，$\nabla^2 = \dfrac{\partial^2}{\partial x^2} + \dfrac{\partial^2}{\partial y^2} + \dfrac{\partial^2}{\partial z^2}$，例如，

$$\nabla^2 u_x = \frac{\partial^2 u_x}{\partial x^2} + \frac{\partial^2 u_x}{\partial y^2} + \frac{\partial^2 u_x}{\partial z^2} \tag{3-30}$$

不可压缩的粘性流体运动方程也称为纳维-斯托克斯方程（简记为 N-S 方程）。这是理论

流体力学的基本方程之一。由于 $\rho$ 是常数，N-S 方程和连续性微分方程可以组成含四个未知数的封闭方程组。结合边界条件，在理论上，N-S 方程可解。然而，要直接描述解决复杂而微妙的粘性流体运动，是非常复杂和困难的，这方面的研究仍是学界的前沿问题。

## 3.5 欧拉运动微分方程的积分

欧拉运动微分方程适用于理想流体沿流线稳定流动的情况。欧拉方程可分别与 $\mathrm{d}x$、$\mathrm{d}y$ 和 $\mathrm{d}z$ 相乘，然后相加得

$$\underbrace{(f_x\mathrm{d}x+f_y\mathrm{d}y+f_z\mathrm{d}z)}_{\langle 1 \rangle}-\underbrace{\frac{1}{\rho}\left(\frac{\partial p}{\partial x}\mathrm{d}x+\frac{\partial p}{\partial y}\mathrm{d}y+\frac{\partial p}{\partial z}\mathrm{d}z\right)}_{\langle 2 \rangle}=\underbrace{\frac{\mathrm{d}u_x}{\mathrm{d}t}\mathrm{d}x+\frac{\mathrm{d}u_y}{\mathrm{d}t}\mathrm{d}y+\frac{\mathrm{d}u_z}{\mathrm{d}t}\mathrm{d}z}_{\langle 3 \rangle} \quad (3-31)$$

### 3.5.1 在恒定势流、仅受重力、不可压缩流体等条件下的积分

**1. 伯努利方程**

因为理想流体只受重力作用，$f_x=f_y=0$，$f_z=-g$，则

$$\langle 1 \rangle = -g\mathrm{d}z \quad (3-31\mathrm{a})$$

因为理想流体是不可压缩的，故

$$\langle 2 \rangle = -\frac{1}{\rho}\mathrm{d}p = -\mathrm{d}\left(\frac{p}{\rho}\right) \quad (3-31\mathrm{b})$$

因为理想的流体为恒定势流 $\dfrac{\partial u_x}{\partial y}=\dfrac{\partial u_y}{\partial x}$，$\dfrac{\partial u_y}{\partial z}=\dfrac{\partial u_z}{\partial y}$，$\dfrac{\partial u_x}{\partial z}=\dfrac{\partial u_z}{\partial x}$，则有

$$\begin{aligned}
\langle 3 \rangle &= \frac{\mathrm{d}u_x}{\mathrm{d}t}\mathrm{d}x+\frac{\mathrm{d}u_y}{\mathrm{d}t}\mathrm{d}y+\frac{\mathrm{d}u_z}{\mathrm{d}t}\mathrm{d}z \\
&= \left(u_x\frac{\partial u_x}{\partial x}+u_y\frac{\partial u_x}{\partial y}+u_z\frac{\partial u_x}{\partial z}\right)\mathrm{d}x + \left(u_x\frac{\partial u_y}{\partial x}+u_y\frac{\partial u_y}{\partial y}+u_z\frac{\partial u_y}{\partial z}\right)\mathrm{d}y \\
&\quad + \left(u_x\frac{\partial u_z}{\partial x}+u_y\frac{\partial u_z}{\partial y}+u_z\frac{\partial u_z}{\partial z}\right)\mathrm{d}z \\
&= \frac{\partial}{\partial x}\left[\frac{1}{2}(u_x^2+u_y^2+u_z^2)\right]\mathrm{d}x+\frac{\partial}{\partial y}\left[\frac{1}{2}(u_x^2+u_y^2+u_z^2)\right]\mathrm{d}y+\frac{\partial}{\partial z}\left[\frac{1}{2}(u_x^2+u_y^2+u_z^2)\right]\mathrm{d}z \\
&= \frac{\partial}{\partial x}\left(\frac{u^2}{2}\right)\mathrm{d}x+\frac{\partial}{\partial y}\left(\frac{u^2}{2}\right)\mathrm{d}y+\frac{\partial}{\partial z}\left(\frac{u^2}{2}\right)\mathrm{d}z \\
&= \mathrm{d}\left(\frac{u^2}{2}\right) \quad (3-31\mathrm{c})
\end{aligned}$$

所以有

$$\mathrm{d}\left(gz+\frac{p}{\rho}+\frac{u^2}{2}\right)=0 \quad (3-32)$$

积分得

$$gz + \frac{p}{\rho} + \frac{u^2}{2} = C' \tag{3-33}$$

或

$$z + \frac{p}{\rho g} + \frac{u^2}{2g} = C \tag{3-34}$$

$$z_1 + \frac{p_1}{\rho g} + \frac{u_1^2}{2g} = z_2 + \frac{p_2}{\rho g} + \frac{u_2^2}{2g} \tag{3-35}$$

可以通过沿流线的积分得到伯努利方程,即式(3-34)。

**2. 伯努利方程的物理意义**

伯努利方程中各项的物理和几何意义如表3-1所示。

表3-1 伯努利方程各项的物理意义和几何意义

| 项 目 | 名 称 | 物 理 意 义 | 几 何 意 义 |
| --- | --- | --- | --- |
| $z$ | 位置水头 | 单位重量流体具有的位置势能 | 位置高度 |
| $\dfrac{p}{\rho g}$ | 压强水头 | 单位重量流体具有的压强势能 | 测压管高度 |
| $\dfrac{u^2}{2g}$ | 流速水头 | 单位重量流体具有的动能 | |
| $z + \dfrac{p}{\rho g}$ | 测压管水头 | 单位重量流体具有的总势能 | |
| $z + \dfrac{p}{\rho g} + \dfrac{u^2}{2g}$ | 总水头 | 单位重量流体具有的机械能 | |

## 3.5.2 在恒定流、仅受重力、不可压缩流体等条件下沿流线的积分

势流条件中的积分结果仅适用于势流,并且有旋流也可沿同一流线进行积分,即

$$\frac{\mathrm{d}x}{\mathrm{d}t} = u_x, \quad \frac{\mathrm{d}y}{\mathrm{d}t} = u_y, \quad \frac{\mathrm{d}z}{\mathrm{d}t} = u_z$$

$$\langle 3 \rangle = u_x \mathrm{d}u_x + u_y \mathrm{d}u_y + u_z \mathrm{d}u_z = \frac{1}{2}\mathrm{d}(u_x^2 + u_y^2 + u_z^2) = \mathrm{d}\left(\frac{u^2}{2}\right)$$

能量方程为

$$z + \frac{p}{\rho g} + \frac{u^2}{2g} = C \tag{3-36}$$

式(3-36)和式(3-34)在形式上是完全一致的,但两者的积分常数存在差别。

## 3.6 恒定平面势流

### 3.6.1 流速势函数与拉普拉斯方程

在平面恒定势流中，$u_x$、$u_y$ 是 $x$ 和 $y$ 的函数，以及 $\omega_z = 0$ 或

$$\frac{\partial u_x}{\partial y} = \frac{\partial u_y}{\partial x}$$

故有

$$d\varphi = u_x dx + u_y dy = \frac{\partial \varphi}{\partial x} dx + \frac{\partial \varphi}{\partial y} dy \tag{3-37}$$

则有

$$u_x = \frac{\partial \varphi}{\partial x},\ u_y = \frac{\partial \varphi}{\partial y} \tag{3-38}$$

式中，$\varphi$ 为流速势函数，定义为无旋流。

流速势函数具有下列几种特点：

(1) 流速的三个分量等于流速势 $\varphi$ 在各自坐标上的偏导，或称流速势 $\varphi$ 的梯度等于速度。

(2) 存在无旋流动，即流度势是无旋转流动。

不可压缩流体的连续性微分方程：

$$\frac{\partial^2 \varphi}{\partial x^2} + \frac{\partial^2 \varphi}{\partial y^2} = 0 \tag{3-39}$$

上式即称为拉普拉斯方程。

### 3.6.2 流函数

**1. 流函数与拉普拉斯方程**

因为流动是恒定的平面流动，并且它们都只是 $x$ 和 $y$ 的函数。不可压缩流的连续性微分方程有

$$\frac{\partial u_x}{\partial x} = -\frac{\partial u_y}{\partial y}$$

故

$$d\psi = -u_y dx + u_x dy = \frac{\partial \psi}{\partial x} dx + \frac{\partial \psi}{\partial y} dy \tag{3-40}$$

即为

$$u_x = \frac{\partial \psi}{\partial y},\ u_y = -\frac{\partial \psi}{\partial x} \tag{3-41}$$

式中，$\psi$ 为流函数，定义为不可压流动。将流函数应用于平面势流，即可得到拉普拉斯方程。

因平面势流有

$$\frac{\partial u_x}{\partial y} = \frac{\partial u_y}{\partial x}$$

则

$$\frac{\partial^2 \psi}{\partial x^2} + \frac{\partial^2 \psi}{\partial y^2} = 0 \tag{3-42}$$

或

$$\nabla^2 \psi = 0 \tag{3-43}$$

式(3-42)的适用条件为不可压缩流体、恒定、平面、势流。

**2. 流函数的物理意义**

(1) 流函数等值线 $\psi(x,y)=C$ 即为流线。

求解：

$$\psi(x,y) = C$$
$$\mathrm{d}\psi = -u_y \mathrm{d}x + u_x \mathrm{d}y = 0$$

得平面流方程：

$$\frac{\mathrm{d}x}{u_x} = \frac{\mathrm{d}y}{u_y}$$

(2) 通过横截面 $AB$ 的单位宽度的流量为

$$Q = \int_{y_1}^{y_2} u \mathrm{d}y = \int_{y_1}^{y_2} \frac{\partial \psi}{\partial y} \mathrm{d}y = \int_{\psi_1}^{\psi_2} \mathrm{d}\psi = \psi_2 - \psi_1$$

在不可压缩的平面流动中，两条流线之间的流函数差值，等于两条流线之间的单位宽度的流量。

**3. 推论——流函数和势函数的关系**

$$\begin{cases} \dfrac{\partial \varphi}{\partial x} = \dfrac{\partial \psi}{\partial y} \\ \dfrac{\partial \varphi}{\partial y} = -\dfrac{\partial \psi}{\partial x} \end{cases} \tag{3-44}$$

上式为柯西-黎曼条件，$\varphi$，$\psi$ 都满足拉普拉斯方程和柯西-黎曼条件。

**例 3-2** 有两种流动：(a) $u_x=1$，$u_y=2$；(b) $u_x=4x$，$u_y=-4y$。试问：

(1) (a)中的流动是否为流函数？

(2) (b)中的流动是否存在势函数？

**解**：(1) 因为

$$\frac{\partial u_x}{\partial x} + \frac{\partial u_y}{\partial y} = 0$$

所以存在流函数

$$\psi = \int u_x \mathrm{d}y - u_y \mathrm{d}x = \int \mathrm{d}y - 2\mathrm{d}x = \int \mathrm{d}(y - 2x) = y - 2x + C_1$$

(2) 因为

$$\frac{\partial u_x}{\partial y} = \frac{\partial (4x)}{\partial y} = 0, \quad \frac{\partial u_y}{\partial x} = \frac{\partial (-4x)}{\partial x} = 0$$

所以

$$\frac{\partial u_x}{\partial y} = \frac{\partial u_y}{\partial x}$$

该流动为无旋流，所以存在势函数

$$\varphi = \int u_x \mathrm{d}x + u_y \mathrm{d}y = \int 4x \mathrm{d}x - 4y \mathrm{d}y = 2\int \mathrm{d}(x^2 - y^2) = 2x^2 - 2y^2 + C_2$$

### 3.6.3 流网

**1. 流网的性质**

（1）其他平面无旋流动的流线与等势线正交。

**证明：** 等势线是流线，

$$\mathrm{d}\psi = u_x \mathrm{d}y - u_y \mathrm{d}x = 0$$

流线上任意一点的斜率为

$$m_1 = \frac{\mathrm{d}y}{\mathrm{d}x} = \frac{u_y}{u_x}$$

$$\mathrm{d}\varphi = u_x \mathrm{d}x + u_y \mathrm{d}y = 0$$

势线上任意一点的斜率为

$$m_2 = \frac{\mathrm{d}y}{\mathrm{d}x} = -\frac{u_x}{u_y}$$

因为 $m_1 m_2 = -1$，所以流线与势线在该点上正交。

（2）流网中各网格边长之比等于 $\dfrac{\Delta \varphi}{\Delta \psi}$。如果取 $\Delta \varphi = \Delta \psi$，流网的网格为正方形网格。

**2. 流网的绘制**

有两种常用的方法绘制流网。一种是图解法，另一种就是电比拟法。

图解法的原则：

（1）等势线与边界正交。

（2）如果垂直于某一液面的自由液面的流速为零，则该自由液面为流线。

（3）按照预先选定好的网格线比例绘制流线和等势线。

**3. 流网的应用**

流网是特定边界内流场的唯一解。因此，流网可以应用于所有相似边界的流场。

$$u_{s_i} \approx \frac{\Delta C}{\Delta s_i} = \frac{\Delta C}{\Delta n_i}$$

流网理论已被广泛用于理想流体势流。

### 3.6.4 势流叠加原理

如果势函数满足拉普拉斯方程,则它们叠加的结果也必定满足拉普拉斯方程。

例如,$\boldsymbol{\nabla}^2 \varphi_1 = 0$ 和 $\boldsymbol{\nabla}^2 \varphi_2 = 0$,它们叠加的结果是 $\varphi = \varphi_1 + \varphi_2$.

$$\boldsymbol{\nabla}^2 \varphi = \boldsymbol{\nabla}^2 (\varphi_1 + \varphi_2) = \boldsymbol{\nabla}^2 \varphi_1 + \boldsymbol{\nabla}^2 \varphi_2 = 0 \tag{3-45}$$

## 习 题

1. 已知二元流,其流速分布为 $u_x = x^2 - t$, $u_y = 2xyt$。试求 $t=1$,点 $(-2, 1)$ 处的流线方程和迹线方程。

2. 已知流函数 $\psi = 2(x^2 - y^2)$,试求速度势函数 $\varphi$。

3. 设 $\psi_1$ 及 $\psi_2$ 均满足拉普拉斯方程,试证明 $\boldsymbol{\nabla}^2 \psi = 0$,$\psi = \psi_1 + \psi_2$。

4. 已知流场 $u_x = -\dfrac{Cy}{r^2}$,$u_y = \dfrac{Cx}{r^2}$。其中 $C$ 为常数,$r^2 = x^2 + y^2$。试绘出流网示意图。

5. 平置矩形断面弯管如图 3-15 所示,其外半径为 $r_1$,内半径为 $r_2$,圆心为 $M$。设直管中流速均匀分布,其值为 $u_0$,0—0 断面上的动压强为 $P_0$,弯管中心对称于 $A$—$A$ 断面,该处流线为以 $M$ 为圆心的圆弧,且符合有势流动的规律,$u_\theta r = $ 常数。试求 $A$—$A$ 断面上流速 $u$ 及动压强 $p$ 的分布。

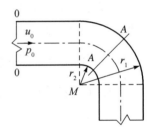

**图 3-15**

# 第 4 章　稳态总流的基本方程

在实际工程中,很多流体问题(如管流、明渠流等)都可以看作一个整体来考虑,将其视为一个整体,不考虑速度、压力等沿主流的横向变化,而仅取其对应截面的平均,从而简化了问题,并获得了较好的计算结果。本章在对一元流的基本定律阐述的基础上,推导分析了总流运动的三个基本方程,即连续性方程、能量方程和动量方程及其应用,这也是本章的学习要点。

## 4.1　总流分析方法

在实际的流体流动中,即使是最简单的流动,其横截面上各点的流速分布是不均匀的。因此,其有关的物理量如动能、动量和压力也是不均匀的。在总流分析中,使用断面平均法来处理基于元流的物理量。

### 4.1.1　元流、总流和总流控制断面

**1. 元流和总流**

元流是指在过流断面上的微元流束。

如果流束的边缘延伸到流场边界,这时边界内的流体流动就是总流。

在分析运动参数如流速、流量、压强等的变化时,可将总流划分为无数个微元流束。由于微元流束的截面非常小,可以认为 dA 段上每一点的运动参数都是均匀的,因此可以用积分法得到总流的运动参数。

**2. 控制断面**

控制断面是一个截面中的每个面积元都与微元流束或流线相交成法线的截面。图 4-1 中给出了两个控制断面 $m—m$ 和 $c—c$,控制断面用 dA 或 A 表示。

### 4.1.2　总流分析方法

**1. 以元流为基础**

总流是由元流构成的。因此,总流的分析方法是先分析元流的运动,然后将其推广到总流上。

**2. 控制断面恒选在渐变流上**

在总流分析法中,其控制断面选择在渐变流处。原因是渐变流的两个特点:一个是横断面是平面的;另一个是各点的流速方向几乎平行。

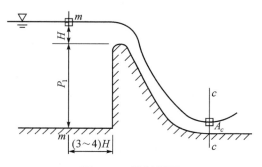

图 4-1　控制断面

## 3. 有关物理量对于断面取平均

(1) 断面平均流速

断面上每一点的流速都不同,其分布曲线是一条抛物线。如图4-2所示,点流速 $u$ 在管轴上达到最大流速 $u_{max}$,在边界上则为零。根据流量的等效性,以平均速度 $v$ 通过断面 $A$ 的流体体积必须等于单位时间内以实际速度通过该断面的流体体积,即有

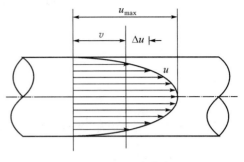

图4-2 平均流速

$$v = \frac{\int_A u \, dA}{A} = \frac{q_v}{A} \tag{4-1}$$

(2) 动量和动量修正系数

动量是物体运动的一种量度,是描述物质机械运动状态的一个重要物理量。

总之,对于任何方向,控制体内动量的变化之和再加上控制体内净流出的动量,等于单位时间内作用在该方向上的控制体上的外力的分量。

对于不可压缩恒定流,其控制面(垂直于切过流体的流速方向),只有两个表面,即流出面和流入面,如果表面的速度是均匀的,则方程中的表面积分就可以去掉。它可简化为

$$d\boldsymbol{K} = dm\boldsymbol{u} = (\rho u \, dA \cdot \Delta t)\boldsymbol{u} = \rho u \boldsymbol{u} \, dA \cdot \Delta t$$

对于总流,则有

$$\sum d\boldsymbol{K} = \int_A \rho u \boldsymbol{u} \, dA \cdot \Delta t = (\int_A \rho u \, dA_u) \boldsymbol{i} \cdot \Delta t \tag{4-2}$$

在此引入动量修正系数 $\beta$,其定义式如下:

$$\beta = \frac{\int_A u^2 \, dA}{q_v v} = \frac{\int_A u^2 \, dA}{A v^2} \tag{4-3}$$

对于不可压缩恒定流,其动量表达式为

$$\sum d\boldsymbol{K} = \left(\rho \int_A u^2 \, dA\right) \boldsymbol{i} \cdot \Delta t = \beta \rho q_v v \cdot \boldsymbol{i} \cdot \Delta t = \beta \rho q_v \boldsymbol{v} \cdot \Delta t \tag{4-4}$$

(3) 动能和动能修正系数

动能是指物体由于机械运动所具有的能量。

在单位时间内通过微元面积 $dA$ 的动能为

$$dE_k = \frac{1}{2} dm \cdot u^2 = \frac{1}{2} \rho u^3 dA$$

对于不可压缩的恒定流,其单位时间所通过的动能为

$$\sum \mathrm{d}E_k = \frac{1}{2}\rho \int_A u^3 \mathrm{d}A = \rho q_v \frac{\alpha v^2}{2} \qquad (4-5)$$

在此引入动能修正系数 $\alpha$,其定义式如下:

$$\alpha = \frac{\int_A u^3 \mathrm{d}A}{q_v v^2} = \frac{\int_A u^3 \mathrm{d}A}{A v^3} \qquad (4-6)$$

## 4.2 不可压缩流体的连续性方程

在流场中选择控制体 1—1、2—2(见图 4-3)。对于不可压缩流体:

$$\int_V \left( \frac{\partial u_x}{\partial x} + \frac{\partial u_y}{\partial y} + \frac{\partial u_z}{\partial z} \right) \mathrm{d}V = \int_A u_n \mathrm{d}A = 0 \qquad (4-7)$$

因为 $\beta=$ 常数,简化得

$$\int_{A_1} u_1 \mathrm{d}A = \int_{A_2} u_2 \mathrm{d}A \qquad (4-8)$$

$$v_1 A_1 = v_2 A_2 \qquad (4-9)$$

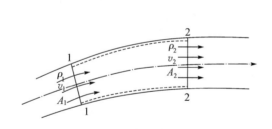

图 4-3 连续性方程总流量

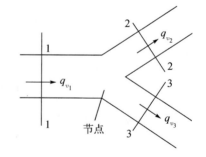

图 4-4 节点处的流量

物理意义是,对于不可压缩流体,平均速度与横截面积成反比。

适用于:不可压缩流体,包括恒定流和非恒定流、理想流体和实际流体。

当两段之间有流体流入或流出时,节点处的流量为

$$\sum_{i=1}^n q_{v_i} = 0 \qquad (4-10)$$

如图 4-4 所示,

$$q_{v_1} - q_{v_2} - q_{v_3} = 0$$

或

$$A_1 v_1 - A_2 v_2 - A_3 v_3 = 0$$

## 4.3 伯努利方程

### 4.3.1 实际流体中的伯努利方程

我们从第 3.5 节得到了元流的伯努利方程:

$$z + \frac{p}{\rho g} + \frac{u^2}{2g} = C$$

但实际流体具有阻力运动的粘性。根据能量方程,可得到实际流动中的元流伯努利方程:

$$z_1 + \frac{p_1}{\rho g} + \frac{u_1^2}{2g} = z_2 + \frac{p_2}{\rho g} + \frac{u_2^2}{2g} + h'_w \quad (4-11)$$

**例 4-1** 如图 4-5 所示,$d_1 = 250$ mm,$d_2 = 100$ mm,U 形水银管压差 $h = 800$ mm,管中的排出量是多少?

**解**:1—1、2—2 段压差:

$$\begin{aligned} p_1 - p_2 &= (\rho_{Hg} - \rho_{H_2O})gh \\ &= (13\,600 - 1\,000) \times 9.8 \times 0.8 \\ &= 98\,784 \text{ Pa} \end{aligned}$$

又因为

$$\frac{A_1}{A_2} = \left(\frac{d_1}{d_2}\right)^2$$

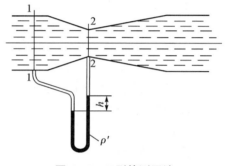

**图 4-5 U 型管测压计**

所以

$$\begin{aligned} Q &= A_1 \sqrt{2(p_1 - p_2)/\rho \left[\left(\frac{A_1}{A_2}\right)^2 - 1\right]} \\ &= \pi \times \frac{0.25^2}{4} \times \sqrt{2 \times 98\,784/[1\,000 \times (2.5^4 - 1)]} \\ &= 0.112 \text{ m}^3/\text{s} \end{aligned}$$

### 4.3.2 沿流线的伯努利方程

**1. 沿流线的伯努利方程的推导**

元流定义为 $dq_v$。将方程(4-11)乘以 $\rho g dq_v$,即可得到元流的能量关系如下:

$$\left(z_1 + \frac{p_1}{\rho g} + \frac{u_1^2}{2g}\right)\rho g dq_v = \left(z_2 + \frac{p_2}{\rho g} + \frac{u_2^2}{2g}\right)\rho g dq_v + h'_w \rho g dq_v \quad (4-12)$$

注意到:$dq_v = u_1 dA_1 = u_2 dA_2$
总流量的能量关系为

$$\int_{A_1} \left(z_1 + \frac{p_1}{\rho g} + \frac{u_1^2}{2g}\right)\rho g u_1 dA_1 = \int_{A_2} \left(z_2 + \frac{p_2}{\rho g} + \frac{u_2^2}{2g}\right)\rho g u_2 dA_2 + \int_{q_v} h'_w \rho g dq_v \quad (4-13)$$

对公式的项进行积分。

（1）势能积分

$$\rho g \int_A \left(z + \frac{p}{\rho g}\right) u \, dA = \rho g \left(z + \frac{p}{\rho g}\right) \int_A u \, dA = \left(z + \frac{p}{\rho g}\right) \cdot \rho g q_v \quad (4-14)$$

（2）动能积分

$$\rho g \int_A \frac{u^3}{2g} dA = \frac{\alpha v^2}{2g} \rho g q_v \quad (4-15)$$

（3）损失部分的积分

$$\int_{q_v} h'_w \rho g \, dq_v = h'_w \rho g q_v \quad (4-16)$$

整理得：$q_{v_1} = q_{v_2} = q_v$，除以 $\rho g q_v$

$$z_1 + \frac{p_1}{\rho g} + \frac{\alpha_1 v_1^2}{2g} = z_2 + \frac{p_2}{\rho g} + \frac{\alpha_2 v_2^2}{2g} + h_w \quad (4-17)$$

这就是实际流体总流的伯努利方程。

**2. 条件及应用**

（1）条件

伯努利方程有几个使用条件：

① 流体为恒定流；
② 流体为不可压缩流体；
③ 流体为理想流体；
④ 流体为沿线流；
⑤ 唯一的质量力是重力。

（2）应用

① 选择平面；
② 选择计算部分；
③ 选择计算点；
④ 列出伯努利方程的解。

**3. 伯努利方程的几何和物理意义**

（1）几何意义

$z$——基准面 0—0 以上的高程，位置水头；

$\dfrac{p}{\rho g}$——在压强 $p$、压差的作用下，单位重量流体的上升高度；

$\dfrac{\alpha v^2}{2g}$——在速度 $v$、动压的作用下，单位重量流体的上升高度；

$h_w$——机械能损失，总水头损失；

$z + \dfrac{p}{\rho g} + \dfrac{\alpha v^2}{2g}$——总水头。

（2）物理意义

$z$——单位重量流体的上升势能；

$\dfrac{p}{\rho g}$——单位重量流体的压力势能；

$\dfrac{\alpha v^2}{2g}$——单位重量流体的动能；

$z + \dfrac{p}{\rho g} + \dfrac{\alpha v^2}{2g}$——单位重量流体的机械能。

在重力作用下不可压缩理想流体的稳定流动中，沿流线单位重量流体的机械能是相同的。通常对于不同的流线，沿流线的单位重量机械能是不同的。

**例 4-2** 如图 4-6 所示，水由弯管虹吸出水箱，顶部截面 $h_1 = 3.6 \, \text{m}$，若此水的蒸发压力为 0.32 atm，则管道内排水量最多时的距离 $h$ 是多少？

**解**：选择 $B$ 截面作为基截面，对于自由面和 $B$ 截面

$$v_B = \sqrt{2gh}$$

由连续性方程：

$$v_A = v_B = \sqrt{2gh} = v$$

写出从 $A$ 点到 $B$ 点的伯努利方程：

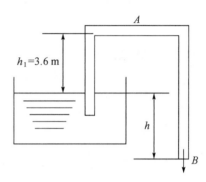

**图 4-6 虹吸**

$$h_1 + h + \dfrac{p_{MA}}{\rho g} = 0$$

$$v = \sqrt{2gh}$$

式中，$h_1$ 保持不变，

$$p_{A\min} = 0.32 \, \text{atm}, \quad p_{MA\min} = (0.32 - 1) \, \text{atm}$$

$$p_{MA} = (0.32 - 1) \, \text{atm} = -0.68 \times 101\,325 = -68\,901 \, \text{Pa}$$

$$h = -h_1 - \dfrac{p_{MA}}{\rho g} = -3.6 - \dfrac{-68\,901}{1\,000 \times 9.81} = 3.42 \, \text{m}$$

### 4.3.3 水头线

位置水头、压力水头和速度水头之和称为伯努利方程中的总水头，用 $H$ 表示。由于伯努利方程的每一组代表一个高程，它们之间的关系可以用几何图形表示。在图 4-7 中，高程基准线为 0—0；线段 $amb$ 是连接每个 $z$ 点的线，称为位置水头线；线段 $enf$ 是连接 $z + p/\rho g$ 每个顶点的线，称为压力计水头线；线段 $gkh$ 为连接 $u^2/2g$ 的每个顶点的线，称为总水头线。几何学中理想流体的伯努利方程的含义是：总水头线是一条水平线，每个水头可以增加或减少，但总水头保持不变。在图 4-7 中，$a$、$m$ 和 $b$ 三个点的总水头为

$$H_1 = z_1 + \dfrac{p_1}{\rho g} + \dfrac{u_1^2}{2g}, \quad H = z + \dfrac{p}{\rho g} + \dfrac{u^2}{2g}$$

$$H_2 = z_2 + \dfrac{p_2}{\rho g} + \dfrac{u_2^2}{2g}$$

图 4-7 水头线

即三个总头都相等。

总水头线坡度为水力坡度：

$$J = \frac{\mathrm{d}h_\mathrm{w}}{\mathrm{d}s} = -\frac{\mathrm{d}H}{\mathrm{d}s} = -\frac{\mathrm{d}\left(z+\dfrac{p}{\rho g}+\dfrac{\alpha v^2}{2g}\right)}{\mathrm{d}s} \tag{4-18}$$

测压水头线的变化可以用测压水头线斜率来表示：

$$J_\mathrm{p} = -\frac{\mathrm{d}\left(z+\dfrac{p}{\rho g}\right)}{\mathrm{d}s} \tag{4-19}$$

由于没有水头损失，理想流体的总水头线是一条水平线。

### 4.3.4 伯努利方程的扩展

**1. 沿程有分流或汇入的伯努利方程**

(1) 分流

流体不可压缩，作稳定流动，流速沿流动路径保持不变。如果两段之间有轴功输出，则可用以下两式进行计算（图 4-8）：

$$z_1 + \frac{p_1}{\rho g} + \frac{\alpha_1 v_1^2}{2g} = z_2 + \frac{p_2}{\rho g} + \frac{\alpha_2 v_2^2}{2g} + h_{\mathrm{w}1,2} \tag{4-20}$$

$$z_1 + \frac{p_1}{\rho g} + \frac{\alpha_1 v_1^2}{2g} = z_3 + \frac{p_3}{\rho g} + \frac{\alpha_3 v_3^2}{2g} + h_{\mathrm{w}1,3} \tag{4-21}$$

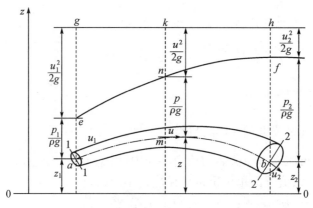

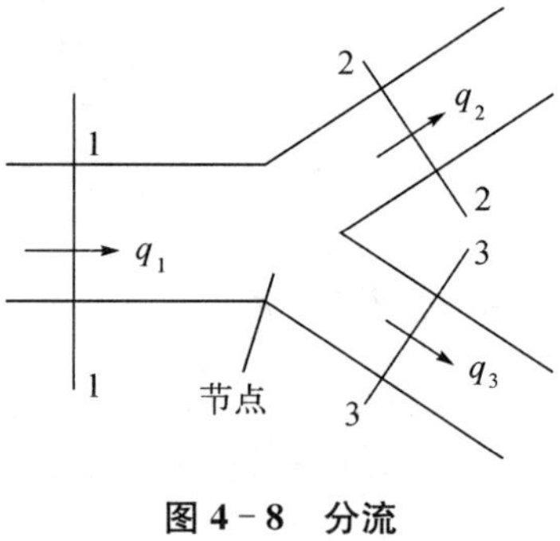

图 4-8 分流

(2) 汇入

当两种流体汇合时，除造成水头损失外，单位重量流体的机械能并不相等，可按下式进行计算：

$$\rho g q_{v1}\left(z_1 + \frac{p_1}{\rho g} + \frac{\alpha_1 v_1^2}{2g}\right) + \rho g q_{v2}\left(z_2 + \frac{p_2}{\rho g} + \frac{\alpha_2 v_2^2}{2g}\right)$$

$$= \rho g q_{v3}\left(z_3 + \frac{p_3}{\rho g} + \frac{\alpha_3 v_3^2}{2g}\right) + \rho g q_{v1} h_{\mathrm{w}1,3} + \rho g q_{v2} h_{\mathrm{w}2,3} \tag{4-22}$$

## 2. 沿程有能量输入或输出的伯努利方程

总的伯努利方程是在没有能量输入或输出的情况下得到的，除引起两截面的水头损失外。当两截面之间有泵等时，流体可以获得额外的能量或失去能量。所以伯努利方程被改变了。

$$z_1 + \frac{p_1}{\rho g} + \frac{\alpha_1 v_1^2}{2g} \pm H = z_2 + \frac{p_2}{\rho g} + \frac{\alpha_2 v_2^2}{2g} + h_w \tag{4-23}$$

## 3. 气体的伯努利方程

总流的伯努利方程是由不可压缩流体导出的，气体是一种可压缩流体，但速度不是很大。所以伯努利方程也适用于气体。

从图 4-9 中，选择截面 1—1 和 2—2 列出伯努利方程。

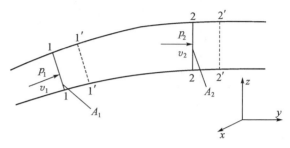

图 4-9 气体的伯努利方程

$$z_1 + \frac{p_{1\text{abs}}}{\rho g} + \frac{v_1^2}{2g} = z_2 + \frac{p_{2\text{abs}}}{\rho g} + \frac{v_2^2}{2g} + h_w$$
$$(\alpha_1 = \alpha_2 = 1)$$

$$\rho g z_1 + p_{1\text{abs}} + \frac{\rho v_1^2}{2} = \rho g z_2 + p_{2\text{abs}} + \frac{\rho v_2^2}{2} + p_w \tag{4-24}$$

因为
$$p_{1\text{abs}} = p_1 + p_a$$
$$p_{2\text{abs}} = p_2 + p_a - \rho_a g(z_2 - z_1)$$

又因为
$$p_1 + \frac{\rho v_1^2}{2} + (\rho_a - \rho)g(z_2 - z_1) = p_2 + \frac{\rho v_2^2}{2} + p_w \tag{4-25}$$

当 $\rho_a = \rho$，$z_2 = z_1$，有

$$p_1 + \frac{\rho v_1^2}{2} = p_2 + \frac{\rho v_2^2}{2} + p_w \tag{4-26}$$

当 $\rho \gg \rho_a$ 时，$\rho_a$ 可以被忽略：

$$p_1 + \frac{\rho v_1^2}{2} - \rho g(z_2 - z_1) = p_2 + \frac{\rho v_2^2}{2} + p_w$$

除以 $\rho g$：

$$z_1 + \frac{p_1}{\rho g} + \frac{v_1^2}{2g} = z_2 + \frac{p_2}{\rho g} + \frac{v_2^2}{2g} + h_w$$

**例 4-3** 一抽水机管路系统如图 4-10 所示。泵输出 $Q = 0.03 \text{ m}^3/\text{s}$，吸管直径 $d = 150 \text{ mm}$，泵所能达到的真空度 $\frac{p_v}{\rho g} = 6.8 \text{ mH}_2\text{O}$。确定从泵轴到池塘水面的最大高程。

**解：** 首先，这里有两个断面：池塘水面 0—0 和泵入口截面 1—1。取截面 0—0 为基准面，$z_0 = 0$，所以

$$z_0 + \frac{p_0}{\rho g} + \frac{\alpha_0 v_0^2}{2g} = z_1 + \frac{p_1}{\rho g} + \frac{\alpha_1 v_1^2}{2g} + h_{w0-1}$$

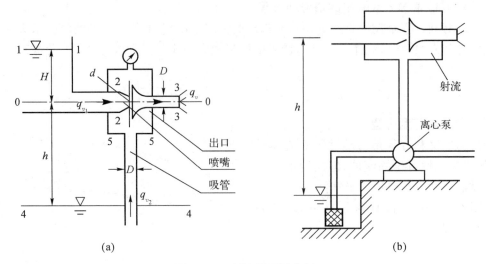

图 4-10 射流原理及应用

其次，伯努利方程中的参数是确定的。

压力 $p_0$ 和 $p_1$ 用相对压力（表压）表示。

$$\frac{p_0}{\rho g}=0, \frac{p_1}{\rho g}=-\frac{p_v}{\rho g}=-6.8 \text{ mH}_2\text{O}$$

由于池塘自由面处的流速 $v_0$ 远小于管中截面 1—1 处的流速 $v_1$，因此 $v_0=0$。速度 $v_1$ 为

$$v_1=\frac{Q}{A}=\frac{0.03}{\pi\times\frac{0.15^2}{4}}=0.17 \text{ m/s}$$

设动能修正因子 $\alpha_1$ 为 1。

流量中的损失：单位重量通过的流体在两段之间的能量损失为

$$h_{w0-1}=1 \text{ mH}_2\text{O}$$

最后，对未知参数进行计算。

将 $v_0=0$，$p_0=0$，$z_0=0$，$p_1=-p_v$，$z_1=h_e$，$\alpha=1$ 和 $v_1=0.17$ 代入伯努利方程，得

$$0+0+0=\frac{v_1^2}{2g}+\frac{p_v}{\rho g}+h_e+h_{w0-1}$$

即：

$$h_e=\frac{p_v}{\rho g}-\frac{v_1^2}{2g}-h_{w0-1}$$

将 $\frac{p_v}{\rho g}$ 和 $v_1$ 值代入上述公式，得

$$h_e=6.8-0.15-1.0=5.65 \text{ m}$$

即从泵井到水面的最大高程为 5.65 m。如果超过，水就会发生沸腾，泵不能正常运转。

## 4.4 动量方程

动量方程是继伯努利方程和连续性方程之后的一维流动的另一个基本方程。我们在工程实践中,总是需要计算流体与固体边界之间的作用力。此外,伯努利方程和连续性方程不能反映流体与固体边界之间的作用力关系,同时,伯努利方程还包括水头损失,但在某些流体中很难确认水头损失的大小。而动量方程可以弥补这些缺点。

### 4.4.1 动量方程的推导

对于任何方向,控制体内动量的变化之和加上动量从控制体中净流出的总和等于单位时间内作用在控制体上的合力的分量。

系统动量关于时间的变化率等于系统上所有作用力的矢量和,即:

$$\sum \boldsymbol{F} = \frac{d\boldsymbol{K}}{dt} = \frac{d(\sum m\boldsymbol{u})}{dt}$$

选取总流量稳定的管流作为控制体,如图 4-11 所示。经过时间 $dt$ 后,流体从 11—22 流动到 $1'1'—2'2'$。

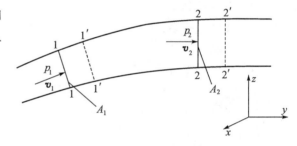

图 4-11 动量方程的推导

($dt$)中的左表面流:$\left(\int_{A_1} \boldsymbol{v}_1 \rho_1 v_1 dA_1\right) dt$

($dt$)中的右表面流:$\left(\int_{A_2} \boldsymbol{v}_2 \rho_2 v_2 dA_2\right) dt$

管内流动的动量变化为

$$d(M\boldsymbol{v}) = \left(\int_{A_2} \boldsymbol{v}_2 \rho_2 v_2 dA_2\right) dt - \left(\int_{A_1} \boldsymbol{v}_1 \rho_1 v_1 dA_1\right) dt$$

动量变化率为:

$$\frac{d(M\boldsymbol{v})}{dt} = \int_{A_2} \boldsymbol{v}_2 \rho_2 v_2 dA_2 - \int_{A_1} \boldsymbol{v}_1 \rho_1 v_1 dA_1$$

稳态总流量的动量方程:

$$\sum \boldsymbol{F} = \int_{A_2} \boldsymbol{v}_2 \rho_2 v_2 dA_2 - \int_{A_1} \boldsymbol{v}_1 \rho_1 v_1 dA_1$$

定义 $\int_A \rho v^2 dA = \beta \rho A v^2$,所以

$$\beta_2 \rho_2 A_2 \boldsymbol{v}_2^2 - \beta_1 \rho_1 A_1 \boldsymbol{v}_1^2 = \sum \boldsymbol{F} \quad (\text{一般假设 } \beta = 1.0)$$

稳态总流中不可压缩动量方程。

$$\rho Q(\boldsymbol{v}_1 - \boldsymbol{v}_2) = \sum \boldsymbol{F}$$

$x$、$y$、$z$ 方向：
$$\begin{cases} \rho Q(u_1 - u_2) = \sum F_x \\ \rho Q(v_1 - v_2) = \sum F_y \\ \rho Q(w_1 - w_2) = \sum F_z \end{cases} \quad (4-27)$$

控制体内流体上所有作用力之和。这种外力包括：
(1) 控制体内流体上的质量力；
(2) 控制体面上的表面力(动压和切应力)；
(3) 周围边界对流体施加的总力。

### 4.4.2 动量方程的应用

动量方程的局限性：
(1) 理想流体，不可压缩稳定流动；
(2) 两个断面必须为渐变流过流断面；
(3) 流量不随沿流变化。

解决问题的步骤：
(1) 选择控制体；
(2) 选择坐标系；
(3) 绘制计算图；
(4) 列出动量方程。

### 4.4.3 推论

修正方程：
$$\sum (\beta \rho q_v \boldsymbol{v})_{\text{流出}} - \sum (\beta \rho q_v \boldsymbol{v})_{\text{流入}} = \sum \boldsymbol{F} \quad (4-28)$$

**例 4-4** 将直径为 10 cm 的流体射流排入空气的喷嘴位于水平 4 cm 直径的右端(图 4-12)。流体的相对密度为 0.85，$p_{M_1} = 7.0 \times 10^5$ Pa，$d_1 = 10$ cm，$d_2 = 4$ cm，流体对锚杆 $S$ 施加的作用力是什么？(忽略摩擦。)

**解：** 选择控制体 1122：

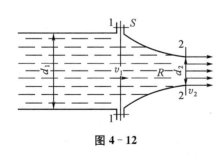

图 4-12

$$v_1 = v_2 \frac{A_2}{A_1} = v_2 \left(\frac{d_2}{d_1}\right)^2$$

断面 1—1 至断面 2—2 之间的伯努利方程：
$$\frac{p_1}{\rho g} + \frac{v_1^2}{2g} = \frac{v_2^2}{2g}$$

有：
$$p_1 = \rho g \left(\frac{v_2^2}{2g} - \frac{v_1^2}{2g}\right) = \rho \frac{v_2^2}{2g} \left[1 - \left(\frac{d_2}{d_1}\right)^4\right]$$

$$v_2 = \sqrt{\frac{2p_1}{\rho\left[1-\left(\frac{d_2}{d_1}\right)^4\right]}} = \sqrt{\frac{2\times 7\times 10^5}{0.85\times 1\,000\times\left[1-\left(\frac{4}{10}\right)^4\right]}} = 41.1 \text{ m/s}$$

有：
$$v_1 = v_2\left(\frac{d_2}{d_1}\right)^4 = 41.1\times\left(\frac{4}{10}\right)^2 = 6.58 \text{ m/s}$$

$$Q = v_1\pi\left(\frac{d_1}{2}\right)^2 = 6.58\times\pi\times\left(\frac{0.1}{2}\right)^2 = 0.051\,6 \text{ m}^3/\text{s}$$

在 $x$ 方向的动量方程：
$$R = p_1 A_1 - \rho Q(v_2 - v_1)$$
$$= 7\times 10^5\times\frac{\pi}{4}\times 0.1^2 - 0.85\times 1\,000\times 0.051\,6\times(41.1-6.58) = 3\,982 \text{ N}$$

## 习　题

1. 一输油管渐变段，两端直径分别为 $d_1=200$ mm，$d_2=60$ mm，油的密度为 $860$ kg/m³。已知断面 1—1 流速 $v_1=2$ m/s，试求断面 2—2 流速及质量流量。

2. 如图 4-13 所示，由两根不同直径的管子与一渐变连接管组成一管路。已知 $d_A=200$ mm，$d_B=400$ mm，$A$ 点相对压强 $p_A=6.86\times 10^4$ N/m²，$B$ 点相对压强 $p_B=3.92\times 10^4$ N/m²；$B$ 点处的断面平均流速 $v_B=1$ m/s。$A$、$B$ 两点的高差 $\Delta z=1$ m。要求判别流动方向，并计算这两断面间的水头损失 $h_\text{w}$。

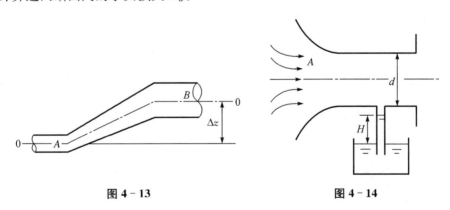

图 4-13　　　　　图 4-14

3. 如图 4-14 所示，离心式通风机借集流器 $A$ 从大气中吸入空气。在直径 $d=200$ mm 的圆柱形管道部分接一根玻璃管，管的下端插入水槽中。若玻璃管中的水上升 $H=150$ mm，求每秒钟所吸取的空气量 $q_v$。（空气的密度 $\rho=1.29$ kg/m³。）

4. 如图 4-15 所示，泵压送水，水泵轴功率 $N=\rho g q_v H=13.3$ kW。效率 $\eta_\text{p}=0.75$。已知 $h=20$ m，管路水头损失 $h_\text{w}=\dfrac{8v^2}{2g}$。试求其流量及泵的扬程 $H_\text{p}(H_\text{p}=h+h_\text{w})$，并定性绘制

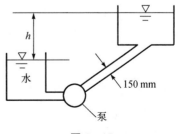

图 4-15

5. 如图 4-16 所示，有一厚度为 50 mm，速度 $v=18$ m/s 的单宽射流水股，在空气中斜向冲击在边长 $l=1.2$ m 的光滑平板上，射流沿平板表面分成两股。已知板与水流方向的夹角 $\theta=30°$。若忽略水流、空气和平板的摩阻，且流动在同一水平面上，试求：(1) 流量分配 $q_{v_1}$ 和 $q_{v_2}$；(2) 射流对平板的冲击力；(3) 若平板与水流方向一致以 $u=8$ m/s 运动时，水流作用在平板上的作用力的大小。

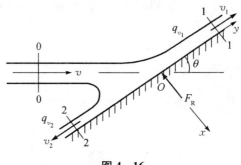

图 4-16

# 第 5 章 相似原理和量纲分析

任何特定问题的无量纲参数都可以通过两种方法来确定,如果这些方程是已知的,则它们可以直接从控制微分方程中推导出来,这里举例说明了这些方法。如果控制方程是未知的,或者在已知的方程中没有出现感兴趣的参数,则可以通过量纲分析来确定无量纲参数。采用前一种方法的好处是,由运动方程确定的无量纲参数更容易解释,并且能与流动中发生的物理现象相联系。因此,无量纲参数的相关知识经常有助于求解过程,特别是在需要假设和近似来求解时。

本节描述并提供了本书其余部分中所需的主要无量纲参数和数字。此处未提及的许多其他内容已在流体力学的广泛领域中被定义和使用。

## 5.1 量纲

### 5.1.1 基本概念

**1. 单位和量纲**

各种物理量由两个因素构成:一个是维度,如长度量纲 $L$ 和质量量纲 $M$;另一个是物理量的单位。一个物理量的量纲是唯一的,但单位有多种表示方法。采用 $\dim q$ 作为物理量 $q$ 的量纲,速度 $v$ 的量纲可表示为:

$$\dim v = LT^{-1}$$

**2. 基本量纲和导出量纲**

彼此独立且不能相互推导的量纲称为基本量纲。由基本量纲推导出来的量纲称为导出量纲。例如,$q$ 是一个导出量纲的物理量,且

$$\dim q = L^{\alpha} T^{\beta} M^{\gamma} \tag{5-1}$$

(1) 若 $\alpha \neq 0, \beta = \gamma = 0$,面积 $A$ 和体积 $V$ 的导出量纲分别为 $\dim A = L^2$,$\dim V = L^3$;

(2) 若 $\alpha \neq 0, \beta \neq 0, \gamma = 0$,速度 $v$ 和加速度 $a$ 的导出量纲分别为 $\dim v = LT^{-1}$,$\dim a = LT^{-2}$;

(3) 若 $\alpha \neq 0, \beta \neq 0, \gamma \neq 0$,力 $F$ 和压强 $p$ 的导出量纲分别为 $\dim F = MLT^{-2}$,$\dim p = ML^{-1}T^{-2}$。

**3. 量纲一的量**

若公式(5-1)中的物理量纲为 $\alpha = \beta = \gamma = 0$,则称该物理量纲为量纲一的量。如角度、π 和应变率,$\dim \theta = 1$,$\dim \pi = 1$,$\dim \varepsilon = 1$。

量纲一的量有三个特点。首先，数值是独立于单位的。其次，量纲一的量是纯数字的，没有尺度效应。最后，量纲一的量可进行超越函数计算。

### 5.1.2 量纲和谐原理

量纲分析的基本原理是量纲和谐。例如，第 4 章中导出的实际流体总伯努利方程：

$$z_1 + \frac{p_1}{\rho g} + \frac{\alpha_1 v_1^2}{2g} = z_2 + \frac{p_2}{\rho g} + \frac{\alpha_2 v_2^2}{2g} + h_w$$

上式中的所有量纲都是 1。恒定总流动量方程式为

$$\sum \boldsymbol{F} = \rho q_v \beta_2 \boldsymbol{v}_2 - \rho q_v \beta_1 \boldsymbol{v}_1$$

上式中的所有量纲为 $MLT^{-2}$。

只有同一类型的物理量才能相加减，不同类型的物理量的加减是没有意义的。

一个方程的量纲是和谐的，方程的形式不会随着单位的变化而变化。如果你用其中一项除以各项，你将得到一个由量纲为一的量构成的方程。因此，量纲和谐原理可以用来检测新方程或经验公式的正确性和完整性。也就是说，量纲和谐原理可以用来确定物理量的量纲指数，或者用来建立物理方程。

## 5.2 量纲分析法

量纲分析法有两种，其中适用于简单问题的称为瑞利法，而另一种通用的方法是 $\pi$ 定理。

### 5.2.1 瑞利法

瑞利法是一种直接以量纲一致原则为基础的量纲分析方法。以下通过举例的方式加以说明。

**例 5-1** 求解管流中的壁面切应力的表达式。

**解**：影响 $\tau_0$ 的物理量有液体的密度 $\rho$、运动粘度 $\mu$、圆管直径 $D$、管壁材料的粗糙度 $\Delta$ 和管中断面的平均流速 $v$。

$$\tau_0 = K \rho^a \mu^b D^c \Delta^d v^e$$
$$(ML^{-1}T^{-2}) = (ML^{-3})^a (ML^{-1}T^{-1})^b (L)^c (L)^d (LT^{-1})^e$$

根据量纲和谐原则：

$$M: 1 = a + b$$
$$L: -1 = -3a - b + c + d + e$$
$$T: -2 = -b - e$$

联立以上三式，得

$$b = 1 - a, \quad c = a - d - 1, \quad e = 1 + a$$

得
$$\tau_0 = K\rho^a\mu^{1-a}D^{a-d-1}\Delta^d v^{1+a}$$

$$\tau_0 = K\left(\frac{\rho D v}{\mu}\right)^a \left(\frac{\Delta}{D}\right)^d \left(\frac{\mu}{\rho D v}\right)\rho v^2 = K(Re)^{a-1}\left(\frac{\Delta}{D}\right)^d \rho v^2 = f\left(Re, \frac{\Delta}{D}\right)\rho v^2$$

当
$$f\left(Re, \frac{\Delta}{D}\right) = \frac{\lambda}{8}$$

得出:
$$\tau_0 = \frac{\lambda}{8}\rho v^2$$

用瑞利法进行量纲分析的步骤一般是:
(1) 确定与物理现象相关的物理量;
(2) 写出各物理量之间的指数乘积的形式;
(3) 用基本量纲表示各物理量的量纲;
(4) 根据量纲和谐原则确定物理量的指数。

### 5.2.2 白金汉定理——$\pi$ 定理

**1. $\pi$ 定理**

将给定过程中的变量关联起来所需的独立无量纲变量组的数量等于 $n-m$,其中 $n$ 为所涉及的变量数,$m$ 为变量中包含的基本量。

$$\varphi(\pi_1, \pi_2, \cdots, \pi_{n-m}) = 0 \qquad (5-2)$$

**2. $\pi$ 定理解题步骤**

(1) 找出与物理过程相关的物理量:

$$f(x_1, x_2, \cdots, x_n) = 0$$

(2) 确定基本量,选择几个基本物理量作为基本量纲;
(3) 确定 $\pi$ 的数量:

$$\pi_i = x_1^{a_i} x_2^{b_i} x_3^{c_i} \cdot x_i$$

(4) 确定量纲一的 $\pi$ 中的各指标;
(5) 将各项 $\pi$ 代入式(5-2):

$$f(\pi_1, \pi_2, \cdots, \pi_{n-m}) = 0$$
$$\pi_4 = f(\pi_1, \pi_2, \cdots, \pi_{n-m})$$

**3. 选择基本量的原则**

(1) 基本量应该与基本量纲相对应;
(2) 在选择基本量时,应选择重要的物理量。

**例 5-2** (1) $f(d, \rho, \mu, v, l, \Delta, \Delta p) = 0$;
(2) 选择 $\rho, d, v$ 作为基本量,$m=3$;
(3) $N(\pi) = n - m = 7 - 3 = 4$

$$\pi_1 = d^{a_1} v^{b_1} \rho^{c_1} \cdot \mu, \quad \pi_2 = d^{a_2} v^{b_2} \rho^{c_2} \cdot l, \quad \pi_3 = d^{a_3} v^{b_3} \rho^{c_3} \cdot \Delta, \quad \pi_4 = d^{a_4} v^{b_4} \rho^{c_4} \cdot \Delta p$$

(4) $\dim \pi_1 = \dim(d^{a_1} v^{b_1} \rho^{c_1} \cdot \mu)$

$$M^0 L^0 T^0 = (L)^{a_1} (LT^{-1})^{b_1} (ML^{-3})^{c_1} \cdot (ML^{-1}T^{-1})$$
$$M: 0 = c_1 + 1$$
$$L: 0 = a_1 + b_1 - 3c_1 - 1$$
$$T: 0 = -b_1 - 1$$
$$a_1 = -1, \ b_1 = -1, \ c_1 = -1, \ \pi_1 = \frac{\mu}{dv\rho} = \frac{1}{Re}$$

同理可得
$$\pi_2 = \frac{l}{d}, \ \pi_3 = \frac{\Delta}{d}, \ \mu_4 = \frac{\Delta p}{\rho v^2}$$

(5)
$$f_1\left(\frac{l}{d}, \frac{1}{Re}, \frac{\Delta}{d}, \frac{\Delta p}{\rho v^2}\right) = 0$$

或者
$$\frac{\Delta p}{\rho v^2} = f_2\left(\frac{l}{d}, Re, \frac{\Delta}{d}\right)$$

$$\Delta p = f_3\left(Re, \frac{\Delta}{d}\right)\frac{l}{d}\rho v^2$$

$$\Delta h = \frac{\Delta p}{\rho g} = f_4\left(Re, \frac{\Delta}{d}\right)\frac{l}{d}\frac{v^2}{2g}$$

设 $\lambda = f_4\left(Re, \frac{\Delta}{d}\right)$，则

$$\Delta h = \lambda \frac{l}{d}\frac{v^2}{2g}$$

## 5.3 流动相似理论基础

根据流体运动方程中获得的无量纲参数，设置了使用小模型进行比例模型测试的条件，这些条件将被证明可用于预测大型设备的性能。具体来说，当两个流场的无量纲参数匹配并且它们的几何形状相似时，它们被认为是动态相似的；即第一个流场中的任何长度量纲都可以通过与单个尺度比的乘积来映射到第二个流场中的对应长度量纲。当两流在动态上相似时，一个流场的分析、模拟或测量可通过比例关系的转化后直接适用于另一个流场。此外，标准无量纲参数的使用减少了在实验或计算中必须改变的参数，并极大地促进了测量或计算结果与先前的在潜在不同条件下进行的工作的对比。

在任何时候，模型和原型的所有参数在整个流场中都具有相同的比例：几何相似、运动相似和动力相似。

### 5.3.1 相似条件

**1. 几何相似**

这个模型是原型的精确几何复制品，模型和原型所有的尺寸比例均相同。

$$\lambda_l = \frac{l_p}{l_m} \tag{5-3}$$

式中，$\lambda_l$ 为模型和原型之间的比例，$l_p$ 为原型的长度，$l_m$ 为模型的长度。

几何相似的结果使得相应的面积比和体积比均保持一定的比例关系。

面积比：
$$\lambda_A = \frac{A_p}{A_m} = \frac{l_p^2}{l_m^2} = \lambda_l^2$$

体积比：
$$\lambda_V = \frac{V_p}{V_m} = \frac{l_p^3}{l_m^3} = \lambda_l^3$$

**2. 运动相似**

运动相似指在整个流场中，原型和模型的对应的各点的速度成一定比例。如：$u_p$ 代表原型某点的速度，$u_m$ 代表模型某点的速度，原型和模型之间的比例为

$$\lambda_u = \frac{u_p}{u_m}$$

所以速度比为
$$\lambda_v = \frac{v_t}{v_m} = \frac{\left(\dfrac{l}{t}\right)_p}{\left(\dfrac{l}{t}\right)_m} = \frac{\lambda_l}{\lambda_t}$$

而加速度比为
$$\lambda_a = \frac{a_p}{a_m} = \frac{\left(\dfrac{v}{t}\right)_p}{\left(\dfrac{v}{t}\right)_m} = \frac{\lambda_v}{\lambda_t} = \frac{\lambda_l}{\lambda_t^2}$$

**3. 动力相似**

动力相似指在整个流场中作用在模型和原型中相对应质点上的所有力具有固定的比例。

力的比例是 $\lambda_F = \dfrac{F_p}{F_m}$（$F_p$ 为原型中的一个力，$F_m$ 为模型中相对应的一个力）

**4. 联系**

几何相似、运动相似和动态相似是相互联系的整体。几何相似是最基本和最明显的要求，动力相似是前提条件，而运动相似是两者的表现形式。

### 5.3.2 相似准数的推导

根据相似原理，相似准数是具有动力相似性的无量纲参数。以纳维-斯托克斯动量方程为起点能推导出不可压缩流的相似准数。对模型来说，为

$$\left[\boldsymbol{f} - \frac{1}{\rho}\boldsymbol{\nabla} p + \nu \boldsymbol{\nabla}^2 \boldsymbol{u} = \frac{\partial \boldsymbol{u}}{\partial t} + (\boldsymbol{u} \cdot \boldsymbol{\nabla})\boldsymbol{u}\right]_m \tag{5-4}$$

对原型来说，为

$$\left[\boldsymbol{f} - \frac{1}{\rho}\boldsymbol{\nabla} p + \nu \boldsymbol{\nabla}^2 \boldsymbol{u} = \frac{\partial \boldsymbol{u}}{\partial t} + (\boldsymbol{u} \cdot \boldsymbol{\nabla})\boldsymbol{u}\right]_p \tag{5-5}$$

如果模型与原型几何相似，边界和初始条件也相似。质量力假设只有重力，其比为

$$\frac{\boldsymbol{f}_p}{\boldsymbol{f}_m} = \frac{g_p}{g_m} = \lambda_g$$

$$\lambda_g \boldsymbol{f}_m - \left(\frac{\lambda_p}{\lambda_\rho \lambda_l}\right)\left(\frac{1}{\rho}\boldsymbol{\nabla} p\right)_m + \left(\frac{\lambda_\nu \lambda_u}{\lambda_l^2}\right)(\nu \boldsymbol{\nabla}^2 u)_m$$
$$= \left(\frac{\lambda_u}{\lambda_t}\right)\left(\frac{\partial \boldsymbol{u}}{\partial t}\right)_m + \left(\frac{\lambda_u^2}{\lambda_l}\right)[(\boldsymbol{u} \cdot \boldsymbol{\nabla})\boldsymbol{u}]_m \qquad (5-6)$$

式(5-4)和式(5-5)都无量纲化:式(5-4)除以 $(\boldsymbol{u} \cdot \boldsymbol{\nabla})\boldsymbol{u}$,得

$$\left[\frac{\boldsymbol{f}}{(\boldsymbol{u} \cdot \boldsymbol{\nabla})\boldsymbol{u}} - \frac{1}{\rho}\frac{\boldsymbol{\nabla} p}{(\boldsymbol{u} \cdot \boldsymbol{\nabla})\boldsymbol{u}} + \frac{\nu \boldsymbol{\nabla}^2 \boldsymbol{u}}{(\boldsymbol{u} \cdot \boldsymbol{\nabla})\boldsymbol{u}} = \frac{\partial \boldsymbol{u}/\partial t}{(\boldsymbol{u} \cdot \boldsymbol{\nabla})\boldsymbol{u}} + 1\right]_m \qquad (5-7)$$

式(5-6)除以 $\left(\frac{\lambda_u^2}{\lambda_l}\right)[(\boldsymbol{u} \cdot \boldsymbol{\nabla})\boldsymbol{u}]$,得

$$\left\{\frac{\lambda_g \lambda_l}{\lambda_u^2}\left[\frac{\boldsymbol{f}}{(\boldsymbol{u} \cdot \boldsymbol{\nabla})\boldsymbol{u}}\right] - \frac{\lambda_p}{\lambda_\rho \lambda_u^2}\left[\frac{1}{\rho}\frac{\boldsymbol{\nabla} p}{(\boldsymbol{u} \cdot \boldsymbol{\nabla})\boldsymbol{u}}\right] + \frac{\lambda_\nu}{\lambda_l \lambda_u}\left[\frac{\nu \boldsymbol{\nabla}^2 \boldsymbol{u}}{(\boldsymbol{u} \cdot \boldsymbol{\nabla})\boldsymbol{u}}\right] = \frac{\lambda_l}{\lambda_t \lambda_u}\left[\frac{\partial \boldsymbol{u}/\partial t}{(\boldsymbol{u} \cdot \boldsymbol{\nabla})\boldsymbol{u}}\right] + 1\right\}_p$$
$$(5-8)$$

方程式(5-7)和方程式(5-8)恒等,则:

$$\frac{\lambda_g \lambda_l}{\lambda_u^2} = \frac{\lambda_p}{\lambda_\rho \lambda_u^2} = \frac{\lambda_\nu}{\lambda_l \lambda_u} = \frac{\lambda_l}{\lambda_t \lambda_u} = 1 \qquad (5-9)$$

用平均流速代替瞬时速度得

$$\frac{v_p^2}{g_p l_p} = \frac{v_m^2}{g_m l_m}, \quad \frac{p_p}{\rho_p v_p^2} = \frac{p_m}{\rho_m v_m^2}, \quad \frac{v_p l_p}{\nu_p} = \frac{v_m l_m}{\nu_m}, \quad \frac{l_p}{v_p t_p} = \frac{l_m}{v_m t_m}$$

或

$$Fr_p = Fr_m, \quad Eu_p = Eu_m, \quad Re_p = Re_m, \quad St_p = St_m \qquad (5-10)$$

$Fr = \frac{v^2}{gl}$ 为弗劳德数,$Eu = \frac{p}{\rho v^2}$ 为欧拉数,$Re = \frac{vl}{\nu}$ 为雷诺数,$St = \frac{l}{vt}$ 为斯特劳哈尔数,因此弗劳德数、欧拉数、雷诺数和斯特劳哈尔数被称为相似准则数。

## 5.3.3 相似准数的物理意义

在模型实验时,识别影响流体运动的主要作用力并选择相似准数对模型设计非常重要。因此,有必要了解相似准数的物理意义。

### 1. N-S 方程的物理意义

作用在流体质点上的力是重力、压力和粘滞力。N-S 方程为

$$\boldsymbol{f} - \frac{1}{\rho}\boldsymbol{\nabla} p + \nu \boldsymbol{\nabla}^2 \boldsymbol{u} = \frac{\partial \boldsymbol{u}}{\partial t} + (\boldsymbol{u} \cdot \boldsymbol{\nabla})\boldsymbol{u} \qquad (3-30)$$
$$\langle 1 \rangle \quad \langle 2 \rangle \quad \langle 3 \rangle \quad \langle 4 \rangle \quad \langle 5 \rangle$$

显然，第⟨1⟩项是重力，第⟨2⟩项是压力，第⟨3⟩项是粘滞力。第⟨4⟩和⟨5⟩项是欧拉加速度。根据公式，⟨4⟩和⟨5⟩项也是加速度惯性力。

需要注意的是：

(1) 这些力对流体运动的影响有强有弱。在类似的模型中，我们需要分析哪些力在流动中起主导作用。

(2) 除惯性力以外的各种力都是用来改变运动状态的。所以流体的改变是惯性力和其他力相互作用的结果。

**2. 相似准数的物理意义**

公式(5-8)的各项分别代表重力、压力、粘滞力和惯性力。弗劳德数表示重力和惯性力的比值。雷诺数表示惯性力和粘滞力的比值。施特劳哈尔数表示时变加速度惯性力和迁移加速度惯性力的比值。

当两个粘性不可压缩流体在完全动力相似时，它们的 $Fr$、$Eu$、$Re$、$St$ 对应相等。

## 5.3.4 动力相似准则

完全动力相似要求每个相似准数均对应且相等。例如，若要使两个流体的弗劳德数和雷诺数相等，则由：

$$(Re)_p = (Re)_m$$

或：

$$\frac{\lambda_v \cdot \lambda_l}{\lambda_\nu} = 1$$

得：

$$(Fr)_p = (Fr)_m$$

或：

$$\frac{\lambda_v}{\lambda_g^{1/2} \cdot \lambda_l^{1/2}} = 1, \ \lambda_g = 1$$

$$\lambda_v = \lambda_l^{1/2}, \ \lambda_l^{1/2} = \lambda_l^{-1}$$

一般情况下，只选取对流体起决定性作用的一种力，使两流体的相似准数相等。

如果流体主要以重力作用为主，则两流体弗劳德数相等，即 $(Fr)_p = (Fr)_m$。

如果流体主要以粘滞力作用为主，则两流体雷诺数相等，即 $(Re)_p = (Re)_m$。

## 5.3.5 推论——可压缩流体中的相似准数

**1. 柯西数**

柯西数代表惯性力与弹性力的比值：$Ca = \dfrac{\rho v^2}{K}$。

$$\left(\frac{f}{f_E}\right)_p = \left(\frac{f}{f_E}\right)_m$$

整理得
$$\left(\frac{\rho v^2}{K}\right)_p = \left(\frac{\rho v^2}{K}\right)_m \quad 或 \quad (Ca)_p = (Ca)_m$$

**2. 马赫数**

由流体在弹性力作用下的相似原理，马赫数表示为 $Ma = \dfrac{v}{c}$。

$$\frac{v_p}{c_p} = \frac{v_m}{c_m} \quad 或 \quad (Ma)_p = (Ma)_m$$

## 5.4 模型实验

### 5.4.1 模型的相似准则

模型和原型在相同的流体介质中很难完全相似，一般只能达到近似相似。因此，重要的是要确定其起主要作用的力，然后选择与主要作用力对应的相似准则。

### 5.4.2 模型设计

模型设计一般执行以下步骤：

（1）确定长度比尺；
（2）设计模型几何尺寸；
（3）选择模型准则；
（4）计算物理量的相似比尺；

① 按雷诺数准则：

速度比尺为：$\dfrac{v_p l_p}{\nu_p} = \dfrac{v_m l_m}{\nu_m}$

若 $\nu_p = \nu_m$，则 $\lambda_v = \dfrac{v_p}{v_m} = \dfrac{l_m}{l_p} = \lambda_l^{-1}$。

② 按弗劳德准则：

速度比尺为：$\dfrac{v_p}{\sqrt{g_p l_p}} = \dfrac{v_m}{\sqrt{g_m l_m}}$

若 $g_p = g_m$，则 $\lambda_v = \dfrac{v_p}{v_m} = \sqrt{\dfrac{l_p}{l_m}} = \lambda_l^{1/2}$。

（5）按比尺计算模型流量、流速等参数，并确定边界条件。

部分流动参数必须由原型给出，再换算为模型的值。比如：$q_{v_m} = \dfrac{q_{v_p}}{\lambda_{q_v}}$。

### 5.4.3 模型实验数据的处理

实验数据分为两类，一类是无量纲的量，另一类是有量纲的量。对于无量纲的量，因为其量纲为一，并且模型和原型相似，其对应的值相等，所以不需要换算。对于有量纲的量，需要按比例关系将模型值转换为原型值。

## 习 题

1. 分别举例说明由重力、粘滞力起主要作用的水流。

2. 试证明 $\tau = \mu \dfrac{\mathrm{d}u}{\mathrm{d}y}$，$z + \dfrac{p}{\rho g} + \dfrac{u^2}{2g} = H$，量纲是和谐的。

3. 用量纲分析法将下列各组物理量组合成量纲一的量：
(1) $\tau$、$v$、$\rho$，其中 $\tau$ 为切应力，$v$ 为流速；
(2) $\Delta p$、$v$、$p$、$g$；
(3) $F$、$\rho$、$l$、$v$，其中 $F$ 为力，$v$ 为流速，$l$ 为某个线性长度；
(4) $v$、$l$、$p$、$\sigma$，其中 $\sigma$ 为表面张力系数。

4. 图 5-1 所示矩形薄壁堰,由实验观察得知,矩形堰的过堰流量 $q_v$、与堰上水头 $H$、堰宽 $b$、重力加速度 $g$ 等有关。试用瑞利法确定堰流流量公式的结构形式。

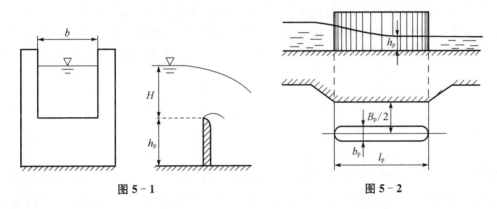

图 5-1　　　　　　　　　　图 5-2

5. 如图 5-2 所示,一桥墩长 $l_p = 24\,\mathrm{m}$,墩宽 $b_p = 4.3\,\mathrm{m}$,水深 $h_p = 8.2\,\mathrm{m}$,桥下水流平均流速 $v_p = 2.3\,\mathrm{m/s}$,两桥台间的距离 $B = 90\,\mathrm{m}$。取 $\lambda_l = 50$ 来设计水工模型实验,试确定模型的几何尺寸和模型实验流量。

# 第6章 流体阻力和能量损失

实际流体与理想流体的区别在于实际流体是粘性的,这反映在第4章实际流体的伯努利方程中的水头损失上。水头损失与流体的物理性质和边界条件密切相关。

本章介绍了水头损失的两种形式(沿程水头损失和局部损失),以及两种流动形态(层流和湍流)中两种流动环境(压力管流和明渠流)下水头损失的计算公式等。最后简要介绍了边界层和阻力。

本章应掌握两种流型和雷诺数的概念,掌握两种流动形态的特点和判别方法,了解圆管层流的运动规律,了解管内湍流的特征、混合长度和湍流速度分布的概念;了解管道沿程水头损失系数的规律,掌握管道沿程水头损失和局部损失的计算方法;了解边界层的概念、边界层的分离和阻力。

## 6.1 简介

粘性流体在运动中必然产生能量损失,在总伯努利方程中,单位重量流体的机械能用 $h_w$ 表示,称为水头损失。粘度是造成水头损失的根本原因,理想流体流动中没有水头损失,因为没有粘度。如果没有特殊说明,本章节和后续章节中的流体均为粘性流体。

### 6.1.1 流动阻力和水头损失类别

根据流体边界条件和产生水头损失的机理不同,水头损失可分为沿程水头损失和局部损失。

**1. 沿程水头损失**

当流体为均匀流动时,由于粘度的原因,横截面上的流速不均匀。相邻两层之间存在相对运动,在相邻层之间以及流动层与边界(摩擦)之间产生切应力,形成流动阻力。均匀流动产生的流动阻力称为摩擦阻力。由摩擦功引起的水头损失称为沿程水头损失,表示为 $h_f$。沿程水头损失沿流动过程均匀分布,大小与过程长度成正比。在侧壁形状、尺寸和方向不变的流动段,如长直通道和等直径压力管,其中水头损失为沿程水头损失。

**2. 局部损失**

流体在局部区域,如弯管、突然膨胀收缩和阀门等(图6-1),由于固体边界的急剧变化而引起的速度分布变化,甚至使主流离开边界,形成称为局部阻力的涡流区。局部阻力功引起的水头损失称为局部损失,用 $h_j$ 表示。局部损失经常发生在突然变化的水流横截面、水流轴急剧弯曲或有突然变化的边界形状或局部障碍。涡流区的形成是局部损失的主要原因。在一个流道中,即使是很长的一段长度,也能产生局部损失。但是为了方便起见,我们通常把它当作一个集中的头部在流体力学的一个部分来处理。

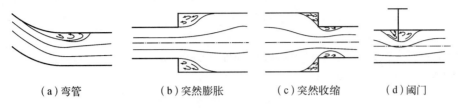

(a)弯管　　　(b)突然膨胀　　　(c)突然收缩　　　(d)阀门

图 6-1　局部损失

### 6.1.2　水头损失计算公式

热能耗散形式的能量损失，不能为其他形式的机械能逆转。如果水流的全程由不同边界的流段组成，且有几个局部损失，则全程水头损失等于沿程水头损失和局部损失之和。即

$$h_w = \sum h_f + \sum h_j \tag{6-1}$$

图 6-2 所示为文丘里管流，出口段 4 连接大气。图中 $H$ 为整个水头线，$H_p$ 为液压坡度线。局部损失包括入口水头损失 $h_{j1}$，文丘里流量计损失 $h_{j2}$ 和阀门损失 $h_{j3}$。沿程水头损失包括 $h_{f1,2}$、$h_{f2,3}$ 和 $h_{f3,4}$，总水头损失为

$$h_w = h_{j1} + h_{j2} + h_{j3} + h_{f1,2} + h_{f2,3} + h_{f3,4}$$

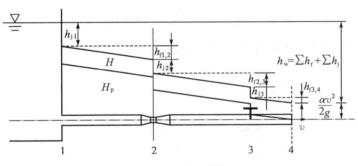

图 6-2　水头损失

气体管道流的机械能损失仍然用第 4 章的气体的伯努利方程来描述，用 $p_w$ 表示。

**1. 沿程水头损失计算公式**

实验表明，圆管流的沿程水头损失 $h_f$ 与流长 $l$ 成正比，与直径 $d$ 成反比，计算公式为

$$h_f = \lambda \frac{l}{d} \frac{v^2}{2g} \tag{6-2}$$

式中，$v$ 为截面的平均速度；$\lambda$ 为沿程水头损失系数。

上述计算公式称为达西-魏斯巴赫公式，也简称达西公式或 D-W 公式。这是一种半经验关系。通过量纲分析（例 5-3）也可以得到相同的形式，因此该公式的结构是合理的。$\lambda$ 的分析和计算是本章的主要内容。

对于非圆管截面，沿程水头损失可表示为

$$h_f = \lambda \frac{l}{4R} \frac{v^2}{2g} \tag{6-3}$$

式中，$R$ 为水力横截面半径，其值为

$$R = \frac{A}{\chi} \tag{6-4}$$

式中，$A$ 为横截面面积；$\chi$ 为流体与固体接触的界面周长，称为湿周。

对于圆管：

$$R = \frac{\pi d^2}{4\pi d} = \frac{d}{4}$$

**2. 局部水头损失计算公式**

$$h_j = \zeta \frac{v^2}{2g} \tag{6-5}$$

式中，$\zeta$ 为局部损耗系数，由实验确定（见第 6.7 节）。

### 6.1.3　沿程水头损失与切应力的关系——均匀流基本方程

如图 6-3 所示，在恒定均匀的管流中，以轴线为对称线，$l$ 为长度，$r$ 为半径，两端 1—1、2—2 为水横截面，均由流体柱组成。对其进行力的分析。假设上下游断面的中心压力分别为 $p_1$ 和 $p_2$，位置高度分别为 $z_1$ 和 $z_2$，重力方向与轴线的夹角为 $\alpha$，沿流动方向作用在柱上的外力包括压力、管壁剪力和重力，力平衡方程为

$$p_1 A' - p_2 A' + \rho g A' l \cos\alpha - \tau \chi' l = 0$$

式中，$\tau$ 为切应力；$\chi'$ 为圆柱体的湿周长；$A'$ 为圆柱的截面面积。

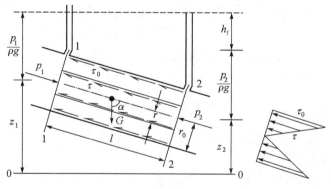

图 6-3　圆管均匀流

将每个单项式除以 $\rho g A'$，$l\cos\alpha = z_1 - z_2$：

$$\left(z_1 + \frac{p_1}{\rho g}\right) - \left(z_2 + \frac{p_2}{\rho g}\right) = \frac{\tau \chi' l}{\rho g A'}$$

根据第 1.1 和 2.2 节中总流量的伯努利方程：

$$\left(z_1 + \frac{p_1}{\rho g}\right) - \left(z_2 + \frac{p_2}{\rho g}\right) = h_f$$

所以

$$h_f = \frac{\tau l}{\rho g R'}$$

或

$$\tau = \frac{\rho g R' h_f}{l} = \rho g R' J \tag{6-6}$$

假设圆管半径为 $r_0$，边界切应力为 $\tau_0$，当 $r=0$，$R'=R$，则

$$h_f = \frac{\tau_0 l}{\rho g R} \tag{6-7}$$

或

$$\tau_0 = \rho g R J \tag{6-8}$$

式(6-7)和(6-8)表示水头损失与切应力之间的关系。对于明渠内的均匀流，用上述类似的方法也可以得到同样的结论。但壁面切应力的分布不如轴对称管流的分布均匀，公式中的 $\tau_0$ 为平均切应力。

上述推导过程与流动特性无关，在均匀流动条件下具有一定的适用性。因此，式(6-8)称为均匀流基本方程。

### 6.1.4 圆管摩擦速度和切应力分布

**1. 均匀流沿程水头损失系数与壁面切应力的关系**

将达西公式(6-2)改为 $J = \frac{\lambda}{4R}\frac{v^2}{2g}$，代入均匀流基本方程(6-8)，得到均匀流沿程水头损失系数与壁面切应力 $\tau_0$ 的关系。公式为

$$\tau_0 = \frac{\lambda}{8}\rho v^2 \tag{6-9}$$

**2. 摩擦速度**

用速度维定义 $v_* = \sqrt{\frac{\tau_0}{\rho}}$，称为摩擦速度。按式(6-9)，有

$$v_* = v\sqrt{\frac{\lambda}{8}} \tag{6-10}$$

后面的内容反复引用了摩擦速度的概念。

**3. 圆管切应力分布**

式(6-6)与式(6-8)之比为

$$\frac{\tau}{\tau_0} = \frac{r}{r_0} \tag{6-11}$$

如图6-3所示，圆管恒定均匀流在水段上的切应力是线性的，在管轴上 $\tau=0$ 和 $\tau_0$ 在管壁处达到最大值。

## 6.2 层流和湍流

经过长期的观察和实践，早在 19 世纪初，科学家们就发现流体的流动具有不同的流动形态。在不同的流动模式下，流体流动时的水头损失与流速之间存在不同的关系。但直到 1883 年，英国物理学家奥斯本·雷诺进行的实验证实了两种流型的存在以及水头损失与流速的关系。

### 6.2.1 雷诺实验、层流和湍流

雷诺实验装置如图 6-4 所示。水箱侧面安装有带喇叭口的水平玻璃管钟，管段下游有一个调节流量的管阀。针形管安装在喇叭口附近，向水中注入颜色。水箱上设有溢流板，使水箱水面不变，管头恒定。在玻璃管的截面 1—2 上还有两个测压管，用于确定截面 1—2 中的水头损失。

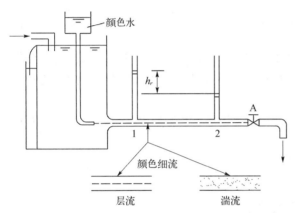

**图 6-4 雷诺实验装置**

实验中，容器内充水，液位保持稳定，管道内流量为稳定流量。稍微打开阀 A，水将以很小的速度流动。

具有颜色的水将呈现出一簇具有不同边界的直线，并且它们不彼此混合。这种现象表明流体颗粒不是相互混合而是有序运动的。流动状态称为层流。如果继续向上转动阀门，管道中的流速将逐渐增加。当流速达到一定值时，有色水开始振动和弯曲，流线在一定断面处逐渐变厚和折断。最后，速度达到 $v_c'$，色水彻底破碎并迅速扩散到整个管道。然后管道中的所有水都被完全着色。这种现象表明流体颗粒不是层流流动，而是流体颗粒的随机运动。这种在不规则流体运动中脉冲的时间和空间中的局部速度、压力和其他物理量称为湍流。层流转化为湍流时的平均流速称为上临界流速，用 $v_c'$ 表示，当实验沿相反方向进行时，即先打开阀门至最大，以确保管道内的流量处于完全展开的湍流状态，然后逐渐关闭阀门。当流速下降到与 $v_c'$ 不同的值时，均匀的颜色又回到一个直线团簇，表明圆管内的流动由湍流烘烤转变为层流。当湍流转变为层流时，管内的平均流速称为下临界流速，用 $v_c$ 表示，$v_c$ 小于 $v_c'$。

### 6.2.2 沿程水头损失与速度的关系

在对数坐标纸上测量了不同流速 $v$ 时的水头损失 $h_f$，并形成了 $v$ 与 $h_f$ 的关系曲线，如图 6-5 所示。

如果实验中流速由慢到快，则实验点落入线 $abcef$ 中。层流可以维持到 $c$ 点，然后变为湍流，且 $c$ 的位置不稳定。如果实验中速度从快到慢，则实验点落在与线 $bce$ 不重合的线 $fedba$ 中。湍流可以维持到 $b$ 点，然后变为层流。在区间 $bce$ 中，层流会因任何意外原因而被破坏，变成湍流，不会回到以前的

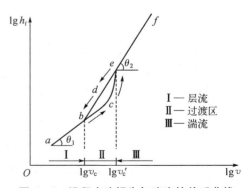

**图 6-5 沿程水头损失与速度的关系曲线**

状态。此区间称为过渡区。$ab$ 为直线，$ef$ 为近似直线，两者均可用下式表示：

$$\lg h_\mathrm{f} = \lg k + m \lg v$$

即：

$$h_\mathrm{f} = k v^m \tag{6-12}$$

实验结果表明，层流时为直线 $ab$ 段，此时 $v < v_\mathrm{c}$，且在 $\theta_1 = 45°$ 时，$m = 1$，即所有测试点在水平轴上呈 $45°$ 的直线分布。沿程水头损失与速度的第一次幂成正比，即 $h_\mathrm{f} \propto v$。湍流时适用直线 $ef$ 段，此时 $v > v_\mathrm{c}'$，$\theta_2 = 60° \sim 63°$，$m = 1.75 \sim 2.0$，即沿程水头损失与速度的第一次幂成正比，即 $h_\mathrm{f} \propto v^{1.75 \sim 2.0}$。

### 6.2.3 层流和湍流的鉴别

雷诺实验结果表明临界流速、流体密度 $\rho$、动力粘度 $\mu$ 和管径 $d$ 关系密切。液体流动形态可用量纲一——雷诺数加以区分：

上限临界雷诺数：

$$Re_\mathrm{c}' = \frac{\rho v_\mathrm{c}' d}{\mu} = \frac{v_\mathrm{c}' d}{\nu}$$

下临界雷诺数：

$$Re_\mathrm{c} = \frac{\rho v_\mathrm{c} d}{\mu} = \frac{v_\mathrm{c} d}{\nu}$$

大量的实验证明了圆管压力流中的下临界雷诺数 $Re_\mathrm{c} \approx 2\,300$，它是一个特别稳定的数值，与外界干扰几乎无关；而上临界雷诺数 $Re_\mathrm{c}' > Re_\mathrm{c}$，是一个不稳定的数值，甚至达到 $12\,000 \sim 20\,000$，这主要与流体进入管道前的稳定性和外界扰动有关。任何小扰动都会使层流进入湍流，因为在上临界雷诺数和下临界雷诺数之间的流动是非常不稳定的。实际工程中总是存在扰动。因此，在实际应用中，介于上临界雷诺数和下临界雷诺数之间的液体流动可视为湍流。因此，下临界雷诺数是判别流体流动状态的判据。

在圆管中：

$$Re_\mathrm{c} = \frac{v_\mathrm{c} d}{\nu} = 2\,300 \tag{6-13}$$

如果 $Re < Re_\mathrm{c}$，层流；当 $Re > Re_\mathrm{c}$ 时，湍流。

明渠水流与天然河道水流：

$$Re_\mathrm{c} = \frac{v_\mathrm{c} R}{\nu} = 575 \tag{6-14}$$

如果 $Re < Re_\mathrm{c}$，层流；当 $Re > Re_\mathrm{c}$ 时，湍流。

**例 6-1** 部分自来水管段，$d = 0.1\,\mathrm{m}$，$v = 1.0\,\mathrm{m/s}$。水温 $10\,℃$，（1）判断管道内的流型；（2）算出保持层流的最大流速。

**解**：（1）水温为 $10\,℃$ 时，水的粘度系数为

$$\nu = \frac{0.017\,75 \times 10^{-4}}{1 + 0.033\,7t + 0.000\,221 t^2} = \frac{0.017\,75}{1.359\,1} \times 10^{-4} = 1.31 \times 10^{-6}\,\mathrm{m^2/s}$$

则：

$$Re = \frac{vd}{\nu} = \frac{1 \times 0.1}{1.31 \times 10^{-6}} = 76\,336 > Re_c = 2\,300$$

圆管内的流动形态为湍流。

(2) $Re_c = \dfrac{v_c d}{\nu}$

$$v_c = \frac{\nu Re_c}{d} = \frac{1.31 \times 10^{-6} \times 2\,300}{0.1} = 0.03 \text{ m/s}$$

因此,为了保持层流,最大流速为 0.03 m/s。

**例 6-2** 实验中矩形明渠水流,底宽 $b=0.2$ m,水深 $h=0.1$ m,流速 $v=0.12$ m/s,水温为 20 ℃,试判断水流形态。

**解**：根据已知条件计算水的运动要素：

$$A = bh = 0.2 \times 0.1 = 0.02 \text{ m}^2$$

$$\chi = b + 2h = 0.2 + 2 \times 0.1 = 0.4 \text{ m}$$

$$R = \frac{A}{\chi} = \frac{0.02}{0.4} = 0.05 \text{ m}$$

当水温为 20 ℃ 时,

$$\nu = 1.0 \times 10^{-6} \text{ m}^2/\text{s}$$

则有：

$$Re = \frac{vR}{\nu} = \frac{0.12 \times 0.05}{1.0 \times 10^{-6}} = 6\,000 > Re_c = 575$$

流动形态为湍流。

## 6.3 圆管中的层流

层流是流体运动的一个简单的例子,也是几种流场之一,可以得出流速分布和水头损失的结论。本章将介绍一些相关的表达式。

### 6.3.1 流速分布

如图 6-6 所示,取圆管层流的均匀流段,分析圆管层流的流速分布。在层流中,切应力满足牛顿内摩擦定律(1-10),考虑 $y$ 与 $r$ 的关系为 $y = r_0 - r$,则 $\mathrm{d}y = -\mathrm{d}r$。所以：

$$\tau = \mu \frac{\mathrm{d}u}{\mathrm{d}y} = -\mu \frac{\mathrm{d}u}{\mathrm{d}r} \quad (6-15)$$

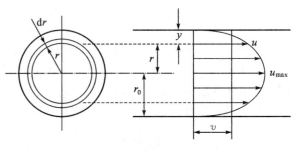

**图 6-6 流速分布**

在式(6-6)中,取 $\tau = \rho g R' J$,其中 $R' = \dfrac{r}{2}$,代入式(6-15),将变量分离,得到:

$$du = -\frac{\rho g J}{2\mu} r \, dr$$

由于均匀流中的 $J$ 相同,上式的不定积分为

$$u = -\frac{\rho g J}{4\mu} r^2 + C \tag{6-16}$$

在管壁处,流体附着在壁上,不满足滑动条件,即当 $r = r_0$ 时,$u = 0$,因此,

$$C = \frac{\rho g J}{4\mu} r_0^2$$

代入式(6-16),有

$$u = \frac{\rho g J r_0^2}{4\mu}\left(1 - \frac{r^2}{r_0^2}\right) \tag{6-17}$$

式(6-17)表明,圆管流段流速分布呈旋转抛物线分布,这是层流的重要特征之一。

根据图6-6,圆管流中的最大流速位于管轴,根据式(6-17):

$$u_{\max} = \frac{\rho g J r_0^2}{4\mu} \tag{6-18}$$

因此,式(6-17) 也可表为

$$u = u_{\max}\left(1 - \frac{r^2}{r_0^2}\right) \tag{6-19}$$

选取宽为 $dr$ 的环形断面,微元面积为 $dA = 2\pi r \, dr$,根据截面平均流速定义,圆管层流平均流速为

$$v = \frac{q_v}{A} = \frac{\int_A u \, dA}{A} = \frac{1}{\pi r_0^2}\int_0^{r_0} \frac{\rho g J}{4\mu}(r_0^2 - r^2) \cdot 2\pi r \, dr = \frac{\rho g J}{8\mu} r_0^2 \tag{6-20}$$

上述方程是哈根-泊肃叶公式。

比较式(6-18) 和(6-20),我们得到:

$$v = \frac{1}{2} u_{\max} \tag{6-21}$$

也就是说,圆管中层流的平均流速是最大流速的一半。与圆管内湍流相比,圆管内湍流流速不均匀,动能修正系数为

$$\alpha = \frac{\int_A u^3 \, dA}{A v^3} = 2.0$$

动量修正系数为

$$\beta = \frac{\int_A u^2 \mathrm{d}A}{Av^2} = \frac{4}{3}$$

### 6.3.2 圆管内层流水头损失的计算

圆管内层流沿程水头损失可用式(6-20)计算。

$$J = \frac{h_f}{l} = \frac{8\mu v}{\rho g r_0^2} = \frac{32\mu v}{\rho g d^2}$$

然后得到：

$$h_f = \frac{32\mu v l}{\rho g d^2} \tag{6-22}$$

计算结果表明，沿程水头损失与层流圆管截面平均速度成正比，与管壁粗糙度无关。此外，式(6-22)也可以重写为达西方程的形式：

$$h_f = \frac{32\mu v l}{\rho g d^2} = \frac{64}{\underbrace{\frac{vd}{\nu}}} \frac{l}{d} \frac{v^2}{2g} = \lambda \frac{l}{d} \frac{v^2}{2g} \tag{6-23}$$

式中，$\lambda = \dfrac{64}{Re}$。

**例 6-3** 某油 $\rho = 850 \text{ kg/m}^3$，$\nu = 0.18 \times 10^{-4} \text{ m}^2/\text{s}$，在直径 $d = 0.1$ m 的管道中以速度 $v = 0.0635$ m/s 移动并形成层流。尝试求出：(1) 管中心的最大流速；(2) 距离管子中心 $r = 20$ mm 处的速度；(3) 沿程水头损失系数 $\lambda$；(4) 管壁切应力 $\tau_0$ 和每千米管长水头损失。

**解：**(1) 管中心最大流速

$$u_{\max} = 2v = 2 \times 0.0635 = 0.127 \text{ m/s}$$

(2) 距离管子中心 20 mm 的速度

$$u = u_{\max}\left[1 - \left(\frac{r}{r_0}\right)^2\right] = 0.127 \times \left[1 - \left(\frac{0.02}{0.05}\right)^2\right] = 0.107 \text{ m/s}$$

(3) 沿程水头损失系数

先解决雷诺数问题：

$$Re = \frac{vd}{\nu} = \frac{0.0635 \times 0.1}{0.18 \times 10^{-4}} = 353 \text{（层流）}$$

所以

$$\lambda = \frac{64}{Re} = \frac{64}{353} = 0.18$$

(4) 切应力和每千米管长水头损失

$$\tau_0 = \frac{\lambda}{8}\rho v^2 = \frac{0.18}{8} \times 850 \times 0.0635^2 = 0.077 \text{ N/m}^2$$

$$h_f = \lambda \frac{l}{d} \frac{v^2}{2g} = 0.18 \times \frac{1\,000}{0.1} \times \frac{0.063\,5^2}{2 \times 9.8} = 0.37 \text{ m}$$

## 6.4 湍流

自然界和工程界的流动问题大多是湍流问题，因此湍流研究具有广泛的意义。与层流不同，湍流是流体粒子相互掺杂的一种无序现象。每个点的速度以及与速度相关的压力、浓度和其他物理量都是在时间和空间上的随机波动（称为湍流脉动）。一个多世纪以来，湍流理论在湍流研究方面可以说是硕果累累。但湍流比层流复杂得多，理论分析难度较大，因此湍流的研究仍处于方兴未艾阶段。下面从湍流的概念、研究方法和主要特点等方面作一个简要的介绍。

### 6.4.1 湍流结构——粘性底层和湍流核心区

**1. 湍流结构**

根据理论分析和实验观察，湍流和层流在墙体上没有滑动条件。在湍流中，有一个非常薄的层流靠近固体边界。由于流体粘度的影响和固体边界的限制，颗粒之间的掺杂被消除，因此流动模式呈现层流，称为粘性底层。

在粘性底层外，流体颗粒被掺杂。流速及其相关物理量的波动明显，称为湍流区。这个区域通常被称为湍流核心区。

粘性核心区和湍流核心区之间有一薄层过渡层，过渡层很薄，边界不清。在进行湍流分析时，将整个截面分为粘性底层和湍流核心区两部分进行讨论（见图 6-7）。

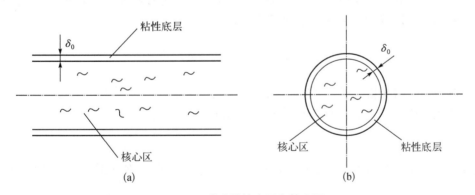

图 6-7 湍流粘性底层和核心区

**2. 粘性底层**

在层流中，壁上的流体速度为零。随着离墙距离的增大，流体速度从零值增大到一定值。速度近似呈线性分布，其梯度值极大。而粘性底层的厚度 $\delta_0$ 通常小于 1 mm，并且随着雷诺数 $Re$ 的增加而减小。

在粘性底层中，切应力服从牛顿的内摩擦定律。取墙切应力 $\tau = \tau_0$，则有

$$\tau_0 = \mu \frac{du}{dy}$$

然后上式的积分：

$$u = \frac{\tau_0}{\mu} y + C$$

根据边界条件,在管壁表面,$y=0$,$u=0$,得到积分常数 $C=0$。所以

$$u = \frac{\tau_0}{\mu} y \tag{6-24}$$

或者 $\mu = \rho \nu$,$v_* = \sqrt{\dfrac{\tau_0}{\rho}}$ 代入上式,变换后我们可以得到

$$\frac{u}{v_*} = \frac{v_*}{\nu} y \tag{6-25}$$

上述公式显示了粘滞底层与切应力的关系。

从后面的内容可以看出,虽然粘性底层很薄,但它对湍流的速度分布和流动阻力都有很大的影响。

**3. 湍流核心区域**

本节中的湍流运动主要针对湍流核心区域中的流体运动,并将在下文中引入。

### 6.4.2 湍流运动的时间平均

忽略流体的随机性和湍流运动元素的时间平均,研究湍流运动规律是研究湍流运动的有效途径。

流体粒子移动到某一点时的速度称为瞬时速度,用 $u_x$ 表示。已知流体粒子的瞬时速度随时间变化。利用样本均值和偏差的方法,我们可以认为瞬时速度由平均速度和波动速度组成。在足够长的时间 $T$ 内,瞬时流速的平均值为

$$\bar{u}_x = \frac{1}{T} \int_0^T u_x(t) \, dt \tag{6-26}$$

式中,$\bar{u}_x$ 称为时间的平均流速,简称时间平均流速。

瞬时速度与平均速度之差称为波动速度 $u'_x$,即

$$u_x = \bar{u}_x + u'_x$$

同样,$y$,$z$ 坐标方向的速度 $u_y$,$u_z$ 和瞬时压力 $p$ 都可以看作是时间平均值和波动两部分:

$$u_y = \bar{u}_y + u'_y$$
$$u_z = \bar{u}_z + u'_z$$
$$p = \bar{p} + p'$$

脉动的每个物理量的平均值总是等于零。

波动速度 $u'_x$ 的平均值为

$$\overline{u'_x} = \frac{1}{T} \int_0^T u'_x \, dt = \frac{1}{T} \int_0^T (u_x - \bar{u}_x) \, dt = \frac{1}{T} (\bar{u}_x - \bar{u}_x) = 0$$

但是每个脉冲的均方值不等于 0。即

$$\overline{u_x'^2} = \frac{1}{T}\int_0^T (u_x - \bar{u}_x)^2 dt \neq 0$$

此外,两个脉动值的乘积的时间平均值也不等于零,即 $\overline{u_x'u_y'}$ 和 $\overline{u_y'u_z'}$ 都不等于零。

在流体运动规律的研究中,经常用脉动速度的均方根来表示脉动幅度的大小。

$$N = \frac{\sqrt{\frac{1}{3}(\overline{u_x'^2} + \overline{u_y'^2} + \overline{u_z'^2})}}{\bar{u}} \tag{6-27}$$

式中,$N$ 称为湍流强度。

当引入时间平均值的概念时,湍流运动可以看作是时间平均值和脉动值的叠加,然后我们可以分别研究它们。因此,在工程实践中,一般流量的计算可以用时间平均值来计算。而湍流中的恒定流也是时间平均的稳定流。

### 6.4.3 湍流中的雷诺应力

根据湍流理论,湍流运动中的应力包括重力作用下的压缩应力和粘性应力。还包括湍流中的附加应力。以平面均匀流为例,可以表示为

$$\tau = \mu \frac{du_x}{dy} - \rho \overline{u_x'u_y'} \tag{6-28}$$

式中,$-\rho \overline{u_x'u_y'}$ 称为湍流中的附加应力,用 $\tau'$ 表示;"$-$"是使附加切应力为正,因为 $u_x'$ 和 $u_y'$ 两者之间的符号相反。

三维流体运动,通过时间平均运算得到 N-S 方程,得到三个附加应力和三个附加的应力正分量,分别是 $-\rho \overline{u_x'u_y'}$,$-\rho \overline{u_x'u_z'}$,$-\rho \overline{u_y'u_z'}$,和 $\rho \overline{u_z'^2}$,$\rho \overline{u_x'^2}$,$\rho \overline{u_y'^2}$,这些附加应力称为雷诺应力。

粘性和附加切应力的比例随湍流条件的不同而不同。实验表明,对于完全展开的湍流,粘性应力主要存在于近壁的薄粘性层中。当附加切应力远大于湍流核心区的粘性应力时,粘性切应力可以忽略。

### 6.4.4 普朗特混合长度湍流半经验理论

建立了雷诺应力与平均流速的本构关系(湍流模型),湍流中的流速和应力分布便可以用理论解释或计算机数值方法求解,这是湍流理论的核心问题。这方面的研究成果很多,其中普朗特混合长度理论是最经典、最简单的湍流模型,具有一定的实用性。

德国学者普朗特利用分子自由程的概念提出了湍流混合长度的假设。他认为湍流速度与平均速度梯度和一定长度(混合长度)的乘积成正比,即:

$$u_x' \propto L_1 \frac{d\bar{u}_x}{dy}, \quad u_y' \propto L_2 \frac{d\bar{u}_x}{dy}$$

普朗特称 $L_1$ 和 $L_2$ 为混合长度,并用 $L^2$ 表示比例常数的乘积。因此,湍流中的附加应力为 $\tau'$:

$$\tau' = -\rho \overline{u_x'u_y'} = \rho L^2 \left(\frac{d\bar{u}_x}{dy}\right)^2 \tag{6-29}$$

式中，L 仍称为混合长度，由实验确定。

### 6.4.5 湍流流速分布

由此可见，湍流的速度分布主要是在湍流的核心区的速度分布。在湍流的核心区域，与湍流的附加应力相比，粘性切应力可以忽略。因此，层流切应力取决于湍流的附加切应力。

$$\tau = \tau' = \rho L^2 \left(\frac{d\bar{u}_x}{dy}\right)^2$$

对于圆管，通过式(6-11)得到 $\tau = \dfrac{\tau_0 r}{r_0}$；根据实验结果我们也可以得到 $L = \kappa y \sqrt{\dfrac{r}{r_0}}$，上面的公式可以表示为

$$\tau_0 = \rho \kappa^2 y^2 \left(\frac{d\bar{u}_x}{dy}\right)^2$$

然后：

$$v_* = \sqrt{\frac{\tau_0}{\rho}} = \kappa y \frac{d\bar{u}_x}{dy}$$

分离变量，积分得

$$\frac{\bar{u}_x}{v_*} = \frac{1}{\kappa} \ln y + C \tag{6-30}$$

上式是由普朗特混合长度理论导出的沿圆管流段法线方向的湍流速度分布，具有普遍意义，可推广到任何管面流、通道流等，但常数 $\kappa$、$C$ 应根据具体实践通过实验确定。

在下面的讨论中，为了简单起见，省略了表示平均速度的线。

## 6.5 圆管内湍流的沿程水头损失

本章主要介绍了圆管内湍流的特点和确定摩擦阻力系数的方法。

### 6.5.1 尼古拉兹实验和湍流阻力分配

1933年，德国力学家、工程师尼古拉兹进行了损失系数分布研究实验，并给出了沿圆管的截面速度和阻力分布图，具有重要意义。这里简要介绍这一经典实验及其主要结果。

**1. 尼古拉兹实验**

尼古拉兹认为有两个因素影响阻力因子：

$$\lambda = f(Re, \Delta/d)$$

式中，$Re$ 为管道流量雷诺数；$\Delta$ 为壁面凸起的粗糙度高度，称为绝对粗糙度；$d$ 为直径；$\Delta/d$ 为相对粗糙度。

采用人工粗糙度法，尼古拉兹将筛出的均匀砂紧密地粘贴在管壁表面，用砂粒直径表示绝对粗糙度。他做了6根长直、直径相同的人造粗管，但相对粗糙度不同。他用这些管子分别做

了流动实验。在实验中,他测量了流速 $v$,水头损失 $h_f$(测压水头差)在长度 $l$ 的流动截面,并用公式 $Re=\dfrac{vd}{\nu}$ 和 $\lambda=\dfrac{d}{l}\dfrac{2g}{v^2}h_f$ 计算了 $Re$ 和 $\lambda$。然后,他用系列数据绘制了 $\lg(100\lambda)$-$\lg Re$ 图(图 6-8),即著名的尼古拉兹图。

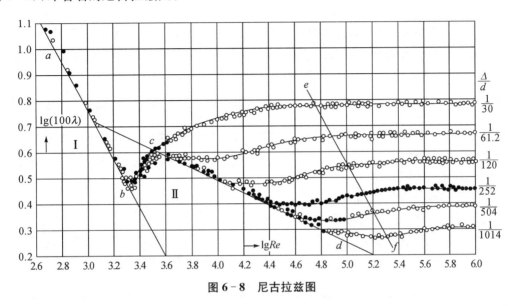

图 6-8 尼古拉兹图

**2. 湍流阻力分区**

根据 $\lambda$ 的变化特征,将尼古拉兹曲线分为五个阻力区。

(1) 层流区,$ab$ 线,$\lg Re<3.36$,$Re<2\,300$。6 种管流的实验点都落在同一条线上,$\lambda$ 与相对粗糙度 $\dfrac{\Delta}{d}$ 无关,从图中可导出相关的相关系数 $\lambda=\dfrac{64}{Re}$。实验结果与理论结果一致。

(2) 从层流到湍流的过渡区,$bc$ 线,$\lg Re=3.36\sim3.6$,$Re=2\,300\sim4\,000$。$\lambda$ 只与 $Re$ 有关。由于流态不稳定,实验结果较为分散。

(3) 湍流的光滑区,$cd$ 线,$\lg Re>3.6$,$Re>4\,000$。流型为湍流,6 种管流均有实验点落在同一条线上,在水力光滑区,$\lambda$ 与 $\dfrac{\Delta}{d}$ 无关,仅与 $Re$ 有关。从图中可以看出,水区的平滑流动在具有较大的 $\dfrac{\Delta}{d}$ 的管中较小;水区域的平滑流动在具有较小的 $\dfrac{\Delta}{d}$ 的管中较大。例如,$\dfrac{\Delta}{d}=\dfrac{1}{120}$ 的管流和 $Re=4\,000\sim12\,600$ 的管流在水力光滑区,$\dfrac{\Delta}{d}=\dfrac{1}{1\,014}$ 的管流和 $Re=4\,000\sim70\,000$ 的管流在水力光滑区。

(4) 湍流过渡区,线 $cd$ 和 $ef$ 之间的区域。6 种不同相对粗糙度的管流有 6 条不同的坡度曲线,这表明湍流过渡区的 $\lambda$ 与 $\dfrac{\Delta}{d}$ 和 $Re$ 有关。

(5) 湍流粗糙区(也称为水力粗糙区)。6 种管流的实验曲线上有 6 条不同的水平线,表明在湍流粗糙区中的 $\lambda$ 仅与 $\dfrac{\Delta}{d}$ 有关,与 $Re$ 无关。例如,当流动在湍流粗糙区时,无论 $Re$ 的值如何，相对粗糙度 $\frac{\Delta}{d} = \frac{1}{120}$ 的管流的 $\lambda$ 为 0.035。在湍流粗糙区，某一管道的 $\lambda\left(\frac{\Delta}{d}\text{为某一值}\right)$ 为常数。由公式 $h_f = \lambda \frac{l}{d} \frac{v^2}{2g}$ 可知，沿程水头损失与流速的平方成正比，所以湍流粗糙区也称为阻力平方区。

当管流处于平滑流区时，该流称为水力平滑管。当管流处于粗糙流区时，管流称为水力粗糙管。水力平滑管和水力管是管道流动特性的名称。对于特定的管道，在低雷诺数时，流量是水力平滑的管道，而在高雷诺数时，流量是水力粗糙的管道。

**推论** 在相似实验中，如果原型和模型处于粗糙区域，它只能在几何相似$\left(\text{等于}\frac{\Delta}{d}\right)$上自动获得阻力相似（等于 $\lambda$），而不等于 $Re$。也就是说，只要模型是按照 Froude 准则设计的，该模型就能自动实现粘性力相似，而不满足雷诺相似准则。因此，阻力平方区判据也被称为相似模型中的自动模型区域。

### 3. 湍流阻力变化原因分析

湍流分为光滑区、过渡区和粗糙区。$\lambda$ 在不同区域的变化是不同的。原因可以通过图 6-9 的粘性亚层的变化来分析。图 6-9a 显示了粘性亚层的厚度大于管壁粗糙度的流动情况，粗糙投影完全隐藏在粘性亚层中，对湍流核心区几乎没有影响。因此，$\lambda$ 仅与 $Re$ 有关，与 $\frac{\Delta}{d}$ 无关，从而形成湍流光滑区的流动阻力特性。图 6-9b 显示，$\delta_0$ 流态与 $\Delta$ 流态接近，其粗糙度对湍流核心区有影响。因此，$\lambda$ 与 $\frac{\Delta}{d}$ 和 $Re$ 均相关，从而形成湍流过渡区的流动阻力特性。图 6-9c 显示了 $\delta_0$ 远小于 $\Delta$，几乎完全进入湍流核心。在这种情况下，$Re$ 的变化对粘性亚层和湍流动力学的影响很小。此外，$\lambda$ 除与 $\frac{\Delta}{d}$ 有关外，与 $Re$ 无关，从而形成湍流粗糙区的流动阻力特性。

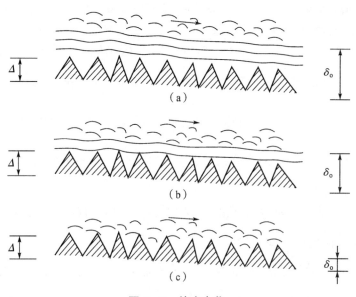

图 6-9 粘度变化

## 6.5.2 圆管内湍流速度分布

**1. 湍流光滑区**

本文介绍了粘性流在底部的速度呈线性分布：

$$u = \frac{\tau_0}{\mu} y \quad (y \leqslant \delta_0)$$

在湍流的核心区域，按式(6-30)

$$\frac{u}{v_*} = \frac{1}{k} \ln y + C$$

参考边界条件 $y = \delta_0$，$u = u_b$，我们确定上述公式的积分常数：

$$C = \frac{u_b}{v_*} - \frac{1}{k} \ln \delta_0, \quad \delta_0 = \frac{\mu}{\tau_0} u_b = \frac{u_b}{v_*^2} v$$

然后得

$$\frac{u}{v_*} = \frac{1}{k} \ln \frac{y v_*}{\nu} + C_1$$

此时 $C_1 = \frac{u_b}{v_*} - \frac{1}{k} \ln \frac{u_b}{v_*}$，然后，通过尼古拉兹实验确定了 $k = 0.4$，$C_1 = 5.5$，得到了水力光滑圆管截面上的流速分布公式。

**2. 湍流粗糙区**

$$\frac{u}{v_*} = 5.75 \lg \frac{y v_*}{\nu} + 5.5 \quad (y > \delta_0) \tag{6-31}$$

圆管内湍流管内的流速分布（省略导数）为

$$\frac{u}{v_*} = 5.75 \lg \frac{y}{\Delta} + 8.48 \quad （核心区） \tag{6-32}$$

式(6-31)和式(6-32)表明，圆管内湍流的流速分布符合对数分布规律。与圆管层流抛物线分布相比，圆管湍流截面上的流速分布更加均匀，如图6-10所示。

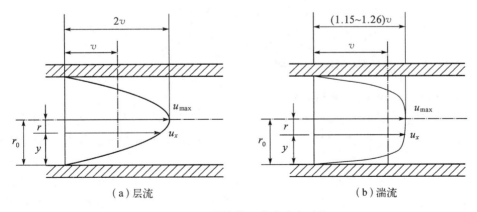

图6-10 圆管截面流速分布对比

上述公式基于普朗特混合长度理论,其未定常数由实验确定,称为半经验公式。

**3. 圆管内湍流速度分布的指数公式**

为了便于工程计算,提出了一个较为简单的计算公式,即圆管内湍流流速指数分布。

$$\frac{u}{u_{\max}} = \left(\frac{y}{r_0}\right)^n \tag{6-33}$$

式中,$u_{\max}$ 为管轴处的最大流速;$r_0$ 为管的半径;$n$ 为指数,可以通过以下经验公式计算。

$$\frac{1}{n} = -2.8\lg\left(\frac{0.034\,55}{Re^{0.1}} - 0.007\,875\right) \tag{6-34}$$

式中,$4 \times 10^3 \leqslant Re < 2 \times 10^6$;当 $Re \geqslant 2 \times 10^6$ 时,$n = \frac{1}{10}$。

公式(6-33)中的 $n$ 值也可以取为常数 $n = \frac{1}{7}$,称为流量分布的 $\frac{1}{7}$ 定律。

**4. 湍流管道流动的几个参数**

根据公式(6-33),平均速度与管中心最大速度的比值为 $\frac{v}{v_{\max}} = 0.79 \sim 0.87$,且随 $Re$ 的增大而增大;流动截面动能修正系数为 $\alpha = 1.077 \sim 1.031$,随 $Re$ 的增大而减小;流动截面动量修正系数为 $\beta = 1.027 \sim 1.011$,随 $Re$ 的增大而减小。

### 6.5.3 圆管湍动阻力因数的半经验公式

用速度分布的半经验公式(6-31)和公式(6-32)计算截面平均速度。然后根据实验数据 $v_* = v\sqrt{\frac{\lambda}{8}}$ 调整常数。最后,我们可以得到沿途阻力因数对应阻力区的尼古拉兹半经验公式。

湍流光滑区:

$$\frac{1}{\sqrt{\lambda}} = -2\lg\left(\frac{2.51}{Re\sqrt{\lambda}}\right) \tag{6-35}$$

湍流粗糙区:

$$\frac{1}{\sqrt{\lambda}} = -2\lg\left(\frac{\Delta}{3.7d}\right) \tag{6-36}$$

### 6.5.4 工业管道、穆迪图和巴尔公式

**1. 工业管道和 Colebrook-White 公式**

没有人工粗糙的实际应用管道称为工业管道。人工粗糙管道与工业管道在粗糙方面有很大的不同。连接两种不同的粗糙形式是在工业管道中如何使用尼古拉兹半经验公式的真正因素。

Colebrook 做了工业管道阻力损失对比实验和尼古拉兹实验。他测量了阻力平方区的工业管道的 $\lambda$。通过对尼古拉兹曲线的比较,确定了在湍流阻力平方区中直径相同、$\lambda$ 相同的人工粗糙管的粗隆起高度 $\Delta$ 为工业管的等效粗糙度。普通工业管道的等效粗糙度如下:焊接钢

管原件为 0.06~1.0 mm,铸铁管原件为 0.3~1.2 mm,水泥管原件为 0.5 mm,混凝土管原件为 0.3~3.0 mm。尼古拉兹半经验公式可应用于具有等效粗糙度的工业管道。

1939 年,Colebrook 和 White 将公式(6-35)和公式(6-36)合并,得到了湍流过渡区的 λ 经验公式。

$$\frac{1}{\sqrt{\lambda}} = -2\lg\left(\frac{2.51}{Re\sqrt{\lambda}} + \frac{\Delta}{3.7d}\right) \tag{6-37}$$

这个公式称为 Colebrook-White 公式。

公式(6-37)与工业管流实验结果一致,可应用于湍流光滑区、过渡区和粗糙区。这是计算沿途湍流因子的通用公式。因此,它在工业管道流量计算中得到了广泛的应用。

**2. 穆迪图**

公式(6-37)是求 λ 值的隐式方程,不易计算。1944 年,美国工程师穆迪绘制了公式的对数曲线,称为穆迪图,如图 6-11 所示。在该图中,我们可以通过 $\frac{\Delta}{d}$ 和 $Re$,直接求得 λ 值。在没有计算机的时代,穆迪图非常实用,是工程流体力学的经典图之一。通过比较穆迪图和尼古拉兹图,我们发现工业管道的流动区域可以分为层流区、临界区(层流湍流过渡区)、湍流光滑区、湍流过渡区和湍流粗糙区。

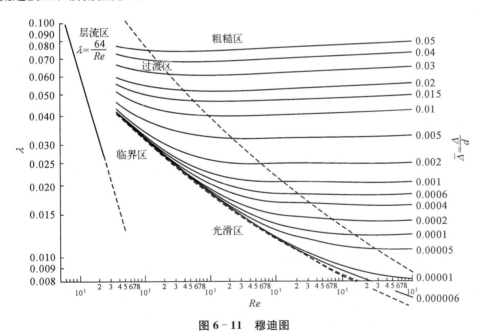

图 6-11 穆迪图

λ 具有相同的变化规律,在湍流过渡带中差别很大。这是一个逐渐突破非均匀粗糙进入工业管道湍流核心区的过程,这不同于颗粒尺寸均匀的人工粗糙度。

**3. 巴尔公式**

除了上面提到的半经验关系外,还有许多经验关系,如 Bula Huge 公式和 Seffrin 松散公式等。巴尔于 1977 年提出的公式如下:

$$\frac{1}{\sqrt{\lambda}} = -2\lg\left(\frac{\Delta}{3.7d} + \frac{5.1286}{Re^{0.89}}\right) \tag{6-38}$$

上述公式为巴尔公式,公式适用范围与公式(6-37)相同,也是一般的阻力面积经验公式。与公式(6-37)相比,最大误差仅为 1%,由于它是 $\lambda$ 的显式公式,便于计算和编程。例如,以前提到的穆迪图是基于巴尔公式的,它与传统的穆迪图完全相同。对于求解 $\lambda$ 的所有公式,优先推荐巴尔公式。

### 6.5.5 湍流阻力区的判别

湍流阻力区判别在两个方面的应用如下:

首先基于湍流阻力区,选择合适的公式计算 $\lambda$ 的值;第二种是基于一般公式,虽然不需要事先确定湍流阻力区域,但在进行流动相似模拟时需要确定管道的湍流阻力区。因此,区分湍流阻力区具有实际意义。

由人工粗糙管导出的湍流阻力区判别半经验公式可供参考。计算结果表明,现有的半经验公式在工业管道湍流阻力区的应用中存在较大误差,拟合得到的管道流动阻力具有较高的精度。

圆管光滑区:

$$\frac{\Delta}{d} \leqslant 0.000\,8 \text{ 且 } 4\,000 < Re \leqslant 10\left(\frac{\Delta}{d}\right)^{-1} \tag{6-39}$$

圆管湍流过渡区:

$$\frac{\Delta}{d} > 0.000\,8, \ 4\,000 < Re \leqslant 576.12\left(\frac{\Delta}{d}\right)^{-1.119}$$

或

$$\frac{\Delta}{d} \leqslant 0.000\,8, \ 10\left(\frac{\Delta}{d}\right)^{-1} < Re \leqslant 576.12\left(\frac{\Delta}{d}\right)^{-1.119} \tag{6-40}$$

圆管湍流粗糙区:

$$Re > 576.12\left(\frac{\Delta}{d}\right)^{-1.119} \tag{6-41}$$

## 6.6 非圆管中湍流的损失

### 6.6.1 带压力管流量的非圆管

如果 $d_e = 4R$ 为非圆管的等效直径,$R$ 为水力半径,可由式(6-3)得到:

$$h_f = \lambda \frac{l}{d_e} \frac{v^2}{2g} \tag{6-42}$$

适用范围:长边 $a$、短边 $b$ 的矩形管适用于 $\frac{a}{b} < 8$ 的湍流;长轴 $d_1$、短轴 $d_2$ 的椭圆管适用于 $\frac{d_1}{d_2} < 3$ 的湍流。

### 6.6.2 通道流量

575 是明渠流的临界雷诺数,当 $Re > 575$ 时,明渠流为湍流。明渠层流的实用价值很小,因此,我们着重研究湍流明渠流动。

法国工程师谢才 1775 年给出了明渠均流平均流速的经验公式,即谢才公式:

$$v = C\sqrt{RJ} \tag{6-43}$$

式中,$C$ 为谢才系数。

$$h_f = \frac{lv^2}{RC^2} = \frac{8g}{C^2} \frac{l}{4R} \frac{v^2}{2g} \tag{6-44}$$

与式(6-3)相比,显然:

$$\lambda = \frac{8g}{C^2}$$

或

$$C = \sqrt{8g/\lambda} \tag{6-45}$$

$C$ 的量纲为 $m^{0.5}/s$。

$C$ 通常由经验公式确定。最著名的经验公式由爱尔兰工程师曼宁于 1890 年提出:

$$C = \frac{1}{n} R^{1/6} \tag{6-46}$$

上式为曼宁公式,其中 $n$ 为粗糙系数,是衡量边墙粗糙度影响的综合系数。一般情况下,它不在本单元中列出。通过测量明渠或管流湍流粗糙区得到数值。例如,铸铁管和钢管的 $n$ 值为 0.011～0.013;混凝土墙 $n$ 值为 0.011～0.014。由上可知,式(6-46)的使用条件为湍流的粗糙区域,当 $n < 0.02$,$R > 0.5$ m 时,计算结果符合实际情况。当式(6-43)被引入式(6-46)计算时,使用条件相同。当消除 $n$ 的单位时,式(6-43)和式(6-46)与该单元不一致。因此,长度单位按规定只能选择米,不能选择厘米等单位。谢才系数仅与 $Re$ 和侧壁粗糙度有关,与横流形状无关。式(6-44)和式(6-46)也可应用于管道的阻力区。

**例 6-4** 一水管长 $l = 500$ m,直径 $d = 0.2$ m,管壁粗糙度高度 $\Delta = 0.1$ mm,水温 $t = 10\ ℃$。沿途水头损失多少?(输送流量 $q_v = 10 \times 10^{-3}$ m³/s。)

**解:**

$$v = \frac{q_v}{\frac{1}{4}\pi d^2} = \frac{10 \times 10^{-3}}{\frac{1}{4} \times 3.14 \times 0.2^2} = 0.318\ 3\ \text{m/s}$$

$t = 10\ ℃$,$\nu = 1.31 \times 10^{-6}$ m²/s,

$$Re = \frac{vd}{\nu} = \frac{0.318\ 3 \times 0.2}{1.31 \times 10^{-6}} = 48\ 595 > 2\ 300$$

因此,管道中存在湍流。

根据巴尔公式:

$$\frac{1}{\sqrt{\lambda}} = -2\lg\left(\frac{\Delta}{3.7d} + \frac{5.128\ 6}{Re^{0.89}}\right)$$

由于
$$\frac{\Delta}{d} = 0.000\,5,\ Re = 48\,595$$

所以
$$\lambda = 0.022\,7$$

$$h_f = \lambda \frac{l}{d} \frac{v^2}{2g} = 0.022\,7 \times \frac{500}{0.2} \times \frac{0.318\,3^2}{2 \times 9.8} = 0.293\text{ m}$$

**例 6-5** 新的水管道，管径 $d = 0.4$ m，管长 $l = 100$ m，粗糙系数 $n = 0.011$，线性损失 $h_f = 0.4$ m，水流为湍流粗糙区。流量多少？

**解：** 管道横断面面积：

$$A = \frac{\pi}{4} d^2 = \frac{3.14}{4} \times 0.4^2 = 0.126\text{ m}^2$$

水力半径：

$$R = \frac{d}{4} = 0.1\text{ m}$$

谢才系数：

$$C = \frac{1}{n} R^{1/6} = \frac{1}{0.011} \times (0.1)^{1/6} = 61.94\text{ m}^{1/2}/\text{s}$$

所以流量为

$$q_v = vA = AC\sqrt{RJ} = 0.126 \times 61.94 \times \sqrt{0.1 \times \frac{0.4}{100}} = 0.156\text{ m}^3/\text{s}$$

## 6.7 局部水头损失

利用公式 $h_j = \zeta \dfrac{v^2}{2g}$ 可以计算局部水头损失，主要是确定局部水头损失系数 $\zeta$。流动雷诺数对 $\zeta$ 的影响很小，可以忽略不计，在正常情况下，$\zeta$ 仅由流道的边界变化决定。除突扩管外的 $\zeta$ 值均可以由试验确定。

### 6.7.1 突扩管

突扩管的 $\zeta$ 可以用总流的三个方程来确定。

假设流体流动是恒定的湍流。如图 6-12 所示，截面 1—1 和 2—2 为控制截面，管壁中流体作用于流动截面 1—2 的切应力忽略不计。实验表明，截面 1—1 的压力分布与静压分布一致。从动量方程(4-30)可以推断（取 $\beta_1 \approx \beta_2 \approx 1.0$）：

$$\sum F = p_1 A_1 - p_2 A_2 = \rho q_v (v_2 - v_1)$$

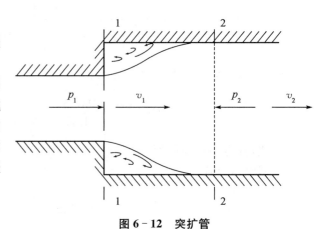

**图 6-12 突扩管**

将动量方程应用于截面 1—1 和截面 2—2,现在 $h_w = h_j$

$$\frac{p_1}{\rho g} + \frac{v_1^2}{2g} = \frac{p_2}{\rho g} + \frac{v_2^2}{2g} + h_j$$

由上述两个公式可知:

$$h_j = \frac{(v_1 - v_2)^2}{2g}$$

由连续性方程,我们可以推出 $v_2 = \frac{v_1 A_1}{A_2}$ 或 $v_1 = \frac{v_2 A_2}{A_1}$,代入上式可得:

$$h_j = \left(1 - \frac{A_1}{A_2}\right)^2 \frac{v_1^2}{2g} = \zeta_1 \frac{v_1^2}{2g} \tag{6-47}$$

或

$$h_j = \left(\frac{A_2}{A_1} - 1\right)^2 \frac{v_2^2}{2g} = \zeta_2 \frac{v_2^2}{2g} \tag{6-48}$$

式中,$\zeta_1$ 和 $\zeta_2$ 分别对应于突扩管前后两个断面的平均速度。

### 6.7.2 突缩管

图 6-13 为突缩管的流动图,从主流断面 1—1 过渡至收缩断面 $c$—$c$,再扩大至截面 2—2,局部损失主要由扩大引起,其值根据经验公式计算

$$h_j = \zeta \frac{v_2^2}{2g}, \quad \zeta = 0.5\left(1 - \frac{A_2}{A_1}\right) \tag{6-49}$$

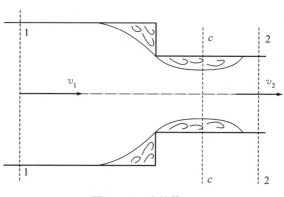

图 6-13 突缩管

### 6.7.3 管道出口和进口

(1) 出口损失系数 $\zeta$。当管道突然扩大到水池(或容器)时,$\frac{A_1}{A_2} \approx 0$(见图 6-14)。由公式 (6-49) 可知:

$$h_j = \zeta \frac{v^2}{2g}, \quad \zeta = 1.0$$

(2) 进口损失系数 $\zeta$。相当于 $A_1 \to \infty$ 突然缩小至 $A_2$,则有

$$h_j = \zeta \frac{v^2}{2g}$$

式中,$\zeta$ 随着靠近流线的入口形状度的增加而减小(见图 6-15)。

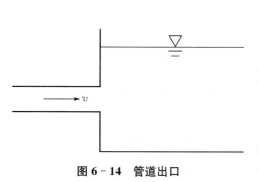

图 6-14 管道出口

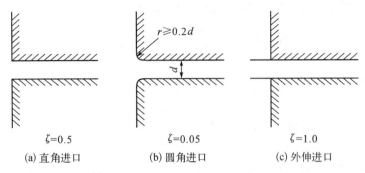

(a) 直角进口 $\zeta=0.5$　　(b) 圆角进口 $\zeta=0.05$　　(c) 外伸进口 $\zeta=1.0$

图 6-15　管道入口

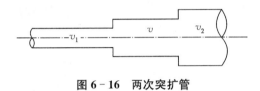

图 6-16　两次突扩管

**例 6-6**　在突扩管中，流速由 $v_1$ 变为 $v_2$，如果中间加一个中等粗细管形成两次突然扩大，忽略局部阻力的相互干扰，采用叠加法计算：(1) 当总局部水头损失最小时，中间管的流速是多少；(2) 计算总局部水头损失，然后与一次扩大时进行比较。

**解：**(1) 两次突扩管的局部水头损失

$$h_j = h_{j1} + h_{j2} = \frac{(v_1-v)^2}{2g} + \frac{(v-v_2)^2}{2g}$$

中间管的流速为 $v$，当总局部水头损失最小时：

$$\frac{\mathrm{d}h_j}{\mathrm{d}v} = 0$$

即

$$\frac{\mathrm{d}h_j}{\mathrm{d}v} = -\frac{2(v_1-v)}{2g} + \frac{2(v-v_2)}{2g} = 0$$

即

$$v = \frac{v_1+v_2}{2}$$

(2) 总局部水头损失

$$h_j = \frac{\left(v_1-\dfrac{v_1+v_2}{2}\right)^2}{2g} + \frac{\left(\dfrac{v_1+v_2}{2}-v_2\right)^2}{2g} = \frac{(v_1-v_2)^2}{4g}$$

一次突然扩大的局部水头损失 $h_j = \dfrac{(v_1-v_2)^2}{2g}$，当两次突然扩大时，总局部水头损失为一次突然扩大的 0.5 倍。

局部水头损失叠加原理只能应用于间隔足够长的局部损失段，如果各损失段间隔非常接近，就会发生相互干扰。则总水头损失约为单个正常局部损失总和的 0.5～3 倍。

## 6.8 边界层的概念

### 6.8.1 边界层

如图 6-17 所示，当均匀流动的流体以速度 $U_0$ 通过平板表面的前缘时，由于粘滞效应，靠近平板表面一层流体粒子将粘附在平板表面，并将速度降至零。外层的流体会受到这一层流体阻碍，使流速降低。距离平板表面越远，则流速降低越小，直到距离平板表面一定距离时，流速将接近原始速度 $U_0$。由于粘性的影响，板表面到未扰动流体之间的流速分布不均匀，流速呈梯度，也存在较大的切应力。这种不可忽视的靠近壁的粘合薄层称为边界层。实际上，边界层内、外区域并没有明显的分界面，在实际应用中规定：沿平板表面外法线速度达到势流速度的 99%（$u_x = 0.99\,U_0$）处的距离为边界层的厚度，用 $\delta$ 代表。边界层厚度 $\delta$ 顺着水流的逐渐增加，因为边界的影响随着边界的长度逐渐向流动区延伸。利用边界层的概念，流场可以分为两个区域来求解。一是边界层内的流动，流体粘度的影响必须包括在这一层内。但由于边界层较薄，可以简化 N-S 方程，利用动量方程得到近似解。第二个是在边界层外的流动，速度梯度为零，没有内摩擦力，因此可以看作是理想流体的流动，因此可以作为势流求解。

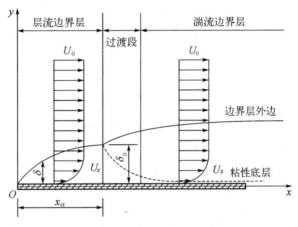

图 6-17 平板绕流

如图 6-17 所示，平板边界层的流动开始处于层流状态，边界层厚度逐渐增加。经过一个过渡段后，层流边界层将转化为湍流边界层。判别边界层内的流动是层流还是湍流的准则为雷诺数，其表达式为：

$$Re_x = \frac{U_0 x}{\nu} \tag{6-50}$$

距离板端越远，则雷诺数越大。当雷诺数达到一定的临界值时，流态则从层流转变为湍流。这种从层流向湍流转变的过渡点称为转捩点。其雷诺数的表达式为

$$Re_{cr} = \frac{U_0 x_{cr}}{\nu} \tag{6-51}$$

$Re_{cr}$ 称为临界雷诺数，其值的大小与流动的波动程度有关，波动较强，$Re_{cr}$ 较小。平板边界层内的流动，由层流转变为紊流的临界雷诺数 $Re_{cr} = 3 \times 10^5 \sim 3 \times 10^6$。

实验结果表明，用下述公式可以计算出板边界层的厚度。

层流边界层：

$$\delta = \frac{5x}{Re_x^{1/2}} \tag{6-52}$$

湍流边界层

$$\delta = \frac{0.377x}{Re_x^{1/5}} \quad (6-53)$$

边界层内的流动全部都是层流的边界层称为层流边界层，边界层内的流动开始是层流，随着 $Re_x$ 的增大逐渐转为湍流的边界层称为混合边界层。在湍流边界层内靠近平板的地方，还有一层薄膜，有较大的流速梯度，粘性切应力仍然起到主导作用，湍流的切应力可以忽略，使得流动形态仍为层流。这一层称为粘性底层。

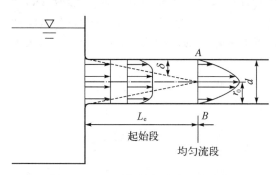

图 6-18 入口初始段边界层的变化

**推论** 根据平板边界层理论，圆管进口段的流速分布沿程发生变化。如图 6-18 所示，流体从水箱中通过光滑的圆形入口进入管道，开始时整个过水断面的流速几乎是均匀的。然而，随着边界层沿流动方向的发展，在边界附近的流速逐渐减小，在管道中心逐渐增大到最大值，流速分布不再发生变化。边界层厚度从入口到管道中心流速达到最大处的发展称为初始段。完整的湍流初始段长度经验值 $L_e = (50 \sim 100)d$，但一般认为，当 $L_e < (20 \sim 40)d$ 时，流速已经接近均匀。

### 6.8.2 边界层的分离

如上所示，当水沿壁面流动时，会产生边界层，厚度沿流动方向增加。在这个过程中，可能会产生边界层和过流壁分离的现象，这种现象称为边界层分离。

对于平板绕流，当压力梯度为 0，即 $\dfrac{dp}{dx}=0$ 时，无论平板多长都不会发生分离现象，边界层厚度只会沿着流动方向不断增加。但是，当边界沿流动方向扩散时，如图 6-19 所示，压力梯度为正，因此，$\dfrac{dp}{dx}>0$，边界层厚度迅速增加，导致边界层分离。边界层内水流的动能一方面要转化为逐渐增加的压力势能，另一方面要用于沿途消耗的粘滞阻力，靠近物体壁面的流体动能被消耗殆尽，这将导致边界层中的流体停滞。上游来流被迫离开固体边壁，从而产生分离。$C$ 点称为边界层的分离点。分离时形成的旋涡，不断地被主流带走，在物体后部形成尾涡区。在分离点 $C$ 处，有

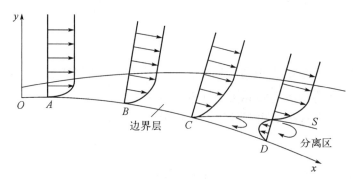

图 6-19 边界层分离

$$\left(\frac{\partial u}{\partial y}\right)_{y=0} = 0$$

$$\tau_0 = \mu\left(\frac{\partial u}{\partial y}\right)_{y=0} = 0$$

从 $C$ 点开始，在下游近壁处形成回流（或漩涡），通常把分离线与边界形成的区域称为尾流。尾流会使能量损失增加，压力降低，从而使绕流体前后形成较大的压降阻力。此外，回流还会引起基础冲刷和泥沙堆积。强烈的涡流紊动，可能会引起绕流体的随机振动破坏流体结构。尾流越大，后果就越严重。尾流减少的主要方法是使绕流体形态尽可能流线化。

边界层的概念最早是由普朗特在 1904 年提出的。边界层理论在流体力学的发展历史上，特别是在航空、船舶和流体机械的研究中具有重要意义。在水利工程中常见的管渠流动，除了入口部分外，几乎所有的流动区域都是边界层流动，因此，它不再分为边界层和外部区域，但仍需要将边界层的概念应用于过坝水流和阻力损失的分析。

## 6.9 阻力

### 6.9.1 绕流的特点

之前讨论过在固体边界内的流体，如管道、明渠的流动阻力和水头损失，都属于所谓的内部流体的问题。而绕流阻力是一个外部流体的问题，如给排水工程、各种闸墩水利工程、铁路和公路桥墩以及运输和国防工业等各种车身和飞行器的绕流问题，都属于外部流体的问题。

当流体与浸入流体中的固体进行相对运动时，流体对固体的作用力根据方向可分为两个分力：一是平行于流动方向的分力 $F_D$，称为绕流阻力，包括边界层粘性摩擦引起的摩擦阻力和边界层分离引起的形体阻力两部分。第二种是垂直作用于流动方向的力 $F_L$，称为升力。该力只能发生在非对称的绕流体中（图 6-20）。

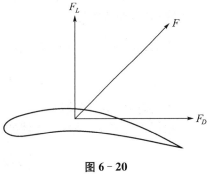

图 6-20

图 6-21a 为流线型绕流体，绕流阻力主要为摩擦力。图 6-21b 为圆柱型绕流体，绕流阻力主要为形体阻力。当比较相同的流量面积时。流线型的绕流阻力为圆柱形的 1/10。因此，流动阻力主要是由形体引起的。

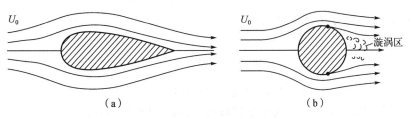

图 6-21 绕流

绕流运动也会使绕流体振动。当 $Re = \dfrac{U_0 d}{\nu}$ 达到一定值时，绕流体形成的边界层分离，形成两侧交替脱落的漩涡，被带到下游，排列成两排，称为卡门涡街。这种周期性的漩涡可以使流体受到横向力,产生绕流体横向振动。如果涡街频率和振动频率相同,就会发生共振,对建筑物造成损害。如拦污栅振动,电线风鸣都由此产生。

### 6.9.2 绕流阻力计算

1726 年,牛顿提出了绕流阻力 $F_D$ 的计算公式:

$$F_D = C_D \rho \frac{U_0^2}{2} A \tag{6-54}$$

式中,$C_D$ 为绕流阻力因数,其值主要由绕流体形态和雷诺数 $Re$ 决定;$C_D$ 可查有关图表;$\rho$ 为流体密度;$U_0$ 为相对速度;$A$ 为投影面积。

上述公式可应用于各种形态的绕流阻力。对于一个小的球体,绕流阻力因数可以由以下公式来确定:

$$C_D = \begin{cases} \dfrac{24}{Re} & (Re < 1.0) \\[2mm] \dfrac{13}{\sqrt{Re}} & (1.0 < Re < 10^3) \\[2mm] 0.44 & (10^3 < Re < 2 \times 10^5) \end{cases} \tag{6-55}$$

### 6.9.3 球体的沉降

我们使用绕流阻力来讨论流体中直径为 $d$ 的球体的沉降现象。

当球体开始下沉到流体中时,由于重力和浮力之差,球体的下沉速度逐渐增加。同时,绕流阻力也增大。当重力、浮力和粘性绕流阻力达到平衡,球体以均匀速度下沉时,这个速度称为沉降速度,简称沉速。

当 $Re = \dfrac{U_0 d}{\nu} < 1$ 时,绕流阻力可由公式(6-54)和(6-55)得到,其中 $U_0 = u_0$,

$$F_D = 3\pi \mu d u_0 \quad (Re < 1.0) \tag{6-56}$$

$U_0$ 为球体的沉降速度。

根据受力平衡的原理,球体的浮力和阻力应等于其重力。即

$$\frac{1}{6}\pi d^3 \cdot \rho g + 3\pi \mu d u_0 = \frac{1}{6}\pi d^3 \cdot \rho_s g$$

求解可以得到球体在流体中的极限沉速

$$u_0 = \frac{d^2}{18\mu}(\rho_s g - \rho g) \tag{6-57}$$

上式的 $\rho$ 和 $\rho_s$ 分别为流体的密度和球体的密度。

利用上述公式的分析可进行沉淀池的设计、疏浚、污水处理等。

# 习 题

1. 在明槽水流某点,激光测速仪(非常灵敏的流速仪)测速,每 0.5 s 取一值,结果如下表。

试求：(1) 湍流强度；(2) 湍流附加切应力。

| 流　速 | 1 | 2 | 3 | 4 | 5 | 6 | 7 | 8 | 9 | 10 |
|---|---|---|---|---|---|---|---|---|---|---|
| $u_x/(\mathrm{m \cdot s^{-1}})$ | 1.88 | 2.05 | 2.34 | 2.30 | 2.17 | 1.74 | 1.91 | 1.91 | 1.98 | 2.19 |
| $u_y/(\mathrm{m \cdot s^{-1}})$ | 0.10 | −0.06 | −0.21 | 0.19 | 0.12 | 0.18 | 0.21 | 0.06 | −0.04 | −0.10 |

2. 由式(6-30)证明，在图 6-22 所示很宽的矩形断面河道中，水深 $y'=0.63h$ 处的流速，等于该断面的平均流速。$\left(\text{提示：}\int_0^h \ln\frac{y}{h}\mathrm{d}\left(\frac{y}{h}\right)=-1\right)$

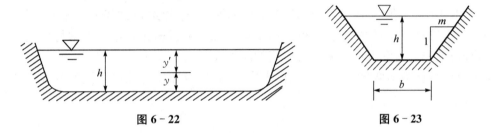

图 6-22　　　　　　　　　　图 6-23

3. 如图 6-23 所示，某梯形断面土渠中发生均匀流动，已知底宽 $b=2$ m，边坡因数 $m=1.5$，水深 $h=1.5$ m，底坡 $i=0.0004$，土壤的糙率 $n=0.0225$。试求：(1) 渠中流速 $v$；(2) 渠中流量 $q_v$。

4. 棱柱形混凝土渠均匀流，断面如图 6-24 所示，水力坡度 $J=1/800$，流量 $q_v=7.05\ \mathrm{m^3/s}$。试求：(1) 湿周上平均摩阻流速；(2) 糙率 $n$。

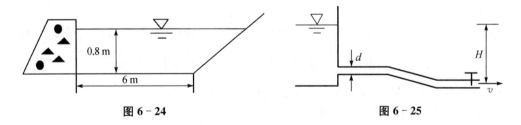

图 6-24　　　　　　　　　　图 6-25

5. 如图 6-25 所示的引水管，管径 $d=500$ mm，总长 $l=1200$ m，沿程阻力因数 $\lambda=0.015$，流量 $q_v=0.2\ \mathrm{m^3/s}$，局部阻力因数：进口 $\zeta_c=0.5$，每个弯头 $\zeta_b=0.8$，阀门 $\zeta_f=0.2$。试求所需的作用水头 $H$。

# 第7章 孔口出流和管嘴出流

液体从位于容器一侧的孔口中流出的现象称为孔口出流。当一根长度约为孔口尺寸3～4倍的管道连接到孔口时,该管道称为圆柱形外管嘴。当液体流过外管嘴并且在出口处形成一个满管流,它被称为管嘴出流。孔口出流和管嘴出流都是工程中常见的现象。给排水工程的取水口、泄水闸孔、孔板流量测量设备均为孔口出流。水流通过路基下的压力涵管、大坝排水管等现象与管嘴出流有关,而管嘴也适用于消防和水力机械化水枪。本章将应用前述的各种理论,分析孔口出流和管嘴出流水力计算的原理和方法。

本章要求掌握基本概念、公式、在工程中的应用,以及非恒定孔口出流和管嘴出流的水力计算方法。

## 7.1 孔口恒定出流

### 7.1.1 薄壁小孔口恒定出流

当孔口有锐缘时,孔壁和流过孔的水仅在轮廓处接触,这意味着孔口的厚度不会影响出流。这种孔被称为薄壁孔口,其形状可以是圆形、矩形、三角形或椭圆形。在这里,我们主要讨论圆形孔口。

根据直径 $d$(或高度 $e$)和孔口形心上的水头高度 $H$ 的比值,孔口可以分为大孔和小孔:$\frac{d}{H} < 0.1$ 为小孔;$\frac{d}{H} \geq 0.1$ 为大孔。对于小孔,由于直径 $d$ 远小于水头高度 $H$,我们可以假设在孔口的收缩断面上的各点的速度和水头高度都是相等的。

**1. 薄壁小孔口自由恒定孔口出流**

如图 7-1 所示,当水从孔口流入大气时,称为自由流出。当孔口形状为圆形时,水箱水位不变,水流自上游从多方向流向孔口,由于惯性作用,流线不能为折角,只能是光滑、连续的曲线,故孔口断面上的各流线不能平行。流出孔口后,水流继续收缩,直到距离孔口 $\frac{d}{2}$ 处,流线趋于平行。该断面称为收缩断面,如图 7-1 中的 $c'—c$ 面。收缩断面面积和孔口截面的比值为收缩系数,用 $\varepsilon$ 表示。$\varepsilon$ 与孔口的形状、孔口边缘情况和孔口位置有关。当孔口在壁面上的位置满足各边到侧壁的距离大于孔口尺寸(或直径)的3倍时,侧壁不能影响水流的收缩,这称为完全收缩。在实验中得出了完全收缩孔口的收缩系

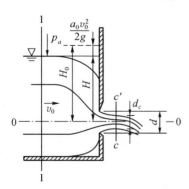

图 7-1 薄壁小孔口恒定自由出流

数，$\varepsilon = \dfrac{A_c}{A} = \left(\dfrac{d_c}{d}\right)^2 = 0.64$。

选择孔口形心位置所在的水平面作为基准水平 0—0，如图 7-1 所示，根据水箱内符合渐变流条件的断面 1—1 和 $c'$—$c$ 得到能量方程：

$$H + \dfrac{p_a}{\rho g} + \dfrac{\alpha_0 v_0^2}{2g} = 0 + \dfrac{p_c}{\rho g} + \dfrac{\alpha_c v_c^2}{2g} + h_w \tag{7-1}$$

当小孔口自由出流时，收缩断面的压强为大气压强，即 $P_c = P_a$。另一方面，水箱内水头的微小损失可以忽略，因此水头损失 $h_w$ 只是水流流经孔口的局部水头损失。

$$h_w = h_j = \zeta_0 \dfrac{v_c^2}{2g}$$

方程中的 $v_c$ 表示收缩断面处的平均速度。

假设 $H_0 = H + \dfrac{\alpha_0 v_0^2}{2g}$（称为孔口自由出流作用水头），$\alpha_c$ 取 1.0，则式(7-1)可以写成

$$H_0 = (1 + \zeta_0) \dfrac{v_c^2}{2g} \tag{7-2}$$

这样就可以得到计算收缩断面处的平均速度的方程

$$v_c = \dfrac{1}{\sqrt{1+\zeta_0}} \sqrt{2gH_0} = \varphi \sqrt{2gH_0} \tag{7-3}$$

式中，$H_0$ 表示作用水头；$\zeta_0$ 为水流流过孔口的局部阻力因数；$\varphi$ 为速度因数，$\varphi = \dfrac{1}{\sqrt{1+\zeta_0}}$。如果忽略水头损失，那么 $\zeta_0 = 0$，$\varphi = 1$，即速度因数 $\varphi$ 为收缩断面的实际流速 $v_c$ 与理想流速 $\sqrt{2gH_0}$ 的比值。根据实验数据，小孔的速度因数 $\varphi = 0.97 \sim 0.98$。所以我们可以得到局部阻力因数，即

$$\zeta_0 = \dfrac{1}{\varphi^2} - 1 = \dfrac{1}{0.97^2} - 1 = 0.06$$

假设孔口断面面积为 $A$，计算自由出流流量 $q_v$ 的方程为

$$q_v = A_c v_c = \varepsilon A \varphi \sqrt{2gH_0} = \mu A \sqrt{2gH_0} \tag{7-4}$$

方程中的 $\mu$ 表示孔口处的流量因数，$\mu = \varepsilon \varphi$。对于薄圆的小孔，$\mu = \varepsilon \varphi = 0.60 \sim 0.62$。如果孔板上游的水箱足够大，$v_0 \approx 0$，则 $H_0 = H$，且流速和流量方程可改写成

$$v_c = \varphi \sqrt{2gH} \tag{7-5}$$

$$q_v = \varepsilon A \varphi \sqrt{2gH} = \mu A \sqrt{2gH} \tag{7-6}$$

对于非完全收缩，如图 7-2 中的孔 1 和孔 2，收缩因数将由实验决定。

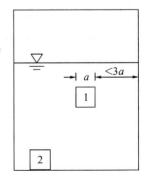

**图 7-2　非完全收缩**

**2. 薄壁小孔口恒定淹没出流**

水从孔口流出,然后进入水体的另一部分,称为淹没出流,如图7-3所示。和自由出流相同,由于惯性的影响,当水流流过孔口时,它会先收缩,然后扩大。假设上下游均匀流断面分别为1—1和2—2,孔口出流的收缩断面为c—c。图中的 $H_0$ 为

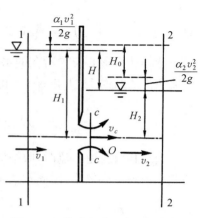

图7-3 薄壁小孔口淹没恒定出流

$$H_0 = \left(H_1 + \frac{\alpha_1 v_1^2}{2g}\right) - \left(H_2 + \frac{\alpha_2 v_2^2}{2g}\right) \tag{7-7}$$

称为孔口淹没出流作用水头,可以看出

$$H_0 = h_w = \sum h_j = (\zeta_0 + \zeta_s)\frac{v_c^2}{2g} \tag{7-8}$$

式中,$\zeta_0$ 为水流进孔口时的局部阻力因数;$\zeta_s$ 为水流出孔口时突然开始扩大时的局部阻力因数,当 $A_2 \gg A_c$,取 $\zeta_s \approx 1$。将它们代入上述的方程中,经过整理,我们可以得到淹没出流的流速和流量的计算方程如下:

$$v_c = \frac{1}{\sqrt{1+\zeta_0}}\sqrt{2gH_0} = \varphi\sqrt{2gH_0} \tag{7-9}$$

$$q_v = \varepsilon A \varphi \sqrt{2gH} = \mu A \sqrt{2gH_0} \tag{7-10}$$

如果孔口上下游的水箱足够大,则 $v_1 \approx v_2 \approx 0$,所以 $H_0 = H$,则方程可以改写为

$$v_c = \varphi\sqrt{2gH} \tag{7-11}$$

$$q_v = \varepsilon A \varphi \sqrt{2gH} = \mu A \sqrt{2gH} \tag{7-12}$$

$\mu$ 为淹没孔口出流的流量因数,其值可以等于自由出流的流量因数,即 $\mu = 0.60 \sim 0.62$。

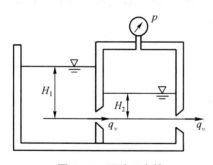

图7-4 双孔口水箱

比较方程(7-6)和(7-12),我们注意到它们的形式是完全相同的,因此它们的流量因数也是等价的。但我们必须注意到,孔口处的作用水头 $H$ 在自由出流时是上游水位与孔口形心之间的高度差,而淹没出流是上下游水位差。另外,淹没出流孔口断面上的每个点都有相同的作用水头,因此淹没出流的流速和流量与孔口的水下深度和孔的大小无关。

**例7-1** 在水箱上有两个相同的孔口,$H_1 = 6$ m,$H_2 = 2$ m,如图7-4所示。试计算出密封容器的表面压强 $p$。

**解:**

$$q_v = \mu A\sqrt{2g\left[(H_1 - H_2) - \frac{p}{\rho g}\right]} = \mu A\sqrt{2g\left(\frac{p}{\rho g} + H_2\right)}$$

$$H_1 - H_2 - \frac{p}{\rho g} = \frac{p}{\rho g} + H_2$$

$$\frac{p}{\rho g} = \frac{1}{2}(H_1 - 2H_2)$$

即

$$p = \frac{\rho g}{2}(H_1 - 2H_2) = \frac{9\,800}{2} \times (6 - 2 \times 2) = 9\,800 \text{ Pa}$$

### 7.1.2 薄壁大孔口自由出流

实验证明,流量方程(7-6)同样适用于大孔口。方程中的 $H$ 应为大孔口形心的水头高度($H_C$),如图 7-5 所示。但是,由于大孔口的收缩系数 $\varepsilon$ 较大,所以流量因数 $\mu$ 也应该较大。在大多数水利工程中,我们可以将闸口视为大孔口,流量因数的值取决于收缩情况,$\mu = 0.66 \sim 0.9$。

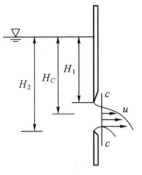

**图 7-5 薄壁大孔口自由出流**

## 7.2 管嘴出流

### 7.2.1 圆柱形外管嘴恒定出流

**1. 自由出流**

如图 7-6 所示,水流进入管嘴后,在入口边缘,水体形成收缩,与壁面分离,形成漩涡区。然后,径流通过收缩截面逐渐扩展到整个横截面。

水箱的水位保持不变,水压为大气压。管嘴流出为自由流出。

列出水箱过水断面 0—0 截面和 2—2 截面的能量方程

$$H + \frac{\alpha_0 v_0^2}{2g} = \frac{\alpha v^2}{2g} + h_w \qquad (7-13)$$

由于管嘴的长度很短,所以沿途的水头损失可以忽略不计。

$$h_w = \sum h_j = \zeta_n \frac{v^2}{2g}$$

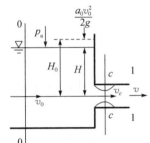

图 7-6 圆柱形外管嘴出流

如果 $H_0 = H + \dfrac{\alpha_0 v_0^2}{2g}$,则流速和流量公式如下:

$$v_c = \frac{1}{\sqrt{\alpha + \zeta_0}} \sqrt{2gH_0} = \varphi_n \sqrt{2gH_0} \qquad (7-14)$$

$$q_v = Av = A\varphi_n \sqrt{2gH_0} = \mu_n A \sqrt{2gH_0} \qquad (7-15)$$

式中,$\zeta_n$ 为管嘴局部阻力因数;$\varphi_n$ 为管嘴流速因数,$\varphi_n = \dfrac{1}{\sqrt{\alpha + \zeta_n}}$;$\mu_n = \varphi_n$。

对直角锐缘管嘴,由实验得

$$\varphi_n = 0.82$$

**2. 淹没出流**

由淹没出流分析可以看出,淹没管嘴出流的公式与公式(7-14)和(7-15)相同。

### 7.2.2 矩形锐缘进口处的圆柱形外管嘴真空度

在孔口外面增加管嘴后,阻力增加了,但流量也增加了。这是因为收缩断面出现了真空,

导致了断面的流速大于孔口自由出流时收缩断面的流速。取 $\alpha_c = \alpha = 1$，则

$$\frac{p_c}{\rho g} + \frac{v_c^2}{2g} = \frac{p_a}{\rho g} + \frac{v^2}{2g} + h_f \quad (7-16)$$

从连续性方程可以得到 $v_c = \dfrac{A}{A_c} v = \dfrac{1}{\varepsilon} v$，局部水头损失主要发生在水流的扩大过程中：

$$h_j = \zeta_{se} \frac{v^2}{2g}$$

$$\frac{p_c}{\rho g} = \frac{p_a}{\rho g} - \frac{v^2}{\varepsilon^2 2g} + \frac{v^2}{2g} + \zeta_{se} \frac{v^2}{2g} \quad (7-17)$$

对于圆柱形外管嘴，有 $v = \varphi_n \sqrt{2gH_0}$；考虑了突然扩大的局部阻力因数可得

$$\zeta_{se} = \left(\frac{A}{A_c} - 1\right)^2 = \left(\frac{1}{\varepsilon} - 1\right)^2$$

$$\frac{p_c}{\rho g} = \frac{p_a}{\rho g} - \left[\frac{1}{\varepsilon^2} - 1 - \left(\frac{1}{\varepsilon} - 1\right)^2\right] \varphi_n^2 H_0 \quad (7-18)$$

对于圆柱形外管嘴，有

$$\varepsilon = 0.64, \quad \varphi_n = 0.82$$

$$\frac{p_c}{\rho g} = \frac{p_a}{\rho g} - 0.75 H_0 \quad (7-19)$$

由式(7-19)可知，圆柱形外管嘴的真空度可以达到作用水头的 0.75 倍。因此，在相同的直径和相同的水头中，圆柱形外管嘴的流量大于孔口。

直角进口圆柱形外管嘴正常工况为：

(1) 水头 $H_0 \leqslant 9$ m；
(2) 管嘴长度 $l = (3 \sim 4) d$；
(3) 管嘴保持满管出流。

### 7.2.3 管嘴的分类

(1) 流线型管嘴(图 7-7b)；
(2) 圆锥形收缩管嘴(图 7-7c)；
(3) 圆锥形扩张管嘴(图 7-7d)。

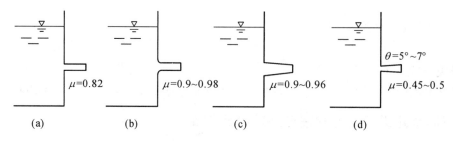

图 7-7 管嘴的分类

管嘴出流的所有基本公式与公式(7-14)和(7-15)相同。

## 7.3 孔口和管嘴的非恒定出流

### 7.3.1 孔口和管嘴非恒定出流容器排水时间计算

图 7-8 是一个面积为 $A_\Omega$ 的圆柱形容器,液体通过壁面自由出流。如果在出流过程中不向容器内添加液体,孔口水头就会逐渐减少。在某一时刻,孔口水头为 $H$,则在微小时间段内流过孔口的流体体积为:

$$q_v \mathrm{d}t = \mu A \sqrt{2gH}\, \mathrm{d}t$$

同时,容器液体在微小时间段 $\mathrm{d}t$ 内降低 $-\mathrm{d}H$。液体体积改变了 $-A_\Omega \mathrm{d}H$,其值等于孔口流出体积。

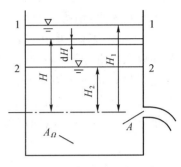

图 7-8 非恒定孔口自由出流

$$-A_\Omega \mathrm{d}H = \mu A \sqrt{2gH}\, \mathrm{d}t$$

$$\mathrm{d}t = -\frac{A_\Omega \mathrm{d}H}{\mu A \sqrt{2gH}}$$

孔口的作用水头由 $H_1$ 降到 $H_2$ 的始末时刻为 $t_1$ 和 $t_2$,则放水时间 $t$ 为

$$t = \int_{t_1}^{t_2} \mathrm{d}t = \int_{H_1}^{H_2} \frac{-A_\Omega}{\mu A \sqrt{2gH}} \mathrm{d}H = \frac{2A_\Omega}{\mu A \sqrt{2g}}(\sqrt{H_1} - \sqrt{H_2})$$

容器的清空($H_2 = 0$)时间 $t$ 为

$$t = \frac{2V}{\mu A \sqrt{2gH_1}} = \frac{2V}{q_{v\max}} \tag{7-20}$$

上述分析也适用于管嘴流出。

### 7.3.2 淹没孔口非恒定出流容器排水时间计算

图 7-9 显示,圆柱形容器内的上游水位固定,下游水位发生变化。流体通过壁面逐渐进入圆柱形容器,即淹没孔口非恒定出流。

该公式也可用前面类似的方法得到。

$$t = \frac{2V}{\mu A \sqrt{2gz_1}} = \frac{2V}{q_{v\max}} \tag{7-21}$$

该公式表明,在水头可变的情况下,孔流充满具有等横截面的圆柱形容器的时间是起始水平在恒定情况下的 2 倍多。

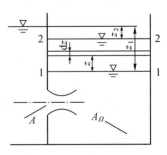

图 7-9 淹没孔口非恒定出流

## 习 题

1. 小孔口自由出流和淹没出流的流量计算公式有何不同?

2. 水位恒定的上、下游水箱,箱内水深为 $H$ 和 $h$。三个直径相等的薄壁孔口1、2、3位于隔板上的不同位置,淹没水中且均为完全收缩。试问:三孔口的流量是否相等?为什么?若下游水箱无水,情况又如何?

3. 薄壁锐缘圆形孔口,直径 $d=10\text{ mm}$,作用水头 $H=2\text{ m}$,不计行近流速。现测得收缩断面处水股直径 $d_c=8\text{ mm}$,在 32 s 时间内,经孔口流出的水量为 $0.01\text{ m}^3$。试求孔口出流的收缩系数 $\varepsilon$、流速因数 $\varphi$ 和流量因数 $\mu$,并计算孔口的局部阻力因数 $\zeta_0$。

4. 如图 7-10 所示,在混凝土坝身设一泄水管,管长 $l=4\text{ m}$,作用水头 $H=6\text{ m}$,现须通过流量 $q_v=10\text{ m}^3/\text{s}$。试确定管径 $d$,并求管中收缩断面处的真空度。可取流量因数 $\mu_n=0.82$。

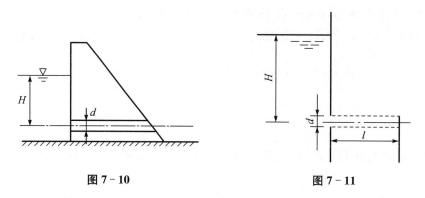

图 7-10    图 7-11

5. 如图 7-11 所示,薄壁容器侧壁上由一小孔口,其直径 $d=20\text{ mm}$,孔口中心线以上水深 $H=5\text{ m}$,试求孔口的出流流速 $v_c$ 和流量 $q_v$。倘若在孔口上外接一长 $l=8d$ 的短管,取短管进口损失系数 $\zeta=0.5$,沿程损失系数 $\lambda=0.02$,试求短管的出流流速 $v'$ 和流量 $q_v'$。

# 第 8 章 有压管道流动

在满管状态下,沿管道流动的流体称为管道中的有压管流。有压管流是生产和生活输水的重要组成部分,其水力性能的计算是实际工程中经常碰到的问题,如工业、农业和市政给水系统中的泵管、配水系统和水电站的引水管和水库的泄水管等。本章利用了前文所述的连续性方程,以及用于水力性能计算的能量方程和动量方程,来解决管道中流体流动的问题。为了方便计算,根据局部水头损失和沿程水头损失占总水头损失的百分比,将管道分为长管和短管。长管是沿程水头损失在管道中占主导地位,而局部水头损失可以忽略不计的管流。短管则是局部水头损失和沿程水头损失均占总水头损失的主要部分,计算时都不可忽略的管流。本章首先介绍了简单管道中不可压缩的恒定管流的水力计算;然后介绍了复杂管道,包括串联管、并联管和沿程均匀泄流管的不可压缩非恒定管流的水力计算。

本章学习要求掌握短管(虹吸管、水泵吸水管、有压涵管等)和长管(串联管和并联管)的水力计算方法及其水头线的绘制方法。

## 8.1 等径短管的水力计算

泵的吸水管、虹吸管、倒虹吸管、坝内排水管和铁路涵管等通常按短管计算。

### 8.1.1 基本公式

对于自由流出的短管,其沿程水头损失和局部水头损失均不可忽略。根据伯努利方程,总水头线 $H$ 和测压管水头线 $H_p$ 如图 8-1 所示。显然,管流的总沿程水头损失是每个流动段的损失之和,表示为 $h_\mathrm{f} = \sum \lambda \dfrac{l}{d} \dfrac{v^2}{2g}$;而总局部水头损失是各局部损失之和,表示为 $h_\mathrm{j} = \sum \zeta \dfrac{v^2}{2g}$。因此,水箱断面 1—1 至管道出口断面 2—2 的伯努利方程如下:

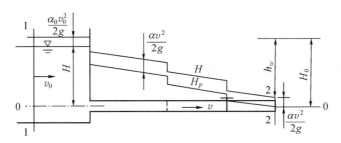

图 8-1 自由出流短管

$$H_0 = \frac{\alpha v^2}{2g} + h_w$$

取 $\alpha = 1$，则有

$$H_0 = \left(1 + \sum \lambda \frac{l}{d} + \sum \zeta\right) \frac{v^2}{2g} \tag{8-1}$$

式中，$H_0$ 为短管自由出流作用水头，$H_0 = H + \frac{\alpha_0 v_0^2}{2g}$，当 $v_0 \approx 0$ 时，$H_0 = H$；$\sum \zeta$ 为管道各局部阻力因数之和。

上述方程是等径短管自由出流的基本公式。

在短管淹没出流的情况下，如图 8-2 所示，同上述可得等径短管淹没出流的基本公式：

$$H_0 = \left(\sum \lambda \frac{l}{d} + \sum \zeta\right) \frac{v^2}{2g} \tag{8-2}$$

式中，$\sum \zeta$ 与自由出流相比，多含管道出口的水头损失因数，$\zeta_{出口} = 1$。

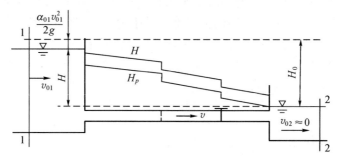

**图 8-2 短管淹没出流**

**推论** 对于非等径短管，容易推得自由出流的公式为

$$H_0 = \sum_{i=1}^{n} \left(\lambda_i \frac{l_i}{d_i} + \zeta_i\right) \frac{v_i^2}{2g} + \frac{\alpha_n v_n^2}{2g} \tag{8-3}$$

淹没出流的公式为

$$H_0 = \sum_{i=1}^{n} \left(\lambda_i \frac{l_i}{d_i} + \zeta_i\right) \frac{v_i^2}{2g} \tag{8-4}$$

式中，$i$ 指的是各管段编号，最后一段编号是 $i = n$。

### 8.1.2 水力计算

**1. 巴尔公式计算 $\lambda$**

替换巴尔公式(6-38)的最后一项为

$$\frac{5.128\,6}{Re^{0.89}} = 5.128\,6 \left(\frac{v}{v_d}\right)^{0.89} = 4.136\,5 \left(\frac{v_d}{q_v}\right)^{0.89}$$

则

$$\frac{1}{\sqrt{\lambda}} = -2\lg\left[\frac{\Delta}{3.7d} + 4.1365\left(\frac{v_d}{q_v}\right)^{0.89}\right] \tag{8-5}$$

上述方程可用于计算光滑区、过渡区和粗糙区。

**2. 曼宁公式计算 $C$**

将 $R = \dfrac{d}{4}$ 和曼宁公式 $C = \dfrac{1}{n} R^{1/6}$ 代入方程式(6-45)，有

$$\lambda = \frac{8g}{C^2} = \frac{12.7gn^2}{d^{1/3}} \tag{8-6}$$

由于引入了经验公式，量纲不和谐，单位以 m、s 计。上述方程仅适用于湍流粗糙区的计算。

大部分等径短管的水力计算是为了解决过流能力、管径或作用水头的问题，这三类问题可以通过上述方程联立或迭代解来解决，详见例 8-1 和例 8-2。如果结合伯努利方程，还可以解决过流断面上的压强问题。

**例 8-1** 堤坝上设置的钢筋混凝土倒虹吸管（表面较光滑），如图 8-3 所示，管长 $l = 200\,\text{m}$，上下游水位差 $H = 8\,\text{m}$，各局部损失因数进口 $\zeta_A = 0.5$，出口 $\zeta_D = 1.0$，弯头 $\zeta_B = \zeta_C = 0.1$，过水流量为 $q_v = 25\,\text{m}^3/\text{s}$，$\nu = 1.011 \times 10^{-6}\,\text{m}^2/\text{s}$，计算所需管径。

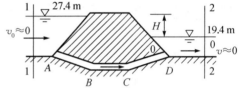

图 8-3 输水短管

**解：** $\sum \zeta = \zeta_A + \zeta_B + \zeta_C + \zeta_D = 0.5 + 0.1 + 0.1 + 1.0 = 1.7$

查表，可取钢筋混凝土管（表面较光滑）$n = 0.013$，由式(8-6)有

$$\sum \lambda \frac{l}{d} = \frac{12.7gn^2}{d^{1/3}} \frac{\sum l}{d} = \frac{12.7 \times 9.8 \times 0.013^2 \times 200}{d^{4/3}} = \frac{4.207}{d^{4/3}}$$

上式量纲为一。引入经验公式，量纲不和谐。

虹吸管为淹没出流，因 $v_0 \approx 0$，有 $H_0 = H$，由式(8-2)得

$$v = \sqrt{\frac{2gH}{\sum \lambda \dfrac{l}{d} + \sum \zeta}} = \sqrt{\frac{2 \times 9.8 \times 8}{4.207/d^{4/3} + 1.7}} = \frac{12.52}{\sqrt{4.207/d^{4/3} + 1.7}}$$

则有

$$d = \left(\frac{4q_v}{\pi v}\right)^{1/2} = \left(\frac{4 \times 25}{\pi} \times \frac{\sqrt{4.207/d^{4/3} + 1.7}}{12.52}\right)^{1/2}$$

$$= 1.594 \times \left(\frac{4.207}{d^{4/3}} + 1.7\right)^{1/4} \,(\text{m})$$

写成迭代解的形式为：

$$d_{(i+1)} = 1.594 \times \left(\frac{4.207}{d_i^{4/3}} + 1.7\right)^{1/4} \,(\text{m}) \quad (i = 0, 1, 2, \cdots)$$

设初值 $d_{(0)} = 1\,\text{m}$，代入上式得 $d_{(1)} = 2.485\,\text{m}$，再往复迭代，依次可得 $d_{(2)} = 2.089\,\text{m}$，

$d_{(3)} = 2.144$ m，$d_{(4)} = 2.135$ m，$d_{(5)} = 2.137$ m，…，可得收敛解 $d = 2.137$ m，取标准管径 $d = 2.25$ m。式(8-6)只适用于湍流粗糙区，故必须对流动所处的阻力区做出判别：

$$Re = \frac{4q_v}{\pi d \nu} = \frac{4 \times 25}{\pi \times 2.137 \times 1.011 \times 10^{-6}} = 1.47 \times 10^7$$

钢筋混凝土管，表面较光滑，查表，可取 $\Delta = 0.001$ m，则

$$\frac{\Delta}{d} = \frac{0.001}{2.139} = 0.000\,47$$

由流区判别式(6-41)，$Re = 1.47 \times 10^7 > 576.12 \left(\frac{\Delta}{d}\right)^{-1.119} = 3.05 \times 10^6$，为湍流粗糙区。说明满足式(8-6)的适用条件。

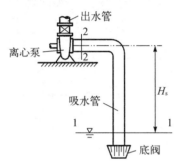

**图 8-4** 离心泵安装高度

**例 8-2** 如图 8-4 所示的离心泵，抽水流量 $q_v = 306$ m³/h，吸水管长度为 $l = 12$ m，直径 $d = 0.3$ m，沿程损失因数 $\lambda = 0.016$，局部损失因数：带底阀吸水口 $\zeta_1 = 5.5$，弯头 $\zeta_2 = 0.3$。允许吸水真空度 $[h_v] = 6$ m，试计算此水泵的允许安装高度 $H_s$。

**解：** 以吸水池水面为基准面，列吸水池水面 1—1 和水泵进口断面 2—2 的伯努利方程。

$$\frac{p_a}{\rho g} = H_s + \frac{p_2}{\rho g} + \frac{\alpha v^2}{2g} + h_w$$

式中

$$v = \frac{4q_v}{\pi d^2} = \frac{4}{\pi \times 0.3^2} \times \frac{360}{3\,600} = 1.20 \text{ m/s}$$

$$h_w = \left(\lambda \frac{l}{d} + \sum \zeta\right) \frac{v^2}{2g} = \left(0.016 \times \frac{12}{0.3} + 5.5 + 0.3\right) \times \frac{1.20^2}{2 \times 9.8} = 0.473 \text{ m}$$

再将 $\dfrac{p_a - p_2}{\rho g} = [h_v] = 6$ m，$\alpha = 1$ 代入上式，可得

$$H_s = \frac{p_a - p_2}{\rho g} - \frac{v^2}{2g} - h_w = 6 - \frac{1.20^2}{2 \times 9.8} - 0.473 = 5.45 \text{ m}$$

## 8.2 长管的水力计算

长管分为简单管路和复杂管路，管径和管内流量沿流程均不变的称为简单管路，其他管属于复杂管路。

### 8.2.1 长管简单管路

复杂管路水力计算的基础就是长管简单管路的计算，如图 8-5 所示，取基准面 0—0，对

断面 1—1 和断面 2—2 构建伯努利方程。忽略流速、水头和局部阻力，只考虑沿程水头损失，因此有

$$H = h_f$$

将 $h_f = \lambda \dfrac{l}{d} \dfrac{v^2}{2g}$，$v = \dfrac{4q_v}{\pi d^2}$ 代入上式，并令

$$S = \dfrac{8\lambda}{g\pi^2 d^5} \quad (8-7)$$

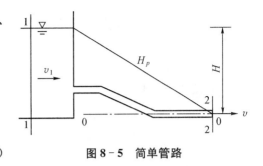

图 8-5　简单管路

则

$$H = Slq_v^2 \quad (8-8)$$

式中，$S$ 为比阻，指单位流量通过单位长度管道所需水头，其量纲为 $[T^2L^{-6}]$。

通过联立方程式(8-5)、(8-7)、(8-8)，得到简单管道的水力计算数学模型：

$$\begin{cases} H = Slq_v^2 \\ S = \dfrac{8\lambda}{g\pi^2 d^5} \\ \dfrac{1}{\sqrt{\lambda}} = -2\lg\left[\dfrac{\Delta}{3.7d} + 4.1365\left(\dfrac{\nu d}{q_v}\right)^{0.89}\right] \end{cases} \quad (8-9)$$

上述公式可用于计算湍流中的各阻力区。

如果管流属湍流粗糙区，可将式 $\lambda = \dfrac{12.7gn^2}{d^{1/3}}$ 代入式(8-7)，因此比阻 $S$ 可用下式计算：

$$S = \dfrac{10.3n^2}{d^{5.33}} \quad (8-10)$$

由于经验公式的引入，量纲不和谐，因此公式中的单位以 m、s 计。

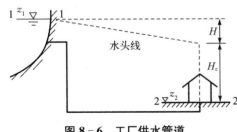

图 8-6　工厂供水管道

**例 8-3**　由水库向工厂供水，如图 8-6，采用铸铁管，$n = 0.013$。已知用水量 $q_v = 300 \text{ m}^3/\text{h}$，管道总长 2 500 m，库水面标高 $z_1 = 87$ m，工厂地面标高 $z_2 = 42$ m，管路末端用户所需要的服务水头（也称自由水头，是给水设计中规定的供水管末端在一定流量下仍保留的压力水头）$H_z = 25$ m，求输水管直径 $d$（流动为紊流粗糙区）。

$$q_v = \dfrac{300}{3\ 600} = 0.083\ 3 \text{ m}^3/\text{s}$$

$$H = z_1 - (z_2 + H_z) = 87 - (42 + 25) = 20 \text{ m}$$

按长管简单管路计算。因流动为湍流粗糙区，按式(8-8)与(8-10)，有

$$S = \dfrac{H}{lq_v^2} = \dfrac{20}{2\ 500 \times 0.083\ 3^2} = 1.153 \text{ s}^2/\text{m}^6$$

$$d = \left(\frac{10.3n^2}{S}\right)^{1/5.33} = \left(\frac{10.3 \times 0.013^2}{1.153}\right)^{1/5.33} = 0.296 \text{ m}$$

取标准管径，$d = 0.3$ m。

若按式(8-9)计算，有

$$d = \left(\frac{8\lambda l q_v^2}{g\pi^2 H}\right)^{1/5} \text{ 及 } \frac{1}{\sqrt{\lambda}} = -2\lg\left[\frac{\Delta}{3.7d} + 4.1365\left(\frac{\nu d}{q_v}\right)^{0.89}\right]$$

查表得，20 ℃时铸铁管当量粗糙度 $\Delta = 1.0 \times 10^{-3}$ m，$\nu = 1.01 \times 10^{-6}$ m²/s。上两式中只有 $\lambda$、$d$ 为未知量，可用迭代法求解。设初值 $d_{(0)} = 0.3$ m，由上两式可迭代求解得收敛解为 $\lambda = 0.0277$，$d = 0.288$ m。两种方法结果基本一致。

### 8.2.2 长管串联管路

如图8-7所示，为不同管径串联而成的沿流管路，沿程的流量可能相同，也可能不同。对于串联管道，有

$$\begin{cases} q_{v_1} = q'_{v_1} + q_{v_2} \\ q_{v_2} = q'_{v_2} + q_{v_3} \\ q_{v_i} = q'_{v_i} + q_{v_{i+1}} \end{cases} \quad (8-11)$$

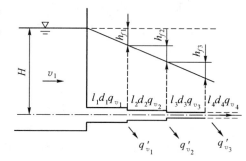

**图8-7 串联管道**

若共有 $k$ 个管段，每段管道较长，均可按长管简单管路计算，则：

$$H = \sum_{i=1}^{k} h_{fi} = \sum_{i=1}^{k} S_i l_i q_{v_i}^2 \quad (8-12)$$

对于流动处在粗糙区的串联管道，水力计算的数学模型由(8-10)、(8-11)和(8-12)联立而得，即：

$$\begin{cases} H = \sum_{i=1}^{k} S_i l_i q_{v_i}^2 \\ S_i = \dfrac{10.3 n_i^2}{d_i^{5.33}} \\ q_{v_i} = q'_{v_i} + q_{v_{i+1}} \end{cases} \quad (8-13)$$

式中，$i$ 为从上游到下游的管段编号，$q'_{v_i}$ 为 $i$ 段和管道的 $(i+1)$ 段之间连接节点的分支流量，$q_{v_{i+1}}$ 为管道的 $(i+1)$ 段中的过流量。

串联管道的计算通常针对水头 $H$、流量 $q_v$ 或管径 $d$ 等问题。在某些情况下，需要补充公式，以获得上述条件的解，详见第8.3.1节。

### 8.2.3 长管并联管道

在两个节点之间设置两个或多个管道的系统是并联管路，如图8-8所示。

并联管路支路的管径、长度和流量不一定相等。但它们都有相同的初始水头和管端保留水头,也就是说,每根管道的水头损失均相等,有

$$h_{f1}=h_{f2}=\cdots=h_{fk} \quad (8-14)$$

以比阻和流量表示即为

$$S_1 l_1 q_{v_1}^2 = S_2 l_2 q_{v_2}^2 = \cdots = S_k l_k q_{v_k}^2$$

任意管段的过流量(下标 $i$ 区分)为

$$q_{v_i} = q_{v_1}\left(\frac{S_1 l_1}{S_i l_i}\right)^{0.5} \quad (8-15)$$

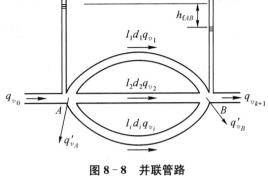

图 8-8 并联管路

假设有 $k$ 根并联管段,结合节点流量平衡条件,得到

$$A:\sum q_{vA}=0 \quad \text{或} \quad q_{v_0}-(q_{v_1}+q_{v_2}+\cdots+q_{v_k}+q'_{v_A})=0 \quad (8-16\text{a})$$

$$B:\sum q_{vB}=0 \quad \text{或} \quad q_{v_1}+q_{v_2}+\cdots+q_{v_k}-(q'_{v_B}+q'_{v_{k+1}})=0 \quad (8-16\text{b})$$

上述公式中减去的为节点的流出流量。

对于粗糙区的并联管路,其水力计算的数学模型是通过联立公式(8-8)、(8-10)、(8-15)和(8-16)得到的,即

$$\begin{cases} q_{v_i}=q_{v_1}\left(\dfrac{S_1 l_1}{S_i l_i}\right)^{0.5} \\ q_{v_0}=q'_{v_A}+q_{v_1}\sum\limits_{i=1}^{k}\left(\dfrac{S_1 l_1}{S_i l_i}\right)^{0.5} \\ S_i=\dfrac{10.3 n_i^2}{d_i^{5.33}} \\ h_{fAB}=S_i l_i q_{v_i}^2 \end{cases} \quad (8-17)$$

式中,$i$ 为并联管道的管段编号;$q_{v_0}$ 为上游节点流入并联管道的流量;$q'_{v_A}$ 为上游节点的分流流量。

该计算模型可以计算并联管道的流量分配等水力问题。

**例 8-4** 有一并、串联输水管路系统,$D$ 端为自由出流,如图 8-9 所示。已知管 $AB$ 流量 $q_{v_0}=0.2 \text{ m}^3/\text{s}$,管长 $l_0=500 \text{ m}$,管径 $d_0=0.35 \text{ m}$,并联管 $BC$ 分别有 $d_1=0.25 \text{ m}$,$d_2=d_3=0.2 \text{ mm}$,$l_1=800 \text{ m}$,$l_2=1\,000 \text{ m}$,$l_3=600 \text{ m}$,$l_4=300 \text{ m}$,$d_4=0.25 \text{ m}$,$B$ 点分流量 $q'_{v_1}=0.029\,5 \text{ m}^3/\text{s}$,$C$ 点分流量 $q'_{v_2}=0.070\,5 \text{ m}^3/\text{s}$。求管路 $AD$ 总作用水头 $H$、并联管的流量分配及 $BC$ 段的水头损失。(流动均处于紊流粗糙区,$n=0.013$)

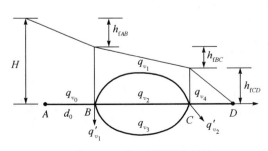

图 8-9 并、串联管路系统

**解**：(1) 先计算并联管路的流量分配

$$S_1 = \frac{10.3 n_1^2}{d_1^{5.33}} = \frac{10.3 \times 0.013^2}{0.25^{5.33}} = 2.82 \text{ s}^2/\text{m}^6$$

同理得

$$S_2 = S_3 = 9.25 \text{ s}^2/\text{m}^6$$

所以

$$q_{v2} = q_{v1} \left(\frac{S_1 l_1}{S_2 l_2}\right)^{0.5} = q_{v1} \left(\frac{2.82 \times 800}{9.25 \times 1\,000}\right)^{0.5} = 0.494 q_{v1}$$

同理得

$$q_{v3} = 0.638 q_{v1}$$

再将 $q_{v0} = 0.2 \text{ m}^3/\text{s}$，$q'_{v1} = 0.029\,5 \text{ m}^3/\text{s}$ 一并代入节点的流量平衡方程式(8-16)，即

$$q_{v0} - (q'_{v1} + q_{v1} + q_{v2} + q_{v3}) = 0$$

有

$$0.2 - 0.029\,5 - q_{v1}(1 + 0.494 + 0.638) = 0$$

解得：

$$q_{v1} = 0.08 \text{ m}^3/\text{s}, \quad q_{v2} = 0.039\,5 \text{ m}^3/\text{s}, \quad q_{v3} = 0.051\,0 \text{ m}^3/\text{s}$$

(2) 由串联管计算作用水头 $H$：

$$H = h_{fAB} + h_{fBC} + h_{fCD} = S_0 l_0 q_{v0}^2 + S_1 l_1 q_{v1}^2 + S_4 l_4 q_{v4}^2$$

式中，$S_0 = \dfrac{10.3 n^2}{d_0^{5.33}} = \dfrac{10.3 \times 0.013^2}{0.35^{5.33}} = 0.469$；$S_4 = S_1 = 2.82$；$l_0 = 500 \text{ m}$，$l_1 = 800 \text{ m}$，$l_4 = 300 \text{ m}$，$q_{v0} = 0.2 \text{ m}^3/\text{s}$，$q_{v1} = 0.08 \text{ m}^3/\text{s}$，$q_{v4} = q_{v0} - q'_{v1} - q'_{v2} = (0.2 - 0.029\,5 - 0.070\,5) = 0.1 \text{ m}^3/\text{s}$，代入上式得

$$h_{fAB} = 9.38 \text{ m}, \quad h_{fBC} = 14.44 \text{ m}, \quad h_{fCD} = 6.84 \text{ m}, \quad H = 30.66 \text{ m}$$

则总作用水头为 30.66 m，并联管路各支管水头损失均为 14.44 m。

### 8.2.4 沿程均匀泄流管路

上文讨论的管道流量在整个管段是恒定且集中在管段末端流出情况，将其称为传输流量。在实际工程中，如灌溉管路、人工降雨管路和滤池冲洗管路中，除了传输流量外，还有沿着管长从侧面不断连续向外泄出的流量 $q_{v_t}$，称为途泄流量(图 8-10)。通常沿程泄出的流量是非均匀的，其中每段单位长度上泄出的流量均相等的管路被称为沿程均匀泄流管路，是较为简单的。为了简化计算，这通常被看作是一个连续进行的过程。

如图 8-10 所示，假设沿途均匀泄流管段的长度为

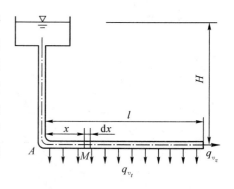

**图 8-10 沿程均匀泄流管路**

$l$,直径为 $d$,则总途泄流量为 $q_{v_t}$,在末端泻出传输流量为 $q_{v_z}$。在离泄流起始点 $A$ 距离为 $x$ 的 $M$ 点断面处,取长度为 $dx$ 的微小管段,且 $dx$ 非常小,可以认为其流量 $q_{v_x}$ 是恒定的,其水头损失可以通过简单管道来计算,即

$$q_{v_x} = q_{v_z} + q_{v_t} - \frac{q_{v_t}}{l} x$$

$$dh_f = S q_{v_x}^2 \cdot dx = S \left( q_{v_z} + q_{v_t} - \frac{q_{v_t}}{l} x \right)^2 dx$$

将上述微小流段的水头损失对整个管道进行积分,可得到整个管段的水头损失:

$$h_f = \int_0^l dh_f = \int_0^l S \left( q_{v_z} + q_{v_t} - \frac{x}{l} q_{v_t} \right)^2 dx$$

当管段的粗糙度和管径保持不变,且流动处于阻力平方区时,比阻 $S$ 是常数,对上式进行积分,可得

$$h_f = Sl \left( q_{v_z}^2 + q_{v_z} q_{v_t} + \frac{1}{3} q_{v_t}^2 \right) \tag{8-18}$$

可以近似写作:

$$h_f = Sl (q_{v_z} + 0.55 q_{v_t})^2 = Sl q_{v_c}^2 \tag{8-19}$$

公式中的 $q_{v_c}$ 称为计算流量:

$$q_{v_c} = q_{v_z} + 0.55 q_{v_t} \tag{8-20}$$

公式(8-19)与简单管道计算公式(8-8) $h_f = Sl q_v^2$ 一致,因此可根据简单管路计算沿途均匀泄流管路。

在流量 $q_{v_z} = 0$ 的特殊情况下,沿程均匀泄流管路的水头损失可通过式(8-18)得到

$$h_f = \frac{1}{3} Sl q_{v_t}^2 \tag{8-21}$$

上面的公式表明,当管道只有沿程均匀途泄流量时,水头损失是相同传输流量通过时的三分之一。

**例 8-5** 由水塔供水的输水塔如图 8-11 所示,由三段铸铁管组成,中段为均匀泄流管段。已知:$l_1 = 500$ m,$d_1 = 0.2$ m,$l_2 = 150$ m,$d_2 = 0.15$ m,$l_3 = 200$ m,$d_3 = 0.125$ m,节点 $B$ 分出流量 $q_v' = 0.01$ m³/s,途泄流量 $q_{v_t} = 0.015$ m³/s,传输流量 $q_{v_z} = 0.02$ m³/s。求需要的水塔高度(作用水头)。

**解**:将第 2 段管道的途泄流量转换为计算流量 $q_{v_c2}$,则各管段的流量为

$$q_{v1} = q_v' + q_{v_t} + q_{v_z} = 0.01 + 0.015 + 0.02$$
$$= 0.045 \text{ m}^3/\text{s}$$

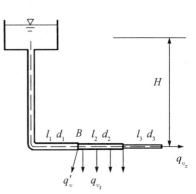

图 8-11 水塔供水管路

$$q_{vc2} = q_{vz} + 0.55 q_{vt} = 0.02 + 0.55 \times 0.015 = 0.028 \text{ m}^3/\text{s}$$

$$q_{v3} = q_{vz} = 0.02 \text{ m}^3/\text{s}$$

整个管路由三管段串联而成,因而作用水头等于各管段水头损失之和,铸铁管 $n = 0.013$,由式(8-13)有

$$H = \sum h_f = 10.3 n^2 \left( \frac{l_1 q_{v1}^2}{d_1^{5.33}} + \frac{l_2 q_{ve2}^2}{d_2^{5.33}} + \frac{l_3 q_{v3}^2}{d_3^{5.33}} \right)$$

$$= 10.3 \times 0.013^2 \times \left( \frac{500 \times 0.045^2}{0.2^{5.33}} + \frac{150 \times 0.028^2}{0.15^{5.33}} + \frac{200 \times 0.02^2}{0.125^{5.33}} \right) = 23.47 \text{ m}$$

## 8.3 管网水力计算的基础

管网是由各类串联管路和并联管路组成的,以城镇中的给水管网为典型代表。将给水管网中多个管段连接起来的管路称为管线;树状网指网络中在任意两个节点之间,只有一条管线;而网络中在任意两个节点之间,至少有两条管的,则称为环状网。而计算相应管网水力,必须满足以下条件:

(1) 连续方程:任一节点流出量和流入量,其代数和为零;
(2) 伯努利方程:任一管网中的闭合环内,其各个管段的水头损失代数和为零;
(3) 管网的流入总流量与所有节点流量总和相同。

以下将具体介绍树状网和环状网的具体水力计算。

### 8.3.1 树状网

计算树状管网的水力,要区分为新建和扩建两种工况。

**1. 新建给水管网的水力计算**

在给水管网设计中,管网布置完成后,各管段的长度、流量、节点流量和标高等均可获取,需要确定管道的直径和管网的水压。方法如下:

(1) 绘制简易管网图。实际管网非常庞大,没有必要甚至有时也不可能做到计算所有管道。因此,可以简化实际的管网。保留主要管线,而省略次要管线。但简化后的管网应该能够反映实际用水情况。

(2) 编号各管网节点和管段。参考如图8-12所示的树状管网,共有10个节点,在每个管段的起止节点上进行编号,例如"9,10"。使各段的流量、管道长度、比阻等参数均可以用编号加以区分。例如 $q_{v5}$ 代表节点5的流量,而 $q_{v9,10}$ 代表管段"9,10"的流量。

(3) 根据连续性方程计算各管段的流量

$$\sum \mp q_{v_{i,j}} - q'_{v_i} = 0 \qquad (8-22)$$

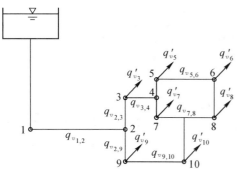

图8-12 树状管网

式中,流入为正,流出为负。根据已知 $q'_{v_i}$,从下游到上游计算 $q_{v_{i,j}}$,进而确定各个管段的流量。

(4) 确定树状管网干管及控制点。控制点通常选择远离水源地、地势高、水头要求大、需水量大的供水点;干管为从水源到控制点的管路;支管为从主管引出的管路。对于新建管网,应首先对干管水力计算。

(5) 按经济流速确定干管管径。在额定供水流量时,控制管道成本最低且技术合理时的流速,称为经济流速。经济流速主要取决于经济和技术的专业需求。在经济效益上,要考虑管道、泵站等的投资和运行成本;在技术上,供水管网防止高压引起的水击,通常限制最大流量小于 2.5~3.0 m/s,同时也为了避免杂质在水中沉积,控制最小流速不低于 0.6 m/s。确定经济流速比较复杂,在给水工程中,通常采用平均经济流速 $v_e$ 作为小型管路的经济流速。当 $d = 0.1 \sim 0.4$ m 时,$v_e = 0.6 \sim 0.9$ m/s;当 $d > 0.4$ m 时,$v_e = 0.9 \sim 1.4$ m/s。

而通过经济流速可以确定管径,公式如下:

$$d_{i,j} = \sqrt{\frac{4q_{v_{i,j}}}{\pi v_e}} \tag{8-23}$$

式中,$q_{v_{i,j}}$ 为 $i$ 和 $j$ 的管段流量。

(6) 计算长管各管段的水头损失,以及干线中按串联管路从起点到控制点的水头损失。

(7) 确定泵的扬程或水塔高度

$$H = \sum h_f + H_z + z_t - z_0 \tag{8-24}$$

式中,$H$ 为泵扬程或水塔距地面的高度;$\sum h_f$ 为管路系统的总水头损失;$H_z$ 为控制点所需的服务水头;$z_t$ 为控制点高程;$z_0$ 为管网起点的水塔标高或泵吸水池标高。

**2. 现有管网扩建水力计算**

与新建管网主管水力计算不同的是,在现有管网扩建(或新建支管)水力计算中已经确定了管道的起点压力,只需再确定管径即可。换言之,计算扩建管网或支管,要根据每个管段的总水头 $H$、管网布置图和流量,进而确定管径。方法如下:

(1) 计算各支管的平均水力梯度。在已知支管起点、终点水头和长度的情况下,可以求出各支管的平均水力梯度 $\bar{J}_{k,n}$:

$$\bar{J}_{k,n} = \frac{H_k - H_n}{l_{k,n}} \tag{8-25}$$

式中,$k$ 为某个支管起点的编号;$n$ 为同一支管终点的编号;$H_k$ 为起点水头;$H_n$ 为终点水头;给水工程中 $H_k$、$H_n$ 应相对于同一基准面,称为水压标高。

(2) 计算支管比阻值

$$S_{i,j} = \frac{\bar{J}_{k,n}}{q_{v_{i,j}}^2} \tag{8-26}$$

式中,$q_{v_{i,j}}$ 为在起点和终点编号为 $k$、$n$ 的支管中,管段编号为 $i$、$j$ 的流量。

(3) 根据各管段比阻值计算管径。计算管径不一定是标准管径,因此部分管段实际选择大于计算管径的标准管径,而另一些标准管径小于实际计算管径,使得总水力性能不变。

**例 8-6** 如图 8-13 为某开发区新建给水管网图。泵站吸水池水面高程 $z_0 = 150$ m,节点 2,4,6 标高分别为 $z_{t2} = 153$ m,$z_{t4} = 155$ m,$z_{t6} = 154.5$ m,节点 4,6 的服务水头分别为 $H_{z4} = 20$ m,$H_{z6} = 12$ m。其他已知值见计算表。试求水泵扬程 $H$ 及各管直径。已知管道 $n = 0.013$(泵吸水管水头损失忽略不计)。

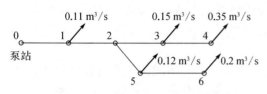

图 8-13 树状管网计算实例

**解:**(1)干管水力计算 以 0—1—2—3—4 的管线为干管,计算各管管径和水头损失。以 0—1 管段为例,选取经济流速 1.2 m/s,按经验,流动处于湍流粗糙区,则

$$d_{0,1} = \sqrt{\frac{4q_{v0,1}}{\pi v_e}} = \sqrt{\frac{4 \times 0.93}{\pi \times 1.2}} = 0.993 \text{ m}$$

取标准管径 $d_{0,1} = 1$ m,求其比阻及水头损失

$$S_{0,1} = \frac{10.3 n_{0,1}^2}{d_{0,1}^{5.33}} = \frac{10.3 \times 0.013^2}{1} = 0.001\ 74 \text{ s}^2/\text{m}^6$$

$$h_{f0,1} = S_{0,1} l_{0,1} q_{v0,1}^2 = 0.001\ 74 \times 1\ 000 \times 0.93^2 = 1.505 \text{ m}$$

其余各管段计算结果列于下表中。

| 管线 | 已知值 | | | 计算值 | | 水头损失 $h_f$/m |
|---|---|---|---|---|---|---|
| | 管号 | 管长 $l$/m | 通过流量 $q_v$/(m³·s⁻¹) | 管径 $d$/mm | 比阻 $S$/(s²·m⁻⁶) | |
| 干管 | 0, 1 | 1 000 | 0.93 | 1 000 | 0.001 74 | 1.505 |
| | 1, 2 | 800 | 0.82 | 900 | 0.003 05 | 1.642 |
| | 2, 3 | 500 | 0.5 | 700 | 0.011 65 | 1.456 |
| | 3, 4 | 1 000 | 0.35 | 600 | 0.026 5 | 3.245 |
| 支管 | 2, 5 | 500 | 0.32 | 450 | 0.117 2 | |
| | 5, 6 | 600 | 0.2 | 400 | 0.30 | |

水泵扬程 $H$ 为

$$H = \sum h_f + H_{z4} + z_{t4} - z_0 = (1.505 + 1.642 + 1.456 + 3.245) + 20 + 155 - 150 = 32.85 \text{ m}$$

(2)求支管管径。以支管"2,5"管段为例:

计算节点 2 的工作水头(水压标高)为

$$H_2 = H_{z4} + z_{t4} + h_{f4,3} + h_{f2,3} = 20 + 155 + 3.245 + 1.456 = 179.701 \text{ m}$$

节点 6 的工作水头为

$$H_6 = H_{z6} + z_{t6} = 12 + 154.5 = 166.5 \text{ m}$$

支管平均水力坡度为

$$\bar{J}_{2,6} = \frac{H_2 - H_6}{l_{2,6}} = \frac{179.701 - 166.5}{500 + 600} = 0.012$$

管段"2,5"的比阻为

$$S_{2,5} = \frac{\bar{J}_{2,6}}{q_{v2,5}^2} = \frac{0.012}{0.32^2} = 0.117\ 2\ \text{s}^2/\text{m}^6$$

管段"2,5"的管径为

$$d_{2,5} = \left(\frac{10.3n^2}{S_{2,5}}\right)^{\frac{1}{5.33}} = \left(\frac{10.3 \times 0.013^2}{0.117\ 2}\right)^{\frac{1}{5.33}} = 0.454 \text{ m}$$

取标准管径为 0.45 m,同理可计算管段"5,6"的直径,已列入表中。

## 8.3.2 环状网

环状管网是并联管路的延伸(如图 8-14 所示),其主要作用是提高给水管网的可靠性。根据求解条件,管网水力计算可分为三类:解环方程、解节点方程、解管段方程。解环方程方法如下。

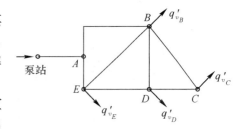

图 8-14 环状管网

环状管网的设计首先根据工程要求,并因地制宜铺设整个管网的管线,确定各段长度和各节点的引出流量。然后通过环状管网水力计算确定各管段流量 $q_v$、管径和相应水头损失,以获得给水系统所需的压力。其管径可由经济流速确定。根据环状管网流动的特点,管网上每个闭环的水流必须满足连续性和能量平衡两个条件。

(1) 节点上满足连续性条件,即流入节点的流量应该等于流出节点的流量:

$$\sum q_{v_i} = 0 \tag{8-27}$$

若设 $n_p$ 为节点数,由此可建立 $n_p - 1$ 个独立方程。其中 $n_p$ 节点的流量平衡方程不是独立方程,可从 $n_p - 1$ 个方程推导出。

(2) 能量平衡。在任何闭环管网中,假设顺时针方向的水头损失为正,逆时针方向的水头损失为负,则在每个环状管网中两者的代数和等于零:

$$\sum h_f = \sum S_i l_i q_{v_i}^2 = 0 \tag{8-28}$$

上面一共有 $n_k$ 个方程,加上方程(8-27),总数为 $n_q = n_k + n_p - 1$,分析管网图可知管网的个数为 $n_q$。如图 8-14 所示,环状管网数 $n_k = 3$,节点数 $n_p = 6$,管段数 $n_q = 3 + 6 - 1 = 8$。方程数与各管段流量未知数相同,可以进一步求解。

上述非线性方程难以直接求解各管段的流量,通常采用渐近法求解。首先,需要初步估计管道中的流量分配,但往往难以满足封闭条件,误差在所难免。因此需要校正初步评估的流量分配。假设环状管网内任何一管段的流量为 $q_{v_i}$,误差为 $\Delta q_{v_i}$,真实值 $h_{fi}$ 为

$$h_{fi} = S_i l_i (q_{v_i} + \Delta q_{v_i})^2 = S_i l_i (q_{v_i}^2 + 2 q_{v_i} \Delta q_{v_i} + \Delta q_{v_i}^2)$$

忽略二次微量 $\Delta q_v^2$，并认为 $\Delta q_v$ 对各个管道都是相同的，则可以得到公式：

$$\sum h_{fi} = \sum S_i l_i q_{v_i}^2 + 2\Delta q_v \sum S_i l_i q_{v_i} = 0$$

因此

$$\Delta q_v = -\frac{\sum S_i l_i q_{v_i}^2}{2\sum S_i l_i q_{v_i}} = -\frac{\sum h_{fi}}{2\sum S_i l_i \dfrac{q_{v_i}}{q_{v_i}^2}} = -\frac{\sum h_{fi}}{2\sum \dfrac{h_{fi}}{q_{v_i}}} \qquad (8-29)$$

如果计算结果为正，说明顺时针方向的管道流量增加 $\Delta q_v$，逆时针方向的流量减少 $\Delta q_v$。如果结果为负，则相反。

环状管网的水力计算方法有很多种，上述方法称为水头平衡法。采用水头平衡法计算环状管网的计算步骤如下：

(1) 初步估计管道流量，使节点满足式(8-27)的要求，即 $\sum q_{v_i} = 0$。

(2) 管径由经济流速 $v_c$ 通过公式 $d_{i,j} = \sqrt{\dfrac{4 q_{v_{i,j}}}{\pi v_c}}$ 确定，并根据计算值选择合适的标准管径。

(3) 根据预估流量计算各管道的水头损失（只计算沿途水头损失）。

(4) 检查环状管网是否满足式(8-28) $\sum h_{wi} = \sum h_{fi} = 0$。如果不满足，则根据公式(8-29)计算修正流量 $\Delta q_v$，修正初始估计流量 $q_v$。重复以上计算，直到误差达到要求的精度。

(5) 各节点水头由各管段水头损失计算确定，然后得到水泵扬程或水塔的水面高度。

### 8.3.3 泵和管路系统的水力特性

给水管路系统的实际水量与设计水量往往不同，而且由于水泵扬程随流量的变化，管路系统的水头损失也随流量而变化。将水直接泵入管路系统时，也存在流量与水压协调平衡的问题。这需要了解水泵的水力特性和管道系统的水力特性。以下作一简单介绍。

**1. 泵的水力特性**

水泵的流量 $q_v$ 与扬程 $H$、效率 $Y$、轴功率 $N$ 的关系曲线，分别称为水泵性能曲线。其中水泵的 $q_v$-$H$ 曲线称为泵的水力性能曲线，给水管网中常用的离心泵水力性能曲线如图8-15所示。在管网设计中，一般选择泵水力参数的最高效率作为管网的设计值。但是不同时期管网的实际用水量会有所不同，由图分析可知，离心泵抽出的水量越小，则压力越高。说明管网实际水头随用水量的变化而变化。

**2. 管路系统的水力特性**

管路系统流量与水头损失的关系曲线称为管道系统性能曲线。对于长管系统，一系列流量的水头损失值可由公式(8-8)计算得出，可画出曲线 $q_v$-$\sum h_f$，也可通过实验得到，如图8-16所示，在图中

$$H_g = H_z + (z_t + z_0)$$

式中，$z_t$ 为控制点标高，$z_0$ 为水泵吸水池水面高程。

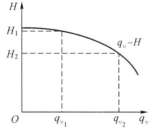

图 8-15 离心泵水力性能曲线

由图分析可知,对于一定的管道系统,流量越大,扬程损失越大,则要求泵提供更大的扬程作用。

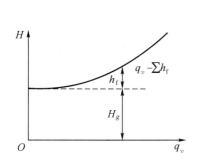

图 8-16 管路性能曲线

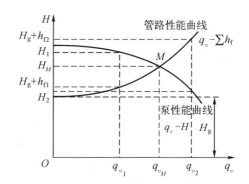

图 8-17 泵与管路系统的水力耦合关系

**3. 泵与管路系统的水力耦合条件**

将泵和管路系统的水力性能曲线以相同的比例绘制在同一坐标图上,曲线 $q_v$-$H$ 和曲线 $q_v$-$\sum h_f$ 会有一个交叉点 $M$,如图 8-17 所示。在图中与 $M$ 点相交的条件下,泵的扬程刚好满足管道系统的损失水头、控制点的服务水头和给水的几何给水高度。这个条件就是泵管路系统的设计工况,$q_{v_M}$ 为设计流量,$H_M$ 为设计水头。

给水管网控制节点一般都设有出口控制阀,可调节用水流量。假设调节阀的水头损失不包含在管路的性能曲线中,即调节阀门时管路的性能曲线不能改变。下面分析调节水量时管内压力的变化。阀门调小时,管道流量减少,泵扬程和泵出口压力增加,而管道损失水头减少,使管道末端控制点的剩余水头提高;如果管道流量增加,则结果相反。

综合以上分析,对于给水管网,考虑到实际流量的较大变化,又需要管网压力稳定,因此选择水力性能曲线比较平坦的水泵就可以满足相应要求。

## 8.4 有压管道中的水击

前文讨论的是与时间 $t$ 无关的恒定流。在本节中将讨论与 $s$ 和 $t$ 相关的有压管的非恒定流。

### 8.4.1 水击现象

**1. 水击传播过程**

如图 8-18 所示,有压管道与水库相连,尾部的阀门用于调节管道流量。管道长度为 $l$,直径为 $d$。正常情况下,阀门开度保持不变,管道内为恒定流,其流量、流速和阀门压力分别为 $q_v$、$v_0$ 和 $p_0$。

当阀门关闭时,阀门内会产生水击升压 $\Delta p$,并以速度 $c$ 在管道中传播。水击波传播的一系列过程可分为四个阶段。由于水击现象,弹性力和惯性力

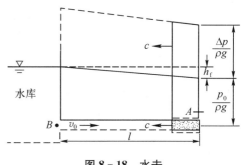

图 8-18 水击

起主要作用,在分析水击现象时,水击波传播循环特性不受影响,水头损失和流速可以忽略不计。

(1) $0 < t \leqslant \dfrac{l}{c}$,增压逆波阶段(图 8-19a)

$\dfrac{l}{c}$ 代表水击波通过管道全长 $l$ 的时间。$0 < t \leqslant \dfrac{l}{c}$ 代表了从阀门前端到进口 $B$ 的水击波阶段。

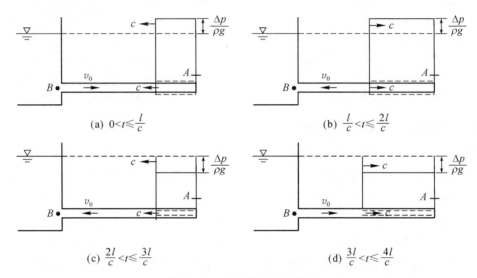

图 8-19 有压管道中的水击

当阀门 $A$ 关闭时,管道中靠近阀门的微小水层停止流动,其长度为 $dl$。流速从 $v_0$ 减小到 0,流体被惯性力压缩,使压力从 $p_0$ 增加到 $p_0 + \Delta p$。在突然增加的压力的作用下,微小水层被压缩并且密度增加,同时管壁会膨胀。上述现象称为水击波的传播现象。

这种水击波称为压力波。

(2) $\dfrac{l}{c} < t \leqslant \dfrac{2l}{c}$,减压顺波阶段(图 8-19b)

由于水库水位不会因管道内水击而变化,当 $t = \dfrac{l}{c}$ 时,管道入口断面 $B$ 左侧的压力维持在恒定流时的 $p_0$,右侧压强为水击冲击后的 $p_0 + \Delta p$,这表明水体不平衡。由于压力差,管网中的水从静止状态流向水库的方向,流速为 $v_0$。所以 $B$ 点的水体先恢复到原来的状态,压力从 $p_0 + \Delta p$ 下降到 $p_0$。然后水和管壁一层层恢复到原来的状态。压力也降低到 $p_0$,这也是一个水击,由管道入口传递到阀门,被称为减压波。

由于水的可压缩性和管壁弹性一致,反射波的速度也为 $c$,即当 $t = \dfrac{2l}{c}$,反射波到达下游阀门。通常这个时间被称为水击相长,用 $T_r$ 表示,也就是 $T_r = \dfrac{2l}{c}$。从 $t = 0$ 到 $t = \dfrac{2l}{c}$ 这段时间称为水击的首相。

(3) $\dfrac{2l}{c} < t \leqslant \dfrac{3l}{c}$,减压逆波阶段(图 8-19c)

第一阶段结束时,所有参数均已恢复。由于惯性作用,阀门 $A$ 附近的水层以速度 $v_0$ 流向水库。但由于阀门完全关闭,水无法补充,所以靠近阀门的水层有远离阀门的趋势。由于连续性的要求,水流停止,流速从 $v_0$ 减小到 0,压力从 $p_0$ 减少了 $\Delta p$,使水体膨胀,密度降低,管壁收缩。这些现象是阀门逐层以相同的方式通过管道入口,即在阀门处产生了降压反射波。在 $t = \dfrac{3l}{c}$ 时,整个水体处于静止状态,管壁收缩,压强从 $p_0$ 减少了 $\Delta p$。

(4) $\dfrac{3l}{c} < t \leqslant \dfrac{4l}{c}$,增压顺波阶段(图 8 - 19d)

在 $t = \dfrac{3l}{c}$ 时,整个水体处于静止、减压和膨胀的状态。管道 $B$ 点左侧压强为 $p_0$,右侧压强为 $p_0 - \Delta p$,受力不平衡。在压力差 $\Delta p$ 的作用下,水以 $v_0$ 的流速流向阀门,膨胀的水体被压缩,压力恢复到 $p_0$,收缩的管壁立即恢复。直到 $t = \dfrac{4l}{c}$ 时,顺波传到阀门上,整个管道的流量、压强和管壁都恢复到水击之前的状态,整个管道的流速为 $v_0$,压力为 $p_0$。

从 $t = \dfrac{2l}{c}$ 到 $t = \dfrac{4l}{c}$ 这段时间被称为水击末相。

综上所述,阀门的快速启闭是水击发生的一个重要原因。由于水的压缩性和管壁的柔韧性,不可能在整个管子内同时产生水击压力,而是一个传播过程,这使得水击压力受到限制。

**2. 直接水击与间接水击**

前文讨论的是,假设阀门突然关闭而发生在简单有压管道上的水击。事实上,阀门的开启和关闭不是瞬时的,需要一段时间 $t_c$。在关闭阀门的过程中,阀门处的压强随之上升,可以看作是阀门不断产生水击增压逆波,并传播到水库,当第一个增压逆波历时 $t = \dfrac{l}{c}$ 后到达水库时,立即反射回一个减压顺波。当减压顺波遇到其他压力波时,必定会降低管内压强。因此,水击分为直接水击和间接水击。

正常情况下,当管道阀门突然关闭时,管道内的流速迅速下降,水击压强明显增加,称之为正水击,其会导致水管爆裂。而管道阀门快速开启,管道内流量迅速增加,水击压强明显降低,这一水击称为负水击。负水击会导致管道中产生真空和气蚀,甚至导致管道凹陷。

## 8.4.2 水击的计算

**1. 水击波的传播速度**

$$c = \dfrac{c_0}{\sqrt{1 + \dfrac{K_0}{K}\dfrac{D}{\delta}}} = \dfrac{1\,435}{\sqrt{1 + \dfrac{K_0}{K}\dfrac{D}{\delta}}} \text{(m/s)} \qquad (8-30)$$

式中,$c_0$ 为声波在水中的速度,水温 5 ℃ 左右,压强在 1~25 个气压之间,$c_0 = 1435$ m/s;$K_0$ 为水的体积模量,水温 5 ℃ 左右时,$K_0 = 2.06 \times 10^3$ MPa;$K$ 为管材的弹性模量;$D$ 为管径,m;$\delta$ 为管壁厚度,m。

**2. 水击压强的计算**

设定阀门瞬时关闭,即 $v = 0$,直接水击压强最大值的计算公式为

$$\Delta p = \rho c v_0 \tag{8-31}$$

或

$$\frac{\Delta p}{\rho g} = \frac{c v_0}{g} \tag{8-32}$$

间接水击是由于正水击波和反射波的相互作用，难以计算，近似公式为

$$\Delta p = \rho c v_0 \frac{T_r}{T_z} \tag{8-33}$$

或

$$\frac{\Delta p}{\rho g} = \frac{c v_0}{g} \frac{T_r}{T_z} = \frac{v_0}{g} \frac{2l}{T_z} \tag{8-34}$$

式中，$T_z$ 为阀门关闭的时间；$v_0$ 表示水击前的平均速度；$T_r = \dfrac{2l}{c}$ 表示水击波相长。

### 8.4.3 水击危害的预防

(1) 设置空气室或安装水击消除阀。
(2) 设置调压塔或调压井。
(3) 延长阀门关闭时间，缩短管道长度，降低管内流速，都是预防水击的有效方法。

## 习 题

1. 如图 8-20 所示，两水池间水位差 $H=8$，如在两水池间布置两条平行且标高相同的管道，其中一条管道直径 $d_1=50$ mm，另一条管道直径 $d_2=100$ mm。两管道长度相等，即 $l_1=l_2=30$ m。(1) 试求每条管道所通过的流量。(2) 改为布置一条管道，长度不变且要求通过的总流量亦不变，求该管道的直径。设每条管道中所有的局部水头损失因数为 $\sum \zeta = 0.5$，管道的沿程阻力因数均为 $\lambda = 0.02$。

2. 一条管道在某处分为 $A$、$B$ 两条分支管道，然后再重新汇合，$A$ 管为镀锌管，$n=0.011$，长 1 500 m，管径 150 mm。$B$ 管为铸铁管，$n=0.013$，管径 200 mm。要求分配 $A$、$B$ 两管所通过的流量相等，试确定 $B$ 管的长度。

3. 如图 8-21 所示，由 $A$、$B$ 两水池，通过管道 $AE$、$BE$ 流经 $E$ 处汇合，再由 $EC$ 管道向 $C$ 处供水，并在该处流入大气中。$A$ 水池水位为 36 m，$B$ 水池水位为 40 m，供水点 $C$ 处高程为 10 m，要求计算各管所通过的流量。

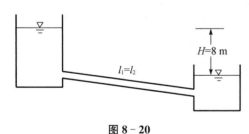

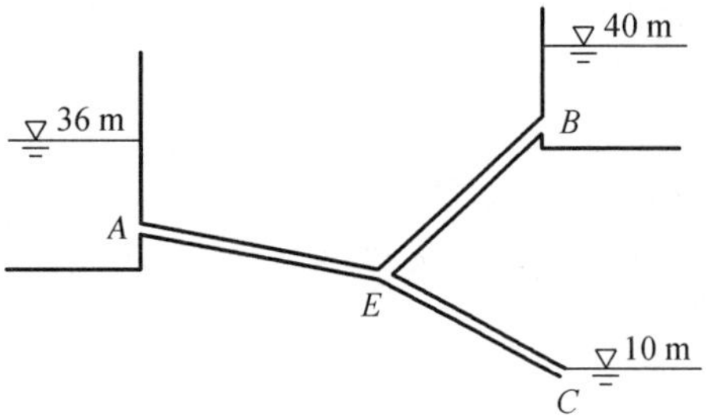

图 8-20

图 8-21

各管道资料如下表所示：

| 管道编号 | 长度/m | 直径/mm | $\lambda$ |
| --- | --- | --- | --- |
| AE | 300 | 300 | 0.024 |
| BE | 450 | 375 | 0.02 |
| CE | 60 | 450 | 0.024 |

4. 枝状供水管网如图 8-22 所示，已知水塔地面标高 $z_A = 15$ m，管网终点 $C$ 和 $D$ 点的标高 $z_C = 20$ m，$z_D = 15$ m，服务水头 $H_z$ 均为 5 m，$q'_{vC} = 0.02$ m³/s，$q'_{vD} = 0.0075$ m³/s，$l_1 = 800$ m，$l_2 = 700$ m，$l_3 = 500$ m，$n = 0.013$。试设计水塔高度和 $AB$、$BC$、$BD$ 段管径。

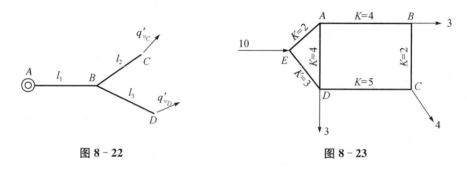

图 8-22　　　　　　　　图 8-23

5. 如图 8-23 所示，有一四边形管网 $ABCD$ 和三角形管网 $ADE$，两管网由 $AD$ 连接。在 $E$ 点有 10 个单位流量流入，在 $B$、$C$、$D$ 三点分别有 3、4、3 个单位流量流出。有关管道的其他资料，均标明于图上。试求各管段所通过的流量，并在图中标明流量的方向（图中 $K = Sl$，即 $h_f = Kq_v^2$）。

# 第 9 章 明 渠 流

人工渠道、天然河道以及未充满水流的管道等统称为明渠。明渠流是一种具有自由表面的流动,自由表面上各点受当地大气压的作用,其相对压强为零,所以又称为无压流动。与有压管流不同,重力是明渠流的主要动力,而压力是有压管流的主要动力。明渠水流根据其水力要素是否随时间变化分为恒定流和非恒定流。明渠恒定流又根据流线是否为平行直线分为均匀流和非均匀流。明渠流动理论将为输水、排水、灌溉渠道的设计和运行控制提供科学依据。

本章主要介绍明渠恒定均匀流的形成条件和特征以及明渠均匀流的水力计算,渠道底坡、允许流速、水力最优断面、水力最优断面条件、明渠均匀流基本公式;无压圆管均匀流水力特征、水力要素、充满度及无压圆管均匀流水力计算。而后讨论明渠恒定非均匀流的三种流态——缓流、临界流和急流,以及三种流态的判别;水跃和水跌的概念以及棱柱形渠道非均匀渐变流水面曲线的分析。教学重点在于明渠恒定均匀流的基本概念和水力计算,明渠恒定非均匀流的三种流态的概念;教学难点在于棱柱形渠道非均匀渐变流水面曲线的分析。

## 9.1 明渠分类

### 9.1.1 棱柱渠和非棱柱渠

明渠可分为棱柱形和非棱柱形。棱柱渠的断面形状和尺寸均保持不变,其断面面积只随深度的变化而变化。非棱柱渠的断面形状和尺寸会不断变化,其断面面积不仅随深度的变化而变化,而且还会随位置改变。

明渠断面形式多样,天然河道的断面一般为不规则形状,而人工渠道的断面均为规则形状,主要为矩形、梯形和圆形等,如图 9-1 所示。

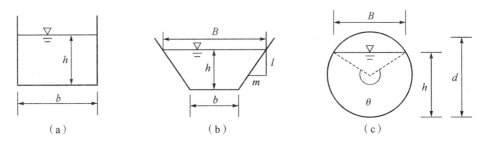

**图 9-1 人工渠道断面形状**

## 9.1.2 顺坡渠、平坡渠和逆坡渠

明渠底部一般会略微向下倾斜，渠道底部与纵剖面间会有一条交线，称为底线。底线沿流程在单位长度内的高程差称为渠道底坡或纵坡，即

$$i = \frac{\Delta z}{l'} = \sin\theta \tag{9-1}$$

式中，$\theta$ 为渠底线与水平线的夹角，如图 9-2 所示。

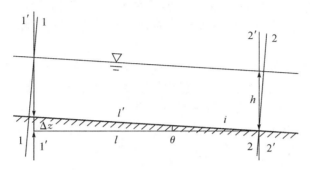

图 9-2 渠道底坡

在实际工程中，渠道底坡一般都很小，有 $\sin\theta \approx \tan\theta$，所以

$$i = \tan\theta = \frac{\Delta z}{l} \tag{9-2}$$

如图 9-3 所示，根据渠道底坡的数值不同，渠道通常可分为顺坡渠、平坡渠和逆坡渠。明渠渠底沿流程降低的称为顺坡渠（$i>0$），明渠渠底高程沿流程不变的称为平坡渠（$i=0$），明渠渠底沿流程升高的称为逆坡渠（$i<0$）。

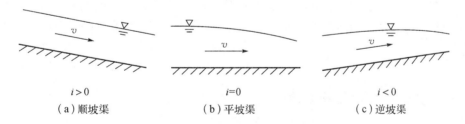

(a) 顺坡渠　　(b) 平坡渠　　(c) 逆坡渠

图 9-3 顺坡渠、平坡渠和逆坡渠

## 9.2 明渠均匀流

### 9.2.1 明渠均匀流的特性和条件

**1. 明渠均匀流的水力特性**

明渠均匀流同时具有均匀流和重力流的特征，其流线是相互平行的直线，所有液体质点都沿着相同的方向作匀速直线运动，在水流方向上所受到的合外力为零。由此可以推知明渠均匀流具有以下特征：

(1) 运动要素沿程不变,即明渠中水深、流速分布、断面平均流速、流量等均沿程保持不变。

(2) 断面压强符合静水压强的分布规律。

(3) 总水头线、水面线(即测压管水头线)和渠底线三线为相互平行的直线,所以水力坡度$J$、水面坡度$J_p$和渠道底坡$i$三者沿流程不变且相等。

(4) 明渠均匀流就是重力在流动方向上的分力与液流阻力相平衡的流动。

**2. 明渠均匀流的形成条件**

由于明渠均匀流具有上述特性,它的形成就需要满足一定的条件。

(1) 渠道必须是长而直的棱柱形渠道,断面形状和尺寸、渠道底坡$i$和粗糙系数$n$要沿程不变。

(2) 渠道必须是顺坡渠($i>0$),平坡渠($i=0$)和逆坡渠($i<0$)中都不能形成均匀流。

(3) 明渠中的水流必须是恒定的,流量保持不变,沿程没有水流流出或汇入。

(4) 渠道中无闸、坝或跌水等建筑物的局部干扰。

上述条件显然只有在人工渠道中才有可能满足,天然河道的断面形状和尺寸、坡度、粗糙系数一般沿程改变,所以不会产生均匀流。

### 9.2.2 明渠均匀流的基本公式

明渠均匀流的水力计算公式为谢才公式,即

$$v = C\sqrt{RJ} = C\sqrt{Ri} \tag{9-3}$$

根据连续性方程和谢才公式,得到明渠均匀流的流量公式:

$$q_v = AC\sqrt{RJ} = AC\sqrt{Ri} = K\sqrt{J} = K\sqrt{i} \tag{9-4}$$

式中,$K = AC\sqrt{R}$,称为流量模数,具有流量的量纲,它表示在一定断面形状和尺寸的棱柱形渠道中,当底坡$i$等于1时通过的流量;谢才系数$C$通常采用曼宁公式来确定,即$C = \dfrac{1}{n} R^{1/6}$。

### 9.2.3 水力最佳断面和允许流速

**1. 水力最佳断面**

从明渠均匀流的计算公式可知,明渠的输水能力取决于明渠断面的形状尺寸、底坡和粗糙系数的大小。在设计渠道时,底坡一般依照地形条件而定,粗糙系数取决于渠道的土质、护面材料及维护情况。从设计的角度考虑,希望在一定的流量下,能得到最小的过流断面面积,以减少工程量,节省投资;或者是在过流断面面积、粗糙系数和底坡一定的条件下,使渠道所通过的流量最大。凡是符合这种条件的断面形式就称为水力最佳断面。

将曼宁公式代入基本公式(9-4),有

$$q_v = AC\sqrt{Ri} = A\frac{1}{n}R^{2/3}i^{1/2} = \frac{1}{n}\frac{A^{5/3}}{\chi^{2/3}}i^{1/2} \tag{9-5}$$

上式表明,当$i$、$n$和$A$一定时,湿周$\chi$越小,其过水能力越大。当几何面积一定时,润周最小,水力半径最大的形状是圆形。对于明渠,半圆形断面是水力最优的,但半圆形断面不易施工,仅在混凝土制作的渡槽等水工建筑物中使用。

在工程中,明渠一般为梯形断面,其边坡系数 $m$ 是由边坡稳定要求和施工条件决定的。根据水力最佳断面的定义,在断面面积一定且湿周最小时,通过的流量最大,即

$$\frac{\mathrm{d}A}{\mathrm{d}h}=0 \text{ 且 } \frac{\mathrm{d}\chi}{\mathrm{d}h}=0$$

对于梯形明渠,其横截面积为

$$A=h(b+mh)$$

$$b=\frac{A}{h}-mh \tag{9-6}$$

梯形明渠的湿周为

$$\chi=b+2h\sqrt{1+m^2}=\frac{A}{h}-mh+2h\sqrt{1+m^2} \tag{9-7}$$

对水深 $h$ 求导有

$$\frac{\mathrm{d}\chi}{\mathrm{d}h}=-\frac{A}{h^2}-m+2\sqrt{1+m^2}=-\frac{b+mh}{h}-m+2\sqrt{1+m^2}=0$$

$$\beta_m=\left(\frac{b}{h}\right)_m=2(\sqrt{1+m^2}-m) \tag{9-8}$$

上式表明,梯形水力最佳断面宽深比只与坡度系数有关。

梯形断面的水力半径为

$$R=\frac{A}{\chi}=\frac{h(b+mh)}{b+2h\sqrt{1+m^2}} \tag{9-9}$$

联立方程(9-8)和(9-9),可得

$$R_m=\frac{h}{2}$$

说明梯形水力最佳断面的水力半径等于水深的一半,且与边坡系数无关。对于矩形断面来说,以 $m=0$ 代入式(9-8),可得

$$\beta_m=2$$

说明矩形水力最佳断面的底宽 $b$ 为水深 $h$ 的 2 倍。

值得注意的是,上述水力最优断面的概念只是从水力学角度提出的,并不完全等同于技术经济最优。在工程实践中还必须依据造价、施工技术、运转要求和养护等各方面的条件来综合考虑和比较,选出最经济合理的断面形式。

**2. 渠道允许流速**

良好的渠道应使设计流速不要太大,以免遭侵蚀。设计流速也不能太小,以防止淤积。因此,在设计渠道时,除考虑上述水力最佳条件及经济因素外,还应使渠道的断面平均流速 $v$ 在允许流速范围内,即

$$v_{\max} > v > v_{\min} \quad 或 \quad v_s > v > v_d$$

式中，$v_{\max}$ 为免遭冲刷的最大允许流速，简称不冲允许流速；$v_{\min}$ 为免受淤积的最小允许流速，简称不淤允许流速。

### 9.2.4 梯形明渠均匀流的数学模型及其解法

联立式（9-5）、（9-6）和式（9-7），可得梯形明渠均匀流的水力计算数学模型如下：

$$q_v = \frac{\sqrt{i}}{n} \frac{[h(b+mh)]^{5/3}}{(b+2h\sqrt{1+m^2})^{2/3}} \tag{9-10a}$$

$$v = \frac{\sqrt{i}}{n} R^{2/3} \quad 或 \quad v = \frac{q_v}{A} \tag{9-10b}$$

$$R = \frac{A}{\chi} \tag{9-10c}$$

上式中有 $q_v$、$n$、$i$、$b$、$h$、$m$ 六个变量，一般情况下，边坡系数 $m$、粗糙系数 $n$ 可根据土质条件确定，其余四个变量再按工程条件预先确定三个变量，然后求解另一个变量。在实际工程中待求变量可能较多，求解比较困难，所以可以根据不同的问题构建不同的方程，利用计算机编程进行计算。

（1）校核明渠输水能力 $q_v$，确定渠道底坡 $i$，或计算渠道粗糙系数 $n$，这些问题已知其他五个量，可以通过式（9-10）直接求解。

（2）设计渠道过流断面尺寸 $b$ 和 $h$。

① 若已知 $q_v$、$n$、$i$、$b$、$m$，需要求解水深 $h$。将公式（9-10a）中 $h^{5/3}$ 的 $h$ 设为待求值 $h_{j+1}$，可得求解 $h$ 的迭代公式如下：

$$h_{j+1} = \left(\frac{nq_v}{\sqrt{i}}\right)^{0.6} \frac{(b+2h_j\sqrt{1+m^2})^{0.4}}{b+mh_j} \quad (j=0,1,2,\cdots) \tag{9-11}$$

若已知 $q_v$、$n$、$i$、$h$、$m$，需要求解渠底宽 $b$。类似可得求解 $b$ 的迭代公式如下：

$$b_{j+1} = \left[\frac{1}{h}\left(\frac{nq_v}{\sqrt{i}}\right)^{0.6}(b_j+2h\sqrt{1+m^2})^{0.4} - mh\right]^{1.3} \times b_j^{-0.3} \quad (j=0,1,2,\cdots) \tag{9-12}$$

上式是加速收敛的加权修正形式，其通式为 $x_{j+1} = x_{j+1}^y \cdot x_j^{1-y}$。若迭代解的收敛性是围绕真解上下振荡型收敛的，则指数 $y$ 取值小于 1，可使振荡振幅减小；若迭代解的收敛性是在真值的单边渐近收敛的，则指数 $y$ 取值大于 1，可使迭代解对真值的偏差减小。加权指数法加速迭代解的收敛性效果显著。

迭代公式（9-11）、（9-12）收敛均较快，一般迭代 3 次或 4 次即可满足工程上的精度要求。

② 若已知 $\beta = b/h$，求解相应的 $h$ 和 $b$。

对于小型渠道的宽深比 $\beta$，一般按水力最优设计，$\beta = \beta_m = 2(\sqrt{1+m^2} - m)$；对于大型渠道的宽深比 $\beta$，则要在考虑经济技术的条件下给出，对通航渠道则应按特殊要求设计。

将 $b = \beta h$ 代入式（9-10a），有

$$q_v = \frac{\sqrt{i}}{n} \frac{[h^2(b+mh)]^{5/3}}{n[h(\beta+2h\sqrt{1+m^2})]^{2/3}} \tag{9-13}$$

$$\begin{cases} h = \left(\frac{nq_v}{\sqrt{i}}\right)^{0.375} \frac{(\beta+2\sqrt{1+m^2})^{0.25}}{(\beta+m)^{0.625}} \\ b = \beta h \end{cases} \tag{9-14}$$

③ 从最大允许流速 $v_{\max}$ 出发,求相应的 $b$ 和 $h$。

解决这类问题的方法是将 $v_{\max}$ 作为设计渠道的实际断面平均流速来考虑,可以确定相应的断面面积 $A$、水力半径 $R$ 和湿周 $\chi$ 如下:

$$A = \frac{q_v}{v_{\max}}$$

$$R = \left(\frac{nv_{\max}}{i^{1/2}}\right)^{3/2}$$

$$\chi = \frac{A}{R}$$

应用梯形断面上水力要素间的几何关系式

$$A = h(b+mh)$$

$$\chi = b + 2h\sqrt{1+m^2}$$

联立求解,可求解出

$$\begin{cases} h = \frac{\chi \pm \sqrt{\chi^2 - 4A(2\sqrt{1+m^2}-m)}}{2(2\sqrt{1+m^2}-m)} \\ b = \chi - 2h\sqrt{1+m^2} \end{cases} \tag{9-15}$$

式(9-11)~(9-15)均引入了曼宁公式,应用时需满足相应条件。

**例 9-1** 一梯形渠道底宽 $b=3.6$ m,水深 $h=1.2$ m,边坡系数 $m=2.0$,粗糙系数 $n=0.0192$,底坡 $i=0.000\ 625$。求通过的流量 $q_v$。

**解:** 水面宽度　　$B = b + 2mh = 3.6 + 2 \times 2 \times 1.2 = 8.4$ m

过水断面面积　$A = (b+mh)h = (3.6+2\times 1.2)\times 1.2 = 7.2$ m²

湿周　　　　　$\chi = b + 2h\sqrt{1+m^2} = 3.6 + 2\times 1.2\sqrt{1+2^2} = 8.97$ m

水力半径　　　$R = \frac{A}{\chi} = \frac{7.2}{8.97} = 0.803$ m

谢才系数　　　$C = \frac{1}{n} R^{1/6} = \frac{1}{0.019\ 2} \times 0.803^{1/6} = 50$ m$^{1/2}$/s

流量　　　　　$q_v = AC\sqrt{Ri} = 7.2 \times 50 \sqrt{0.803 \times 0.000\ 625} = 8.07$ m³/s

**例 9-2** 一梯形断面土渠,通过流量 $q_v = 1$ m³/s,底坡 $i = 0.005$,边坡系数 $m = 1.5$,粗糙

系数 $n=0.025$，不冲允许流速 $v_{\max}=1.2$ m/s。试按允许流速及水力最佳条件，分别设计断面尺寸。

**解：**（1）按不冲允许流速 $v_{\max}=1.2$ m/s 设计

$$A=\frac{q_v}{v_{\max}}=\frac{1}{1.2}=0.83 \text{ m}^2$$

由

$$v_{\max}=\frac{1}{n}i^{1/2}A^{2/3}\chi^{-2/3}=\frac{1}{0.025}0.005^{1/2}\,0.83^{2/3}\,\chi^{-2/3}=1.20 \text{ m/s}$$

解得

$$\chi\approx 3.0 \text{ m}$$

由梯形断面条件，得

$$A=(b+mh)h=bh+1.5h^2=0.83 \text{ m}^2$$
$$\chi=b+2h\sqrt{1+m^2}=b+3.61h=3.0 \text{ m}^2$$

联立上述两式，解得

$$b_1=-0.79 \text{ m},\ h_1=1.05 \text{ m（舍去）}$$
$$b_2=1.63 \text{ m},\ h_2=0.38 \text{ m}$$

（2）按水力最佳条件设计

水力最佳宽深比为 $\quad \beta=\dfrac{b}{h}=2(\sqrt{1+m^2}-m)=0.61$

则 $\quad b=0.61h$

因 $\quad A=(b+mh)h=(0.61h+1.5h)h=2.11h^2$

$$C=\frac{1}{n}R^{1/6}$$
$$R=0.5h$$

故 $\quad q_v=AC\sqrt{Ri}=\dfrac{1}{n}AR^{2/3}i^{1/2}=3.76\,h^{8/3}=1 \text{ m}^3/\text{s}$

解得 $\quad h=0.61 \text{ m},\ b=0.37 \text{ m}$

校核 $\quad A=2.11h^2=0.79 \text{ m}^2,\ v=\dfrac{q_v}{A}=\dfrac{1}{0.79}=1.27 \text{ m/s}$

故需要采取适当的加固措施，否则会造成冲刷。

## 9.3 无压圆管均匀流

### 9.3.1 水力数学模型和求解方法

无压圆管是指圆形断面不满流的长管道，管道内的流动具有自由表面，且表面压强为大气压强。无压管流均匀流属于明渠均匀流的特殊断面形式，它的形成条件、水力特征以及基本公式都和前述明渠均匀流相同。

无压圆管均匀流水力计算数学模型依旧由断面几何关系式及前述明渠均匀流基本公式联立给出。无压圆管均匀流过流断面的几何要素如图 9-4 所示,基本量有圆管直径 $d$、水深 $h$、充满度 $\alpha$ 或充满角 $\theta$,充满度定义为 $\alpha = \dfrac{h}{d}$。

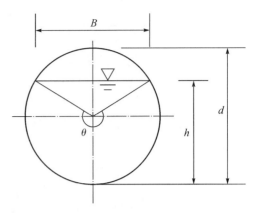

图 9-4　无压圆管过流断面

断面面积　　　$A = d^2(\theta - \sin\theta)/8$　　　(9-16a)

湿周　　　　　$\chi = \dfrac{\theta d}{2}$　　　(9-16b)

水力半径　　　$R = \dfrac{d}{4}\left(1 - \dfrac{\sin\theta}{\theta}\right)$　　　(9-16c)

流速　　　　　$v = 0.397 \dfrac{\sqrt{i}}{n}\left(d\dfrac{\theta - \sin\theta}{\theta}\right)^{2/3}$　　　(9-16d)

流量　　　　　$q_v = 0.0496 \dfrac{d^{8/3} i^{1/2}}{n}\dfrac{(\theta - \sin\theta)^{5/3}}{\theta^{2/3}}$　　　(9-16e)

充满度　　　　$\alpha = \dfrac{h}{d} = \sin^2(\theta/4)$　　　(9-16f)

水面宽度　　　$B = d \cdot \sin(\theta/2)$　　　(9-16g)

在其他量均已知的情况下,可以直接求解另一个未知量 $q_v$、$i$ 或 $d$。而直接求解 $\theta$ 比较困难,可迭代求解。式(9-16e)改写成下面的形式:

$$\left(\dfrac{q_v n}{0.0496 d^{8/3} i^{1/2}}\right)^{5/3} = \dfrac{\theta_{j+1} - \sin\theta_j}{\theta_j^{2/5}}$$

迭代解为

$$\theta_{j+1} = k\theta_j^{0.4} + \sin\theta_j \qquad (9-17)$$

## 9.3.2　无压圆管均匀流的水力特性

无压圆管均匀流除了具有明渠均匀流的特征外,另外还具有如下特征:

(1) 在未达到满管流之前流量达到最大值,或流速达到最大值,相应的充满度是最优充满度。现在引入满流时的流量 $q_0$ 和流速 $v_0$,与不满流时的流量 $q$ 和流速 $v$ 进行比较。不同的充满度对应一个流量和流速,采用无量纲的结合量,表示充满度与流量、流速的关系,如图 9-5 所示。

① 当 $\alpha = \dfrac{h}{d} \approx 0.95$ 时,$\dfrac{q}{q_0} \approx 1.08$,即 $q$ 达到最大值,此时管中通过的流量是满流的 1.08 倍;

② 当 $\alpha = \dfrac{h}{d} \approx 0.81$ 时,$\dfrac{v}{v_0} \approx 1.16$,即 $v$ 达到最大值,此时管中流速是满流的 1.16 倍。

分析可知,无压圆管的最大流量和最大流速均不发生于满管流时,这是由于圆形断面上部充水时,超过某一水深后,其湿周比过流断面积增长得快,水力半径开始减少,从而导致流量和流速的减少。

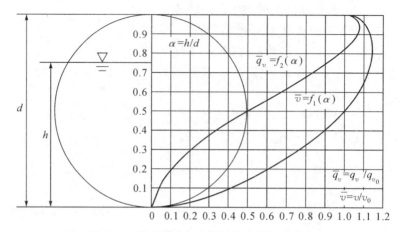

图 9-5 管道满流与不满流的无量纲参数图

(2) 当圆管流充满度接近 1 时,均匀流很难保持稳定。一旦受到外界波动的干扰,很容易形成有压流和无压流的交替流动。

因此,下水道的水力计算需符合工程中的有关规定,不允许超过最大设计充满度。

## 9.4 明渠恒定非均匀流

### 9.4.1 明渠非均匀流

明渠非均匀流是指通过明渠的流速和水深沿程变化的流动。天然明渠和人工明渠中的水流多为非均匀流。因为天然明渠不存在棱柱形渠道,即使是人工明渠,其断面形状、尺寸、粗糙度和底坡也可能沿程改变,或在明渠中修建水工建筑物(涵洞、闸、桥),使得明渠水流发生非均匀流。

与明渠均匀流相比,明渠非均匀流的特征如下:
① 断面平均流速、水深沿程改变;
② 流线不是相互平行的直线,同一条流线上的流速大小和方向不同。即总水头线 $J$、水面线 $J_p$ 和底坡 $i$ 三者不相等,$J \neq J_p \neq i$。

为了便于区分均匀流和非均匀流,后面将均匀流对应的参数以下标"0"表示。

### 9.4.2 断面单位能量

明渠水流的流动状态,可以从能量的角度分析判断。如图 9-6 所示,设明渠非均匀渐变流,基准面 0—0 受单位重力作用的液体的机械能为

$$E = z + \frac{p}{\rho g} + \frac{\alpha v^2}{2g} \tag{9-18}$$

若将该断面基准提高 $z_1$,使其通过该断面的最低点,受单位重力作用的液体相对于新基准面 $0_1—0_1$ 的机械能为

$$E_s = E - z_1 = h + \frac{\alpha v^2}{2g} \tag{9-19}$$

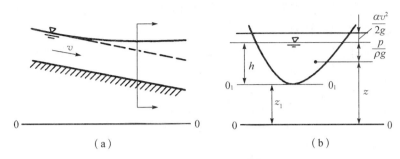

图 9-6 断面单位能量

或
$$E_s = h + \frac{\alpha q_v^2}{2gA^2} \tag{9-20}$$

式中，$E_s$ 称为断面单位能量，或断面比能，是相对于通过该断面最低点的基准面、受单位重力作用的液体所具有的机械能。

断面单位能量 $E_s$ 和受单位重力作用的流体的机械能 $E$ 是两个不同的能量概念。受单位重力作用的流体的机械能 $E$ 是全部水流相对于沿程同一基准面的机械能，其值沿程减少。而断面单位能量 $E_s$ 是按通过各自断面最低点的基准面计算的机械能，其值沿程可能增加，可能减少，只有在均匀流中，沿程不变。

明渠非均匀流水深是可变的，一定的流量，可能以不同的水深通过某一过流断面而有不同的断面单位能量。对于棱柱形渠道，流量一定时，断面单位能量只随水深的变化而变化，即

$$E_s = h + \frac{\alpha q_v^2}{2gA^2} = f(h)$$

从上式可以看出：在断面形状、尺寸以及流量一定时，当 $h \to 0$ 时，$A \to 0$，$v \to \infty$，则 $E_s \to \infty$，若以水深 $h$ 为纵坐标，断面单位能量 $E_s$ 为横坐标，则横坐标轴就是函数曲线 $E_s = f(h)$ 的渐近线；当 $h \to \infty$ 时，$A \to \infty$，$v \to 0$，则 $E_s \approx h \to \infty$，曲线以通过坐标原点与横轴成 45°角的直线为渐近线，其间存在断面单位能量最小值 $E_{s\min}$，它对应着临界水深 $h_c$。

### 9.4.3 临界水深

**1. 临界水深的含义和计算公式**

临界水深是指在断面形式及流量一定的条件下，相应于断面单位能量为最小值时的水深。亦即 $E_s = E_{s\min}$ 时所对应的水深 $h_c$，相应于 $h_c$ 的各水力要素均以下标"c"表示。令 $\dfrac{dE_s}{dh} = 0$，并考虑到 $\dfrac{dA}{dh} \approx B$，有

$$\frac{dE_s}{dh} = \frac{d}{dh}\left(h + \frac{\alpha q_v^2}{2gA^2}\right) = 1 - \frac{\alpha q_v^2}{gA^3}\frac{dA}{dh} = 1 - \frac{\alpha q_v^2}{gA^3}B = 0$$

可得
$$\frac{\alpha q_v^2}{g} = \frac{A_c^3}{B_c} \tag{9-21}$$

当给定渠道流量、断面形状和尺寸时，就可由上式求得 $h_c$ 值。由上式可知，临界水深仅与

断面形状、尺寸和流量有关，而与渠底坡度 $i$ 及壁面粗糙系数 $n$ 无关。

**2. 矩形断面临界水深的计算**

对于矩形断面明渠，水面宽度 $B_c = b$，单宽流量 $q_l = q_v/b$，代入式(9-21)中，有

$$\frac{\alpha q_v^2}{g} = \frac{A_c^3}{B_c} = \frac{b_c^3 h_c^3}{b_c}$$

得
$$h_c = \sqrt[3]{\frac{\alpha q_l^2}{g}} \qquad (9-22)$$

**3. 梯形断面临界水深的计算**

对于梯形断面，可将 $A_c = h_c(b + mh_c)$ 和 $B_c = b + 2mh_c$ 代入式(9-21)中，有

$$\frac{[h_c(b + mh_c)]^3}{b + 2mh_c} = \frac{\alpha q_v^2}{g}$$

该式是高次隐式方程，不易求解，可写成迭代形式：

$$h_{c(j+1)} = \left[\frac{\alpha q_v^2}{g} \frac{b + 2mh_{cj}}{b + mh_{cj}}\right]^{1/3} \quad (j = 0, 1, 2, \cdots) \qquad (9-23)$$

**4. 圆形断面临界水深的计算**

将式(9-16a)代入公式(9-21)中，可导出求解 $\theta$ 的迭代公式：

$$\theta_{j+1} = \left[8\left(\frac{\alpha q_v^2}{gd^5}\sin\frac{\theta_j}{2}\right)^{1/3} + \sin\theta_j\right]^{0.7} \cdot \theta_j^{0.3} \quad (j = 0, 1, 2, \cdots) \qquad (9-24)$$

临界水深为
$$h_c = d \cdot \sin^2\frac{\theta}{4}$$

式(9-24)中迭代过程加速收敛的加权指数为 0.7，表明很快就会收敛。

### 9.4.4 临界底坡

在棱柱形渠道中，断面形状、尺寸和流量一定时，若水流的正常水深 $h_0$ 恰好等于临界水深 $h_c$ 时，相应的渠底底坡称为临界底坡，用 $i_c$ 表示。临界底坡 $i_c$ 应用明渠均匀流基本方程 $q_v = A_c C_c \sqrt{R_c i_c}$ 和临界水深关系方程 $\frac{\alpha q_v^2}{g} = \frac{A_c^3}{B_c}$ 联立求解，可得

$$i_c = \frac{q_v^2}{A_c^2 C_c^2 R_c} = \frac{g}{\alpha C_c^2} \frac{\chi_c}{B_c} \qquad (9-25)$$

临界底坡 $i_c$ 并不是实际存在的渠道底坡，它只是为了便于分析和计算非均匀流动而引入的一个假想均匀流 ($h_0 = h_c$) 的假想底坡。如果实际的明渠底坡小于某一流量下的临界坡度，即 $i < i_c (h_0 > h_c)$，此时渠底坡度称为缓坡；如果 $i > i_c (h_0 < h_c)$，此时渠底坡度称为急坡或陡坡；如果 $i = i_c (h_0 = h_c)$，此时渠底坡度称为临界坡。值得注意的是，上述关于渠底坡度的缓、急之称，是对应于一定流量而言的。对于某一渠道，底坡是一定的，但当流量增大或变小时，所对应的 $h_c$ (或 $i_c$) 要发生变化，从而该渠道的缓坡或急坡之称也要随之改变。

### 9.4.5 缓流、急流、临界流及其判别

明渠水流在临界水深时的流速称为临界流速,以 $v_c$ 表示。这样的明渠水流状态称为临界流。当明渠水流流速小于临界流速时,称为缓流,多发生在平原和近海河流。大于临界流速时,称为急流,多发生在山区河流和陡槽中。常用的判别方式如下。

**1. 临界水深法**

将渠道中的水深 $h$ 与临界水深 $h_c$ 相比较,可以判别明渠水流的流态:$h>h_c$,$v<v_c$,水流为缓流;$h=h_c$,$v=v_c$,水流为临界流;$h<h_c$,$v>v_c$,水流为急流。

**2. 波速法**

将石子投入平静湖水中,会产生微小的干扰波,干扰波以石子落点为中心,以一定的速度 $c$ 向四周传播,湖面上形成一连串同心圆的波形,将这种在静水中传播的微波速度称为相对波速,用 $c$ 表示:

$$c=\sqrt{g\bar{h}}$$

可以通过比较水流的断面平均流速 $v$ 和相对波速 $c$ 的大小,判别水流是属于哪一种流态。当 $v<c$ 时,干扰波能向上游传播,水流为缓流;当 $v=c$ 时,干扰波恰好不能向上游传播,水流为临界流;当 $v>c$ 时,干扰波不能向上游传播,水流为急流。

**3. 弗劳德数法**

定义明渠流速 $v$ 与相对波速 $c$ 之比为弗劳德数,用 $Fr$ 表示,即

$$Fr=\frac{v}{c}=\frac{v}{\sqrt{g\bar{h}}}$$

可作为水流流态的判别数:$Fr<1$,水流为缓流;$Fr=1$,水流为临界流;$Fr>1$,水流为急流。

弗劳德数在流体力学中是一个重要的判别数,为了加深对它的物理意义的理解,可把它的形式改写为

$$Fr=\frac{v}{\sqrt{g\bar{h}}}=\sqrt{\frac{2v^2/2g}{\bar{h}}} \tag{9-26}$$

以上方法对于均匀流和非均匀流的流态判别均适用。对于明渠均匀流,还可根据底坡来判别,当底坡 $i<i_c$,则为缓流;当 $i=i_c$,则为临界流;当底坡 $i>i_c$,则为急流。

上述各种判别方法是等效的,可依工程条件,采用其中一种方法。

**例 9-3** 长直的矩形断面渠道,底宽 $b=1\text{ m}$,粗糙系数 $n=0.014$,底坡 $i=0.0004$,渠内均匀流正常水深 $h_0=0.6\text{ m}$。试判别水流的流动状态。

**解:** 断面平均流速为
$$v=C\sqrt{Ri}$$

式中
$$R=\frac{bh_0}{b+2h_0}=0.273\text{ m}$$

$$C=\frac{1}{n}R^{1/6}=57.5\text{ m}^{1/2}/\text{s}$$

于是得
$$v=0.6\text{ m/s}$$

(1) 用波速法判别

$$c = \sqrt{g\bar{h}} = \sqrt{gh_0} = 2.43 \text{ m/s} > v$$

流动为缓流。

(2) 用弗劳德数判别

$$Fr = \frac{v}{\sqrt{gh_0}} = 0.25 < 1$$

流动为缓流。

(3) 用临界水深判别

$$q_v = vh_0 = 0.36 \text{ m}^2/\text{s}$$

$$h_c = \sqrt[3]{\frac{\alpha q_v^2}{g}} = 0.24 \text{ m} < h_0$$

流动为缓流。

(4) 用临界底坡判别

由于流动是均匀流,还可以用临界底坡来判别水流状态,计算相应量得

$$B_c = b = 1 \text{ m}$$

$$\chi_c = b + 2h_c = 1.48 \text{ m}$$

$$R_c = \frac{bh_c}{\chi_c} = 0.16 \text{ m}$$

$$C_c = \frac{1}{n} R_c^{1/6} = 52.7 \text{ m}^{1/2}/\text{s}$$

$$i_c = \frac{g}{\alpha C_c^2} \frac{\chi_c}{B_c} = 0.005\,2 > i = 0.000\,4$$

此渠道为缓坡,均匀流为缓流。

### 9.4.6 水跃与消能

明渠中的水流由急流过渡到缓流时,水流的自由表面会突然跃起,并在表面形成旋滚,这种急变流现象称为水跃。它可以在溢洪道下、泄水闸下、跌水下形成,也可以在平坡渠道中闸下出流时形成。

水跃区上部分都是急流冲入缓流激起的表面旋流,翻腾滚动,饱掺空气,称为"表面旋滚"。旋滚下面是断面向前扩张的主流。由于形成表面旋滚其底部为主流,水流紊动,流体质点互相碰撞,掺混强烈。旋滚与主流间质量不断交换,致使水跃段内有较大的能量损失,可达跃前断面急流能量的 60%~70%。因此,常用水跃来消除泄水建筑物下游高速水流的巨大能量,即水跃常用于泄水建筑物下游的消能,是一种有效的消能手段。

**1. 水跃基本方程**

现以平坡渠道上的完整水跃为例,建立水跃方程。在推导过程中,根据水跃发生的实际情况,作下列一些假设:

(1) 水跃段长度不大,可忽略渠道边壁的摩擦阻力;
(2) 跃前、跃后两个过流断面为渐变流过流断面,符合静水压强分布规律;
(3) 跃前、跃后两个过流断面的动量修正系数相等,即 $\beta_1=\beta_2=\beta$。

取跃前断面和跃后断面之间的水跃空间为控制体,列出流动方向总流的动量方程如下:

$$\rho q_v(\beta_2 v_2 - \beta_1 v_1) = \rho g h_{c1} A_1 - \rho g h_{c2} A_2 \quad (9-27)$$

式中,$h_{c1}$、$h_{c2}$ 分别为跃前、跃后断面形心点的水深;$\rho g h_{c1} A_1$、$\rho g h_{c2} A_2$ 分别为作用在跃前、跃后断面上的动水压力。

由连续性方程可知,跃前、跃后断面的平均流速分别为

$$v_1 = \frac{q_v}{A_1}, \ v_2 = \frac{q_v}{A_2}$$

则式(9-27)可写为

$$\frac{q_v^2 \beta}{g A_1} + h_{c1} A_1 = \frac{q_v^2 \beta}{g A_2} + h_{c2} A_2 \quad (9-28)$$

式(9-28)就是棱柱形平坡渠中水跃的基本方程。它说明水跃区单位时间内,流入跃前断面的动量与该断面动水总压力之和,同流出跃后断面的动量与该断面动水总压力之和相等。式中,$A$ 和 $h_c$ 都是水深的函数,其余量均为常量,所以可写成如下形式:

$$J(h) = \frac{q_v^2 \beta}{gA} + h_c A \quad (9-29)$$

水跃基本方程也可以写成如下形式:

$$J(h') = J(h'') \quad (9-30)$$

上述水跃基本方程表明,对于某一流量 $q_v$,具有相同的水跃函数 $J(h)$ 的两个水深,这一对水深即为共轭水深。

**2. 水跃基本计算**

(1) 共轭水深计算

对于矩形断面的棱柱渠道,令 $\alpha=\beta=1.0$,并将 $A_1=bh'$,$A_2=bh''$,$h_{c1}=\dfrac{h'}{2}$,$h_{c2}=\dfrac{h''}{2}$,$q_l=\dfrac{q_v}{b}$,$h_c^3=\dfrac{\alpha q^2}{g}$ 代入式(9-28),消去 $b$ 得

$$\frac{q_l^2}{gh'} + \frac{h'^2}{2} = \frac{q_l^2}{gh''} + \frac{h''^2}{2} \quad (9-31)$$

经过整理,得二次方程:

$$h'h''(h'+h'') = \frac{2q_l^2}{g}$$

分别以跃前水深 $h'$ 和跃后水深 $h''$ 为未知量,解得

$$h' = \frac{h''}{2}\left(\sqrt{1+8\frac{q_l^2}{gh''^3}}-1\right) = \frac{h''}{2}\left[\sqrt{1+8\left(\frac{h_c}{h''}\right)^3}-1\right] \quad (9-32)$$

$$h'' = \frac{h'}{2}\left(\sqrt{1+8\frac{q_l^2}{gh'^3}}-1\right) = \frac{h'}{2}\left[\sqrt{1+8\left(\frac{h_c}{h'}\right)^3}-1\right] \quad (9-33)$$

由于
$$Fr_1^2 = \frac{v_1^2}{gh'} = \frac{q_l^2}{gh'^3}, \ Fr_2^2 = \frac{v_2^2}{gh''} = \frac{q_l^2}{gh''^3}$$

所以式(9-32)、式(9-33)可写成：

$$h' = \frac{h''}{2}(\sqrt{1+8Fr_2^2}-1) \quad (9-34)$$

$$h'' = \frac{h'}{2}(\sqrt{1+8Fr_1^2}-1) \quad (9-35)$$

共轭水深之比为
$$\eta = \frac{h'}{2}(\sqrt{1+8Fr_1^2}-1) \quad (9-36)$$

由上式可以看出，$\eta$ 是随着跃前断面的弗劳德数 $Fr_1$ 的增大而增大。

(2) 水跃长度计算

水跃长度是泄水建筑物消能设计的主要依据之一。由于水跃现象的复杂性，目前理论研究尚不成熟，水跃长度的确定仍以实验研究为主。现介绍用于计算平底坡矩形渠道水跃长度的经验公式。

① 以跃后水深表示，适用于 $4.5 < Fr_1 < 10$ 的情形：$l_j = 6.1 h''$；
② 以水跃高度表示，$l_j = 6.9(h'' - h')$；
③ 以弗劳德数表示，$l_j = 9.4(Fr_1 - 1) h'$。

(3) 水跃消能计算

研究发现，水跃造成的能量损失主要集中在水跃段，仅有极少部分发生在跃后段。对平底矩形渠道，对跃前、跃后断面列总能量方程，可得水跃的能量损失为

$$\Delta E_j = \left(h' + \frac{\alpha_1 v_1^2}{2g}\right) - \left(h'' + \frac{\alpha_2 v_2^2}{2g}\right)$$

将 $v_1 = \frac{q_l}{h'}$，$v_2 = \frac{q_l}{h''}$ 和 $h'h''(h'+h'') = \frac{2q_l^2}{g}$ 代入上式，并取 $\alpha_1 = \alpha_2 = 1.0$，得

$$\Delta E_j = \frac{(h''-h')^3}{4h'h''} \quad (9-37)$$

式(9-37)说明，在给定流量下，跃前与跃后水深相差越大，水跃消除的能量值就越大。

### 9.4.7 水跌

处于缓流状态的明渠水流，因渠底突然变为陡坡或下游渠道断面形状突然扩大，引起水面降落，转变为急流，这种从缓流向急流过渡的局部水力现象称为水跌，或称跌水。了解跌水现象对分析和计算明渠恒定非均匀流的水面曲线具有重要的意义。例如缓坡渠道后接一急坡渠道，水流经过连接断面时的水深可认为是临界水深，这一断面称为控制断面，其水深称为控制水深。在进行水面曲线分析和计算时可作为已知水深，从而给分析、计算提供了一个已知条件。

**例 9-4** 某泄水建筑物下游矩形断面渠道,泄流单宽流量 $q_l = 1.5 \text{ m}^2/\text{s}$。产生水跃,跃前水深 $h' = 0.8 \text{ m}$。试求:(1) 跃后水深 $h''$;(2) 水跃长度 $l_j$;(3) 水跃消能率 $\Delta E_j / E_1$。

**解:** (1) 求跃后水深 $h''$

$$Fr_1^2 = \frac{q^2}{gh'^3} = \frac{15^2}{9.8 \times 0.8^3} = 44.84$$

$$h'' = \frac{h'}{2}(\sqrt{1 + 8Fr_1^2} - 1) = \frac{0.8}{2} \times (\sqrt{1 + 8 \times 44.84} - 1) = 7.19 \text{ m}$$

(2) 求水跃长度 $l_j$

$$l_j = 6.1 h'' = 6.1 \times 7.19 = 43.86 \text{ m}$$

$$l_j = 6.9(h'' - h') = 6.9 \times 6.39 = 44.09 \text{ m}$$

$$l_j = 9.4(Fr_1 - 1)h' = 42.83 \text{ m}$$

(3) 求水跃消能率

$$\Delta E_j = \frac{(h'' - h')^3}{4 h' h''} = \frac{(7.19 - 0.8)^3}{4 \times 0.8 \times 7.19} = 11.34 \text{ m}$$

$$\frac{\Delta E_j}{E_1} = \frac{\Delta E_j}{h' + \dfrac{q^2}{2gh'^2}} = \frac{11.34}{0.8 + \dfrac{15^2}{2 \times 9.8 \times 0.8^2}} = 61\%$$

## 9.5 明渠渐变流水面曲线分析

明渠非均匀渐变流的水深 $h$ 随着流程 $s$ 而变化,自由水面线称为水面曲线 $h = f(s)$,是与渠底不平行的曲线。明渠恒定非均匀渐变流水面曲线分析的主要任务,就是根据渠道的槽身条件、来流条件以及水工建筑物情况等确定水面曲线的沿程变化趋势和变化范围,定性地绘出水面曲线。

### 9.5.1 棱柱形明渠非均匀渐变流微分方程

图 9-7 为一明渠非均匀渐变流段,沿水流方向取一微小流段 $ds$,其上游 1—1 断面相对于 0—0 基准面的渠底高程为 $z$,水深为 $h$,断面平均流速为 $v$;下游 2—2 断面相应参数为 $z + dz$,$h + dh$,$v + dv$。列出 1—1、2—2 断面的能量方程如下:

$$(z + h) + \frac{\alpha v^2}{2g} = (z + dz) + (h + dh) + \frac{\alpha (v + dv)^2}{2g} + dh_w \quad (9-38)$$

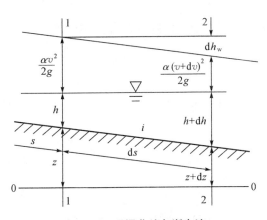

**图 9-7** 明渠非均匀渐变流

式(9-38)中的 $(v+\mathrm{d}v)^2$ 按二项式展开,并且忽略高阶微项 $(\mathrm{d}v)^2$,整理后可得

$$\mathrm{d}z + \mathrm{d}h + \mathrm{d}\left(\frac{\alpha v^2}{2g}\right) + \mathrm{d}h_w = 0$$

用 $\mathrm{d}s$ 除上式,得

$$\frac{\mathrm{d}z}{\mathrm{d}s} + \frac{\mathrm{d}h}{\mathrm{d}s} + \frac{\mathrm{d}}{\mathrm{d}s}\left(\frac{\alpha v^2}{2g}\right) + \frac{\mathrm{d}h_w}{\mathrm{d}s} = 0 \tag{9-39}$$

式中第 1 项

$$\frac{\mathrm{d}z}{\mathrm{d}s} = -\frac{z_1 - z_2}{\mathrm{d}s} = -i$$

式中第 3 项

$$\frac{\mathrm{d}}{\mathrm{d}s}\left(\frac{\alpha v^2}{2g}\right) = \frac{\mathrm{d}}{\mathrm{d}s}\left(\frac{\alpha q_v^2}{2gA^2}\right) = -\frac{\alpha q_v^2}{gA^3}\frac{\mathrm{d}A}{\mathrm{d}s}$$

而棱柱形渠道过流断面面积只随水深变化,即 $A = f(h)$,则

$$\frac{\mathrm{d}A}{\mathrm{d}s} = \frac{\mathrm{d}A}{\mathrm{d}h}\frac{\mathrm{d}h}{\mathrm{d}s} = B\frac{\mathrm{d}h}{\mathrm{d}s}$$

式中第 4 项,渐变流段局部水头损失很小,可忽略不计,有 $\mathrm{d}h_w = \mathrm{d}h_f$,得

$$\frac{\mathrm{d}h_w}{\mathrm{d}s} = \frac{\mathrm{d}h_f}{\mathrm{d}s} = J$$

将上面各项代入公式(9-39),得

$$\frac{\mathrm{d}h}{\mathrm{d}s} - \frac{\alpha q_v^2}{gA^3}\left(B\frac{\mathrm{d}h}{\mathrm{d}s}\right) = i - J$$

整理得

$$\frac{\mathrm{d}h}{\mathrm{d}s}\left(1 - \frac{\alpha q_v^2 B}{gA^3}\right) = i - J$$

渐变流过流断面沿程变化缓慢,水头损失可按均匀流计算,即

$$J = \frac{q_v^2}{A^2 C^2 R} = \frac{q_v^2}{K^2} \tag{9-40}$$

最后得

$$\frac{\mathrm{d}h}{\mathrm{d}s} = \frac{i - J}{1 - \left(\frac{\alpha q_v^2 B}{gA^3}\right)} = \frac{i - J}{1 - F_r^2} = \frac{i - \left(\frac{q_v}{K}\right)^2}{1 - F_r^2} \tag{9-41}$$

上式即为棱柱形明渠恒定非均匀渐变流得微分方程。

## 9.5.2 明渠渐变流的流动现象和空间划分

### 1. 非均匀渐变流流动现象

明渠非均匀渐变流的微分方程中包含了 $h$、$h_0$、$h_c$ 及 $i$ 的相互关系。由于在不同渠道底坡下，上述三个水深值有不同的组合，从而形成了明渠非均匀流水面曲线的各种变化：$\dfrac{\mathrm{d}h}{\mathrm{d}s}>0$、$\dfrac{\mathrm{d}h}{\mathrm{d}s}=0$、$\dfrac{\mathrm{d}h}{\mathrm{d}s}<0$、$\dfrac{\mathrm{d}h}{\mathrm{d}s}\to i$ 及 $\dfrac{\mathrm{d}h}{\mathrm{d}s}\to\pm\infty$ 等。

(1) $\dfrac{\mathrm{d}h}{\mathrm{d}s}>0$，水深沿程增加，水面曲线为壅水曲线，如图 9-8 中的断面 1—1 至断面 2—2 间的水流。

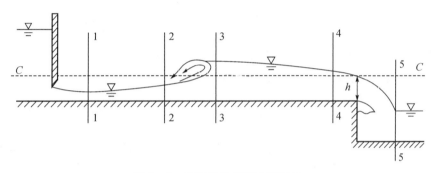

图 9-8　明渠非均匀流水流现象

(2) $\dfrac{\mathrm{d}h}{\mathrm{d}s}<0$，水深沿程减小，水面曲线为降水曲线，如图 9-8 中的断面 3—3 至断面 4—4 间的水流。

(3) $\dfrac{\mathrm{d}h}{\mathrm{d}s}\to+\infty$，水面线不连续，为水跃现象，如图 9-8 中的断面 2—2 至断面 3—3 间的水流。

(4) $\dfrac{\mathrm{d}h}{\mathrm{d}s}\to-\infty$，水面线不连续，为水跌现象，如图 9-8 中的断面 4—4 至断面 5—5 间的水流。

(5) $\dfrac{\mathrm{d}h}{\mathrm{d}s}=0$，水深沿程不变，水流为均匀流。

(6) $\dfrac{\mathrm{d}h}{\mathrm{d}s}=i$，则 $\dfrac{\mathrm{d}h}{\mathrm{d}s}=-\dfrac{\mathrm{d}z}{\mathrm{d}s}$，水面线是水平线（图 9-9）。

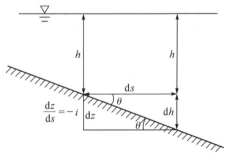

图 9-9　明渠流水平线

### 2. 流动空间分区

为了便于分析水面曲线沿程变化的情况，一般在水面曲线的分析图上作出两条平行于渠底的直线。其中一条距渠底 $h_0$，为正常水深线 $N$—$N$；而另一条距渠底 $h_c$，为临界水深线 $C$—$C$。在渠底以上画出的这两条辅助线（$N$—$N$ 和 $C$—$C$）把渠道水流划分成三个

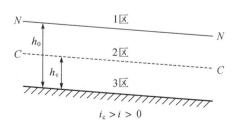

图 9-10　明渠空间划分

不同的区域。这三个区分别称为1区、2区和3区,各区的特点如下:

1区:$h > h_0$ 且 $h > h_c$;

2区:$h_0 > h > h_c$(缓坡)或 $h_0 < h < h_c$(急坡);

3区:$h < h_0$ 且 $h < h_c$。

### 9.5.3 明渠渐变流12种水面曲线

明渠渐变流实际所在区域不同、底坡不同,其水面曲线有不同形式,下面将分别进行分析。

**1. 缓坡渠道($i_c > i > 0$)**

缓坡渠道中正常水深 $h_0$ 大于临界水深 $h_c$,由 N—N 和 C—C 两条辅助线将流动空间分为1、2和3三个区域,出现在各区的水面曲线不同,如图9-11a所示。

(1) 1区($h > h_0 > h_c$)

在式(9-43)中,分子 $h > h_0$,$J < i$,$i - J > 0$;分母 $h > h_c$,$Fr < 1$,$1 - Fr^2 > 0$,所以 $\frac{dh}{ds} > 0$,水深沿程增加,水面线是壅水曲线,称为 $M_1$ 型壅水曲线。

如图9-11a所示,上游 $h \to h_0$,$J \to i$,$i - J \to 0$;$h > h_c$,$Fr < 1$,$1 - Fr^2 > 0$,所以 $\frac{dh}{ds} \to 0$,水深沿程不变,水面线以 N—N 线为渐近线。下游 $h \to \infty$,$J \to 0$,$i - J \to i$;$h \to \infty$,$Fr \to 0$,$1 - Fr^2 \to 1$,所以 $\frac{dh}{ds} \to i$,单位距离水深的增加等于渠底高程的降低,水面线为水平线。

在缓坡渠道上修建挡水建筑物,抬高控制水深 $h$,超过该流量的正常水深 $h_0$,挡水建筑物上游将出现 $M_1$ 型水面曲线,如图9-11b所示。

(2) 2区($h_0 > h > h_c$)

在式(9-43)中,分子 $h < h_0$,$J > i$,$i - J < 0$;分母 $h > h_c$,$Fr < 1$,$1 - Fr^2 > 0$,所以 $\frac{dh}{ds} < 0$,水深沿程减小,水面线是降水曲线,称为 $M_2$ 型降水曲线。

如图9-11a所示,上游 $h \to h_0$,类似于 $M_1$ 型水面线的分析,$\frac{dh}{ds} \to 0$,水深沿程不变,水面线以 N—N 线为渐近线。下游 $h \to h_c < h_0$,$J > i$,$i - J < 0$;$h \to h_c$,$Fr \to 1$,$1 - Fr^2 \to 0$,所以 $\frac{dh}{ds} \to -\infty$,水面线与 C—C 线正交,此处水深急剧降低,已不再是渐变流,而是水跌情形。

缓坡渠道末端为跌坎,渠道内为 $M_2$ 型水面线,跌坎断面通过临界水深,形成水跌,如图9-11b所示。

(3) 3区($h_0 > h_c > h$)

在式(9-43)中,分子 $h < h_0$,$J > i$,$i - J < 0$;分母 $h < h_c$,$Fr > 1$,$1 - Fr^2 < 0$,所以 $\frac{dh}{ds} > 0$,水深沿程增加,水面线为壅水曲线,称为 $M_3$ 型壅水曲线。

如图9-11a所示,上游水深由出流条件控制,下游 $h \to h_c$,$Fr \to 1$,$1 - Fr^2 \to 0$,所以 $\frac{dh}{ds} \to \infty$,发生水跃,形成下凹的壅水曲线。

在缓坡渠道上闸门部分开启时,门后水深小于临界水深,形成的急流,由于阻力作用,流速

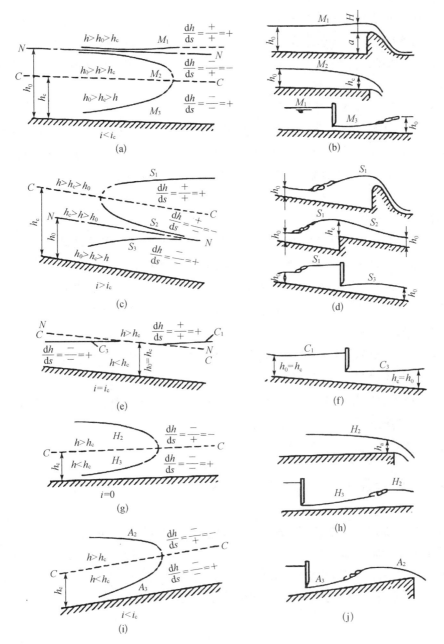

**图 9-11  各类水面曲线及工程实例**

沿程减小,水深增加,形成 $M_3$ 型水面曲线,如图 9-11b 所示。

**2. 陡坡渠道 ($i > i_c > 0$)**

陡坡渠道中正常水深 $h_0$ 小于临界水深 $h_c$,由 $N—N$ 线和 $C—C$ 线将流动空间分为 1、2 和 3 三个区域,出现在各区的水面曲线与缓坡不同,如图 9-11c 所示。

1 区水面曲线为 $S_1$ 型壅水曲线,上游发生水跃,下游水面曲线趋于水平。在陡坡渠道中修建挡水建筑物,上游形成 $S_1$ 型水面曲线,如图 9-11d 所示。

2 区水面曲线为 $S_2$ 型降水曲线,上游发生水跌,下游水面线以 $N—N$ 线为渐近线。水流从陡坡渠道流入另一段渠底抬高的陡坡渠道时,在上游渠道中形成 $S_1$ 型水面曲线,在下游陡坡渠

道中形成 $S_2$ 型水面曲线,而在变坡断面立坎顶上通过临界水深,形成水跌,如图 9-11d 所示。

3 区水面曲线为 $S_3$ 型壅水曲线,上游水深由出游条件控制,下游水面线以 $N—N$ 线为渐近线。在陡坡渠道中修建挡水建筑物,下泄水流的收缩水深小于正常水深,下游形成 $S_3$ 型水面曲线,如图 9-11d 所示。

**3. 临界坡渠道($i=i_c$)**

在临界坡渠道中,正常水深 $h_0$ 等于临界水深 $h_c$。$N—N$ 线与 $C—C$ 线重合,流动区间分为 1、3 两个区域,没有 2 区。水面曲线都是壅水曲线,分别称为 $C_1$ 型壅水曲线和 $C_3$ 型壅水曲线,且在接近 $N—N(C—C)$ 线时都近于水平,如图 9-11e 所示。

在临界坡渠道中,泄水闸门上、下游将形成 $C_1$、$C_3$ 型水面曲线,如图 9-11f 所示。

**4. 平坡渠道($i=0$)**

在平坡渠道中,不可能发生均匀流。只有临界水深线 $C—C$,将流动分为 2、3 区,无 1 区。2 区为 $H_2$ 型降水曲线,3 区为 $H_3$ 型壅水曲线,如图 9-11g 所示。

在平坡渠道末端跌坎上游将形成 $H_2$ 型水面曲线,平坡渠道中泄水闸门开启高度小于临界水深,闸门下游将形成 $H_3$ 型水面曲线,如图 9-11h 所示。

**5. 逆坡渠道($i<0$)**

在逆坡渠道中,也只有临界水深线存在,将流动分为 2、3 区。水面线分别为 $A_2$ 型降水曲线与 $A_3$ 型壅水曲线,如图 9-11i 所示。

在逆坡渠道末端跌坎上游将形成 $A_2$ 型水面曲线,逆坡渠道中泄水闸门开启高度小于临界水深,闸门下游将形成 $A_3$ 型水面曲线,如图 9-11j 所示。

### 9.5.4 水面曲线分析步骤

(1) 绘出 $N—N$ 线和 $C—C$ 线(平坡和逆坡渠道无 $N—N$ 线),将流动空间分区,每个区域只相应一种水面曲线。

(2) 选择控制断面。控制断面应选在水深为已知,且位置确定的断面上,然后以控制断面为起点进行分析计算,确定水面曲线的类型,并参照其增深、减深的形状和边界情况,进行描绘。

(3) 如果水面曲线中断,出现了不连续而产生跌水或水跃时,要作具体分析,一般情况下,水流至跌坎处便形成跌水现象,水流从急流到缓流,便发生水跃现象。

## 9.6 明渠渐变流水面曲线计算

定性分析了水面曲线的变化后,须对它进行定量计算。对棱柱形渠槽,通常是采用数学上一般的近似积分法,称为数值积分法。另一类方法是分段求和法,该方法是将明渠整个流程分成若干段,将微分方程式用有限差分式代替,进行分段直接求和。分段求和法,对棱柱形、非棱柱形渠槽均适用,现分述如下。

### 9.6.1 数值积分法

对基本微分方程式(9-41)分离变量,得

$$ds = \frac{1 - \alpha q_v^2 \dfrac{B}{g} A^3}{i - J} dh \tag{9-42}$$

积分上式,有

$$l = \int_{h_1}^{h_2} \frac{1 - \alpha q_v^2 \dfrac{B}{g} A^3}{i - J} dh \tag{9-43}$$

对于粗糙系数 $n$ 和渠道底坡 $i$ 不变的棱柱形渠道,式(9-45)的被积项仅是水深 $h$ 的函数,即有

$$F(h) = \frac{1 - \alpha q_v^2 \dfrac{B}{g} A^3}{i - J}$$

则

$$l = \int_{h_1}^{h_2} \frac{1 - \alpha q_v^2 \dfrac{B}{g} A^3}{i - J} dh = \int_{h_1}^{h_2} F(h) dh \tag{9-44}$$

倘若按照梯形法积分,可得

$$l \approx \sum \Delta l = \sum_{n=1}^{m} \frac{F(h_n) + F(h_{n+1})}{2} (h_{n+1} - h_n) \tag{9-45}$$

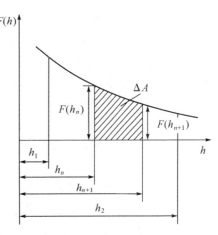

图 9-12 数值积分法

绘出 $F(h)-h$ 曲线,如图 9-12 所示,位于两个水深 $h_n$ 和 $h_{n+1}$ 之间曲线下面的面积 $\Delta A$,就是两水深断面间的长度 $\Delta l$。积分区间 $(h_1, h_2)$ 可分成 $m$ 个小区间,由某一已知水深可以计算出一系列 $\Delta h_i$ 和 $\Delta l_i$ 值,于是可以定量绘制水面曲线。

### 9.6.2 分段求和法

根据基本微分方程式(9-41):

$$\frac{dh}{ds} = \frac{i - J}{1 - \left(\alpha q_v^2 \dfrac{B}{g} A^3\right)}$$

又有断面比能微分方程:

$$\frac{dE_s}{dh} = 1 - \frac{\alpha q_v^2 B}{g A^3}$$

则

$$\frac{dh}{ds} = \frac{i - J}{\dfrac{dE_s}{dh}}$$

$$ds = \frac{dE_s}{i - J}$$

将上式写成有限差分式：

$$\Delta l = \Delta s = \frac{\Delta E_s}{i - \bar{J}} = \frac{E_{sd} - E_{su}}{i - \bar{J}} \qquad (9-46)$$

则流程总长度为

$$l = \sum \Delta l = \sum \Delta s = \sum \frac{\Delta E_s}{i - \bar{J}} = \sum \frac{E_{sd} - E_{su}}{i - \bar{J}} \qquad (9-47)$$

式中，$\Delta E_s$ 为计算流段两端断面上的断面比能之差；$E_{su}$、$E_{sd}$ 分别为计算流段上、下游断面的断面比能；$\bar{J}$ 为计算流段的平均水力坡度，一般 $\bar{J} = \frac{1}{2}(J_u + J_d)$ 或 $\bar{J} = \frac{\bar{v}^2}{\bar{C}^2 \bar{R}}$，其中 $\bar{v} = \frac{v_u + v_d}{2}$，$\bar{C} = \frac{C_u + C_d}{2}$，$\bar{R} = \frac{R_u + R_d}{2}$。

式(9-47)即为分段求和法计算水面曲线的基本公式。将整个流段分成若干段 $\Delta l$，应用式(9-47)，逐步计算明渠非均匀流各断面的水深 $h$ 及相隔的距离 $\Delta l$，从而可求得整个流段 $l$ 上的水面曲线。分段求和法对于棱柱形渠道和非棱柱形渠道的恒定非均匀梯度流都适用。

**例 9-5** 矩形排水长渠道，底宽 $b = 2$ m，粗糙系数 $n = 0.025$，底坡 $i = 0.000\,2$，排水流量 $q_v = 2.0$ m³/s，渠道末端排入河中(图 9-13)。试绘制水面曲线。

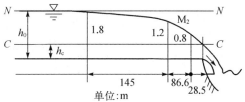

图 9-13 水面线绘制

**解**：(1) 判别渠道底坡性质及水面线类型

正常水深由式(9-6)试算得 $h_0 = 2.26$ m；临界水深由式(9-24)算得 $h_c = 0.467$ m。按 $h_0$、$h_c$ 计算值，在图中标出 $N$—$N$ 线和 $C$—$C$ 线。$h_0 > h_c$ 为缓坡渠道，末端(跌坎)水深为 $h_c$，渠内水流在缓坡渠道 2 区流动，水面线为 $M_2$ 型降水曲线。

(2) 水面线计算

渠道内为缓流，末端水深 $h_c$ 为控制水深，向上游推算。

取 $h_2 = h_c = 0.467$ m，$A_2 = b h_2 = 0.943$ m²，$v_2 = \frac{q_v}{A_2} = 2.14$ m/s，$\frac{v_2^2}{2g} = 0.234$ m，$E_2 = h_2 + \frac{v_2^2}{2g} = 0.7$ m，$R_2 = \frac{A_2}{\chi_2} = 0.32$ m，$C_2 = \frac{1}{n} R_2^{1/6} = 33.07$ m$^{1/2}$/s。

设 $h_1 = 0.8$ m，$A_1 = b h_1 = 1.6$ m²，$v_1 = \frac{q_v}{A_1} = 1.25$ m/s，$\frac{v_1^2}{2g} = 0.08$ m，$E_1 = h_1 + \frac{v_1^2}{2g} = 0.88$ m，$R_1 = \frac{A_1}{\chi_1} = 0.44$ m，$C_1 = \frac{1}{n} R_1^{1/6} = 34.94$ m$^{1/2}$/s。

平均值 $\bar{v} = \frac{v_1 + v_2}{2} = 1.695$ m/s，$\bar{R} = \frac{R_1 + R_2}{2} = 0.38$ m，$\bar{C} = \frac{C_1 + C_2}{2} = 34$ m$^{1/2}$/s，$\bar{J} = \frac{\bar{v}^2}{\bar{C}^2 \bar{R}} = 0.006\,5$，$\Delta l_{1-2} = \frac{\Delta l}{i - \bar{J}} = \frac{-0.18}{-0.006\,3} = 28.57$ m

继续按 $h = 1.2$ m、1.8 m、2.1 m，重复以上步骤计算各段长度，各段计算结果见表 9-1。

根据计算值,便可绘制泄水渠内水面线。

表 9-1 水面曲线计算表

| 断面 | $h$/m | $A$/m² | $v$/(m·s⁻¹) | $\bar{v}$/(m·s⁻¹) | $\dfrac{v^2}{2g}$/m | $E$/m | $\Delta E$/m |
|---|---|---|---|---|---|---|---|
| 1 | 0.476 | 0.934 | 2.140 |       | 0.234 | 0.70  |        |
| 2 | 0.800 | 1.600 | 1.250 | 1.695 | 0.080 | 0.880 | −0.180 |
| 3 | 1.200 | 2.400 | 0.833 | 1.640 | 0.035 | 1.235 | −0.355 |
| 4 | 1.800 | 3.600 | 0.556 | 0.694 | 0.016 | 1.816 | 0.581  |
| 5 | 2.100 | 4.200 | 0.476 | 0.516 | 0.012 | 2.112 | −0.296 |

| 断面 | $R$/m | $\bar{R}$/m | $C$/(m^{1/2}·s⁻¹) | $\bar{C}$/(m^{1/2}·s⁻¹) | $\bar{J}$ | $i-\bar{J}$ | $\Delta l$/m | $\sum\Delta l$/m |
|---|---|---|---|---|---|---|---|---|
| 1 | 0.320 |       | 33.07 |        |        |         |       |         |
| 2 | 0.440 | 0.380 | 33.94 | 33.505 | 0.006 5 | −0.006 3 | 28.57  | 28.57   |
| 3 | 0.545 | 0.492 | 36.15 | 35.045 | 0.004 3 | −0.004 1 | 86.59  | 115.16  |
| 4 | 0.643 | 0.594 | 37.16 | 36.655 | 0.000 6 | −0.000 4 | 1 452  | 1 567.16 |
| 5 | 0.677 | 0.660 | 37.48 | 37.320 | 0.000 3 | −0.000 1 | 3 288  | 4 855   |

# 习　　题

1. 明渠均匀流有哪些水力特征？在什么条件下可能产生明渠均匀流？

2. 两条明渠的断面形状和尺寸均一样,而底坡和糙率不一样,当通过的流量相等时,问两明渠的临界水深是否相等？

3. 已修成的某梯形长直棱柱体排水渠道,长 $l=1.0$ km,底宽 $b=3$ m,边坡系数 $m=2.5$,底部落差为 0.5 m,设计流量 $q_v=9.0$ m³/s。试校核当实际水深 $h=1.5$ m 时,渠道能否满足设计流量的要求(粗糙系数 $n$ 取 0.025)。

4. 梯形断面壤土渠道通过流量 $q_v=10.5$ m³/s,底宽 $b=8.9$ m,边坡系数 $m=1.5$,正常水深 $h_0=1.25$ m,粗糙系数 $n=0.025$,求底坡 $i$,并判断渠道底坡是陡坡还是缓坡？流态是急流还是缓流？

5. 一矩形断面平坡渠道,底宽 $b=2.0$ m,流量 $q_v=10$ m³/s,当渠中发生水跃时,跃前水深 $h'=0.65$ m,求跃后水深 $h''$ 及水跃长度 $l_j$。

# 第10章 堰 流

堰是水务工程中主要的引水和泄水建筑物,常用来控制河渠中的流量和水位,同时也是一种重要的量水设备,被广泛地应用于各类实验室、给水工程及有关工业企业。研究堰流的主要目的是探求流经堰的流量 $q_v$ 与堰基本特征量之间的关系,为堰工程设计和过流能力的计算提供科学依据。

本章主要介绍了各类堰流的水力特征、基本公式、应用特点及水力计算方法等堰流的基础知识和桥孔的设计应用。学习中,要求掌握堰流的水力特征和基本公式,掌握薄壁堰、实用堰、宽顶堰的形式和分类;掌握流量计算和影响因素;了解小桥孔径的设计方法。

## 10.1 堰流的定义和分类

### 10.1.1 堰和堰流

**1. 定义**

明渠中设置堰墙或两边束窄的边墙等建筑物形成障壁,当无压缓流经障壁顶部下泄,溢流上表面不受约束的开敞水流称为堰流,此障壁称为堰(图10-1)。

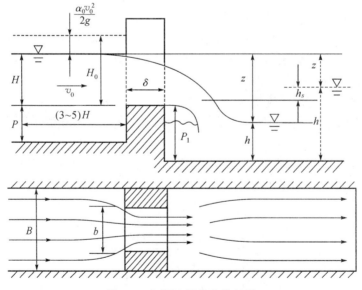

图10-1 堰流及基本特征量

**2. 堰流的基本特征量**

研究的主要目的是探讨流经堰的流量 $q_v$ 和堰流相关的基本特征量之间的关系。堰流的基本特征量包括(图 10-1)：

(1) 相关水头 $H$、$H_0$、$H_d$

$H$ 为堰顶水头(堰上作用水头)，距堰壁 $L=(3\sim5)H$ 处(不受堰顶水面降落的影响)为上游水位高程与堰顶高程之差。

$H_0$ 为堰顶的总水头(堰上总作用水头)，考虑了近流速影响时的堰上作用水头。即

$$H_0 = H + \frac{\alpha_0 v_0^2}{2g}$$

$H_d$ 为堰的设计水头，是堰体形设计时的参数。当实际水头 $H_0$ 和 $H_d$ 相等时，堰的设计流量就是堰的过流量。

(2) 相关宽度 $b$、$B$

$b$ 为堰宽，即水流溢过堰顶部的宽度。

$B$ 为渠宽，即溢流堰上游的渠道宽度。

(3) 堰高 $P$、$P_1$

上游堰高度 $P$ 和下游堰高度 $P_1$，分别为堰顶高程和上、下游的河床高差。

(4) 堰顶宽度 $\delta$

堰顶的厚度也称为堰壁厚度，堰顶宽度，上下游堰壁之间的厚度。

(5) 上游和下游的水位差 $z$

上下游水位差为 $(3\sim5)H$ 处的水位与堰壁下游正常水位之差。

(6) 堰下游水深深度为 $h$ 且下游水位高于堰顶高度 $h_s$(当下游水位比较高时，如图 10-1 中虚线所示)，堰前流速为 $v_0$ 等。

### 10.1.2 堰和堰流的分类

根据堰壁厚度 $\delta$ 和水头 $H$ 之间的关系，可将堰分为薄壁堰、实用堰和宽顶堰。

(1) 薄壁堰，$\dfrac{\delta}{H}<0.67$，如图 10-2a 所示。过堰水流形成"水舌"，水舌下缘先上弯后回落，落至堰顶高程时，距上游壁面约 $0.67H$，堰顶厚度 $\delta<0.67H$ 时，堰顶对水舌无干扰，这样的堰被称为薄壁堰。

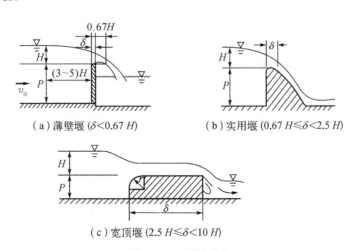

图 10-2 堰流分类

(2) 实用堰，$0.67 \leqslant \dfrac{\delta}{H} < 2.5$，如图 10-2b 所示。过堰水流受到堰顶的约束和顶托，但堰上水流仍有一次连续跌落，这样的堰称为实用堰。

(3) 宽顶堰，$2.5 \leqslant \dfrac{\delta}{H} < 10$，如图 10-2c 所示。堰顶出现近似水平流动，堰顶厚度对水流有顶托作用，能形成二次水跌，这样的堰称为宽顶堰。

当 $\dfrac{\delta}{H} > 10$ 时，因沿程水头损失不能忽略，故是明渠，而不是堰。

堰的分类还有其他方法，如按堰宽 $b$ 与渠宽 $B$ 是否相等，可分为侧收缩堰 ($b<B$) 和无侧收缩堰 ($b=B$)。还可以根据下游水位是否影响堰的过流能力将堰流分为自由式堰流（自由出流）和淹没式堰流（淹没出流），如图 10-3 所示。

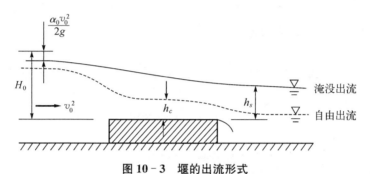

**图 10-3　堰的出流形式**

## 10.2　堰流的基本公式

### 10.2.1　无侧收缩堰自由出流基本公式

下面以薄壁矩形无侧收缩堰自由出流为例，推导堰流基本公式。

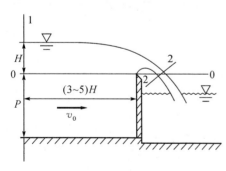

**图 10-4　薄壁堰的出流方式**

如图 10-4 所示，过堰顶取基准面 0—0，距上游堰面 $(3\sim5)H$ 处取渐变流断面 1—1，过基准面与水舌中心线的交点取过水断面 2—2，列伯努利方程，有

$$H + \frac{P_a}{\lambda g} + \frac{\alpha_0 v_0^2}{2g} = \frac{P_2}{\lambda g} + \frac{\alpha_2 v_2^2}{2g} + \zeta \frac{v_2^2}{2g}$$

因水舌上下游与大气接触，故式中 $P_2 = P_a$，令 $H_0 = H + \dfrac{\alpha_0 v_0^2}{2g}$，$\alpha_2 = \alpha$，$v_2 = v$，则有

$$v = \frac{1}{\sqrt{\alpha + \zeta}} \sqrt{2gH_0} = \varphi \sqrt{2gH_0}$$

故

$$q_v = Av = \varphi be \sqrt{2gH_0}$$

式中，$\varphi = \dfrac{1}{\sqrt{\alpha+\zeta}}$ 为流速因数，$B=b$ 为溢流宽度，$e$ 为断面 2—2 上水舌厚度。

若令 $e = kH_0$，则上式为

$$q_v = \varphi k e \sqrt{2g} H_0^{3/2} = mb\sqrt{2g} H_0^{3/2} \tag{10-1}$$

式中，$m = k\varphi$，称为堰流流量因数。上式是无侧收缩堰自由出流的基本公式。

如果将行近流速 $v_0$ 的影响计入流量因数中，则式（10-1）可写为

$$q_v = m_0 b \sqrt{2g} H_0^{3/2} \tag{10-2}$$

式中，$m_0$ 为计入行近流速的堰流流量因数。

### 10.2.2 淹没流和侧缩影响的堰流公式

当下游水位超过堰的一定高度时，就会发生淹没式溢流，当流速不变时，上游水位被下游水位的回水抬高，堰的泄流能力开始下降。式（10-1）的右边需要乘以淹没因数 $\sigma(\sigma<1)$。

当实用堰的宽度小于渠道的宽度时，受水闸闸墩和边墩的影响，水流进入堰后，边墙收缩。因此，流经该段的有效流量减少，水头损失增加，堰的流通能力下降。式（10-1）的右边需要乘以侧缩系数 $\varepsilon(\varepsilon<1)$。侧缩系数与水闸闸墩和侧墩的形状有关。

考虑到潜流和侧缩的影响，式（10-1）可以写成

$$q_v = \sigma \varepsilon m b \sqrt{2g} H_0^{3/2} \tag{10-3}$$

上式适用于各种类型的围堰和围堰的各种影响因素。

上述各式就是适用于各种堰型及各种影响因素的堰流普遍公式。以上各式虽然是由薄壁堰推导而得，但由于式中因数 $\delta$、$\varepsilon$、$m(m_0)$ 都是根据不同堰型、不同影响因素及影响程度由试验得出的，不同堰型、不同流动情况下有不同值。因此，以上各式对其他堰型也普遍适用。

## 10.3 薄壁堰

### 10.3.1 矩形薄壁堰

如图 10-5 所示，堰口形状为矩形的薄壁堰，称为矩形堰。常用作量水设备，通常是根据实测的水头 $H$ 来计算流量，而将行近流速 $v_0$ 的影响计入流量因数中。且常将堰宽 $b$ 与渠宽 $B$ 做成相等，使其无侧收缩影响。故流量计算公式为式（10-2）。

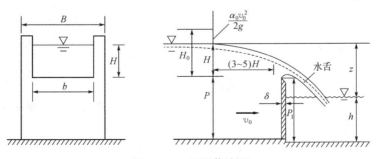

图 10-5 矩形薄壁堰

$$q_v = m_0 b\sqrt{2g} H^{3/2}$$

$$m_0 = \left(0.405 + \frac{0.0027}{H}\right)\left[1 + 0.55\left(\frac{H}{H+P}\right)^2\right] \tag{10-4}$$

式中，$H$ 为堰上水头；$P$ 为上游堰高；$b$ 为堰宽，均以米计。适用条件：$0.1\text{ m} < H < 1.24\text{ m}$，$0.2\text{ m} < b < 2\text{ m}$，$0.2\text{ m} < P < 1.13\text{ m}$。

如果有侧缩，流量因数 $m_0$ 包括在侧缩中。它应该根据以下经验公式来计算：

$$m_0 = \left(0.405 + \frac{0.0027}{H} - 0.03\frac{B-b}{B}\right)\left[1 + 0.55\left(\frac{H}{H+P}\right)^2\left(\frac{b}{B}\right)\right]^2 \tag{10-5}$$

式中，$B$ 为渠道宽度；$B$、$H$、$P$ 的单位是米。

### 10.3.2 三角形薄壁堰

堰口呈三角形的薄壁堰，称为三角形堰，如图 10-6 所示。它通常用于实验室作小流量（$q_v < 0.1\text{ m}^3/\text{s}$）的测量设备。当堰口作成直角（$\theta = 90°$）时叫直角三角形堰，它的流量计算公式常用汤普森经验公式

$$q_v = 1.4 H^{5/2} \tag{10-6}$$

**图 10-6 三角形薄壁堰**

式中，$q_v$ 的单位为 $\text{m}^3/\text{s}$，$H$ 的单位以 m 计，适用范围 $0.05\text{ m} < H < 0.25\text{ m}$，$P \geqslant 2H$，$B \geqslant (3 \sim 4)H$。

实验表明，薄壁堰的淹没条件是：
(1) 下游水位超过堰顶。
(2) 下游发生淹没水跃。

用薄壁堰进行流量量测时还需注意：
(1) 因淹没水跃水面波动大，不宜在淹没条件下工作。
(2) 水舌的下方应保持与大气相通，否则影响测量精度。

## 10.4 实用堰

实用堰根据其剖面形式，可分为曲线形（图 10-2b）和折线形（图 10-7）两类。这里只讨论曲线形实用堰。

### 10.4.1 堰面水力特性

如图 10-8 所示，曲线形实用堰的剖面形状，由下列几部分组成：
(1) 上游直线段 $AB$；
(2) 堰顶曲线段 $BOC$，是影响堰泄流能力的最主要部位；
(3) 下游坝坡段 $CD$ 和反弧段 $DE$。

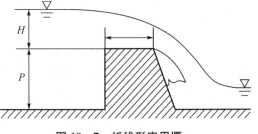

**图 10-7 折线形实用堰**

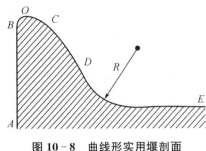

图 10-8  曲线形实用堰剖面

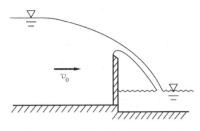

图 10-9  薄壁堰水舌下缘曲线

对照图 10-9 可知,实用堰堰顶曲线实际上是根据薄壁堰水舌下缘曲线设计的。所设计的堰顶曲线,在同样设计水头下,与薄壁堰自由出流时的水舌下缘相比,如果稍微高胖一点,则堰顶水流对堰面有正压作用,形成"非真空剖面堰"。如略为低瘦一点,则在堰面上产生真空,形成"真空剖面堰"。实验表明,堰面压力减小,过流能力增大。真空堰能增大泄水能力,但当真空值过大时,常导致堰体振动,引起堰面发生空蚀破坏,且水流也不稳定。因此,真空堰在实际工程中应尽量避免。

从实验得知,流量因数 $m$ 值随作用水头 $H_0$ 的增大而增大,堰面压强随作用水头增大而减小。因此,对同一个堰体而言,即使设计水头下是非真空堰,但当作用水头 $H_0$ 大于堰的设计水头 $H_d$ 时,也可能出现堰面真空,形成真空堰。

曲线形实用堰有多种成熟的堰形,其中美国水道实验站给出的 WES 剖面堰,目前工程界采用最广泛。

### 10.4.2  流量计算公式

实用堰的流量计算公式仍为式(10-3),但式中实用堰的流量因数 $m$,对不同堰形有不同值。上游面为垂直的 WES 剖面堰,流量因数 $m$ 公式为:

$$m = -0.024\frac{H_0}{H_d} + 0.185\,03\sqrt{\frac{H_0}{H_d}} + 0.340\,6 \tag{10-7}$$

式(10-7)的适用条件是 WES 剖面与堰上游面垂直,且 $\frac{P}{H_d} \geqslant 1.33$,$\frac{H_0}{H_d} < 1.5$。式中 $H_0$ 为实际过堰水流的总水头,$H_d$ 为堰剖面的设计水头,$P$ 为上游堰高。WES 堰的坝面压力,当 $\frac{H_0}{H_d} \leqslant 1.0$ 时,无负压;当 $\frac{H_0}{H_d} > 1.0$ 时,产生负压;当 $\frac{H_0}{H_d} = 1.33$ 时,负压可达 $0.5H_d$。由式(10-7)知,当 $\frac{H_0}{H_d} = 1.0$ 时,其流量因数可取 $m = m_d = 0.502$。

实用堰的淹没条件与薄壁堰相同,即:
(1) 下游水位超过堰顶。
(2) 下游发生淹没水跃。

## 10.5  宽顶堰

### 10.5.1  宽顶堰的水力特性

宽顶堰可分为有坎宽顶堰和无坎宽顶堰,如小桥桥孔和短涵洞过水,有槛或无槛(平底)的节制闸当闸门全开时的过流等。它们自由出流时的过流特点都具有如图 10-10 所示的宽顶堰流典型特征,即堰顶水面有两次降落。在堰进口不远处水面发生第一次降落,堰顶形成一收缩水深,此收缩水深略小于堰顶断面的临界水深。以后形成流线近似平行于堰顶的渐变流,接着在堰尾产生水面二次降落。工程上,对明渠缓流中的过障壁水流,凡符合宽顶堰的水流特征,且满足 $2.5 < \dfrac{\delta}{H} < 10$ 的条件,则一般都按宽顶堰考虑。

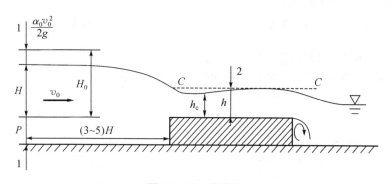

图 10-10 宽顶堰

### 10.5.2 流量计算公式

宽顶堰的流量计算公式仍为式(10-3),式中各因数介绍如下。

**1. 宽顶堰流量因数 m**

如图 10-10 所示,取断面 1—1 和 2—2,建立能量方程,忽略水头损失 $h_w$ 并取 $\alpha = 1.0$ 时,有

$$v = \sqrt{2g(H_0 - h)}$$

则

$$q_l = h\sqrt{2g(H_0 - h)}$$

因不计损失,此时过堰流量最大,有

$$\frac{\partial q_l}{\partial h} = \frac{\partial (h\sqrt{2g(H_0 - h)})}{\partial h} = 0 \tag{10-8}$$

$$H_0 = \frac{3}{2}h \tag{10-9}$$

由式(10-8)与式(10-9)可得

$$h = \sqrt[3]{\frac{q_l^2}{g}} = h_c \tag{10-10}$$

将式(10-9)及式(10-10)代入式(10-1),整理得

$$m = 0.385$$

表明不计水流阻力,$m = 0.385$ 为宽顶堰流量因数的最大值。一般情况下,$m$ 值在 0.32~

0.385 之间。

对直角进口宽顶堰,当 $0 < \dfrac{P}{H} \leqslant 3$ 时,

$$m = 0.32 + 0.01 \dfrac{3 - \dfrac{P}{H}}{0.46 + 0.75 \dfrac{P}{H}} \qquad (10\text{-}11)$$

当 $\dfrac{P}{H} > 3$ 时,$m = 0.32$。

对堰顶圆弧进口,当 $\dfrac{r}{H} \geqslant 0.2$ 时($r$ 为进口圆弧半径),

$$m = 0.36 + 0.01 \dfrac{3 - \dfrac{P}{H}}{1.2 + 1.5 \dfrac{P}{H}} \qquad (10\text{-}12)$$

当 $\dfrac{P}{H} > 3$ 时,$m = 0.36$。

**2. 宽顶堰的侧收缩系数 ε**

宽顶堰的侧收缩系数 ε 用单孔宽顶堰的经验公式计算,为

$$\varepsilon = 1 - \dfrac{a}{\sqrt[3]{0.2 + \dfrac{P}{H}}} \sqrt[4]{\dfrac{b}{B}} \left(1 - \dfrac{b}{B}\right) \qquad (10\text{-}13)$$

式中,$a$ 为墩形系数,矩形边墩 $a = 0.19$,圆形边墩 $a = 0.1$。

**3. 宽顶堰的淹没因数 σ**

宽顶堰的淹没因数 σ 使用经验公式计算,为

$$\sigma = -96.01 \left(\dfrac{h_s}{H_0}\right)^3 + 235.32 \left(\dfrac{h_s}{H_0}\right)^2 - 193.23 \left(\dfrac{h_s}{H_0}\right) + 54.14 \quad \left(0.8 \leqslant \dfrac{h_s}{H_0} \leqslant 0.98\right)$$

$$(10\text{-}14)$$

由试验得出宽顶堰的淹没判别标准是:

$$\dfrac{h_s}{H_0} \geqslant 0.8 \qquad (10\text{-}15)$$

**例 10-1** 某矩形断面渠道,为引水灌溉修筑了单孔宽顶堰,如图 10-11 所示。已知:堰前渠中流速 $v_0 = 1.48 \,\text{m/s}$,渠道宽 $B = 3 \,\text{m}$,堰顶宽 $b = 2 \,\text{m}$,堰高 $h_1 = h_2 = 1 \,\text{m}$,堰顶水头 $H = 2 \,\text{m}$,堰顶进口为直角进口,墩头形状为矩形,下游水深 $h = 2 \,\text{m}$。试求过堰流量。

**解:**(1)判别出流形式。因为

$$h_s = h - h_2 = 2 - 1 = 1 \,\text{m} > 0$$

$$H_0 = H + \dfrac{\alpha_0 v_0^2}{2g} = 2 + \dfrac{1.48^2}{2 \times 9.8} = 2.11 \,\text{m}$$

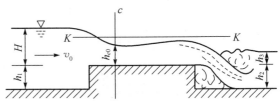

图 10-11 单孔宽顶堰

所以 $\dfrac{h_s}{H_0} < 0.8$，不满足淹没出流条件，为自由出流，$\sigma_s = 1$。

(2) 计算流量因数

因为 $b < B$，所以有收缩现象。堰口为直角进口，$\dfrac{h_1}{H} = 0.5 < 3$，由式得

$$m = 0.32 + 0.01 \dfrac{3 - \dfrac{h_1}{H}}{0.46 + 0.75 \dfrac{h_1}{H}} = 0.35$$

(3) 计算收缩系数

该堰为单孔宽顶堰，故由式可得

$$\varepsilon = 1 - \dfrac{a}{\sqrt[3]{0.2 + \dfrac{h_1}{H}}} \sqrt[4]{\dfrac{b}{B}} \left(1 - \dfrac{b}{B}\right)$$

$$= 1 - \dfrac{0.19}{\sqrt[3]{0.2 + 0.5}} \times \sqrt[4]{\dfrac{2}{3}} \times \left(1 - \dfrac{2}{3}\right) = 0.936$$

(4) 计算流量

$$q_v = \sigma \varepsilon m b \sqrt{2g}\, H_0^{3/2} = 1 \times 0.936 \times 0.35 \times 2 \times \sqrt{2 \times 9.8} \times 2.11^{1.5} = 8.88 \text{ m}^3/\text{s}$$

## 10.6 小桥孔径水力计算

### 10.6.1 小桥孔过流的水力计算

**1. 小桥孔过流的淹没标准**

与宽顶堰溢流一样，小桥过流也分为自由出流和淹没出流两种情况。由实验得知，当桥的下游水深 $h < 1.3 h_c$（$h_c$ 为桥孔水流的临界水深）时，桥孔水流为急流，下游水位不影响过桥水流，小桥过流为自由式，如图 10-12 所示；当桥下游水深 $h \geqslant 1.3 h_c$ 时，桥孔水流为缓流，下游水位影响到上游，小桥过流为淹没式，如图 10-13 所示。

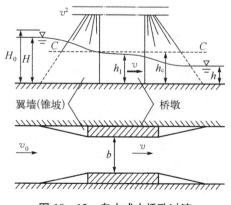

图 10-12 自由式小桥孔过流

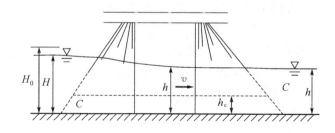

图 10-13 淹没式小桥孔过流

**2. 水力计算公式**

(1) 桥孔水深 $h_1$

工程设计中，$h_1$ 常用下列关系确定：

$$h_1 = \begin{cases} h & \text{（淹没式）} \\ \psi h_c & \text{（自由式）} \end{cases} \qquad (10-16)$$

式中，$\psi$ 为垂向收缩系数，视小乔进口形式而定，非平滑进口 $\psi = 0.75 \sim 0.80$，平滑进口 $\psi = 0.8 \sim 0.85$，有的设计方法认为 $\psi = 1.0$。

(2) 水力计算公式

小桥过流水力计算公式可由恒定总流的伯努利方程和连续性方程导出，为

$$v = \varphi \sqrt{2g(H_0 - h_1)} \qquad (10-17)$$

$$q_v = \varepsilon b h_1 \varphi \sqrt{2g(H_0 - h_1)} \qquad (10-18)$$

### 10.6.2 水力计算要求

为了设计一座安全、经济的桥梁，水力计算应满足以下要求：

(1) 孔径能够使桥孔过流能力达到水文计算确定的设计流量。

(2) 桥孔处的流速不能超过河床土壤或铺砌材料不冲刷的允许流速。

(3) 桥前水深不超过路肩及桥梁底部至水面的超高决定的允许水深。

### 10.6.3 水力计算方法

**1. 计算桥孔的临界深度 $h_c$**

因侧收缩影响，有效宽度为 $\varepsilon b$，则临界水深为

$$h_c = \sqrt[3]{\frac{\alpha q_v^2}{g(\varepsilon b)^2}} \qquad (10-19)$$

由连续性方程，考虑到 $h_1 = \psi h_c$，有

$$q_v = \varepsilon b h_1 v_{\max} = \varepsilon b \psi h_c v_{\max}$$

代入式(10-19)中，得到允许速度 $v_{\max}$ 和临界水深 $h_c$ 之间的关系：

$$h_c = \frac{\alpha \psi^2 v_{\max}^2}{g} \qquad (10-20)$$

**2. 计算桥孔孔径 $b$**

根据 $q_v = \varepsilon b h_1 v_{\max}$，小桥孔径大小为

$$b = \frac{q_v}{\varepsilon h_1 v_{\max}} \qquad (10-21)$$

在工程上，桥梁开孔一般采用标准孔径 $B(B > b)$。实际工程中常采用标准孔径，铁路、公路桥梁的标准孔径有 4 m、5 m、6 m、8 m、10 m、12 m、16 m、20 m 等多种。

## 3. 校核桥前壅水水深 $H$

桥前壅水水深可以通过以下公式检查：

$$H \approx H_0 = h_1 + \frac{q_v^2}{2g\varphi^2(\varepsilon B h_1)^2} \leqslant H' \tag{10-22}$$

式中，$H'$ 为允许壅水水深；$h_1$ 为桥孔水深，如果是自由出流，$h_1 = \psi h_c$；如果是淹没出流，$h_1 = h$。

**例 10-2** 由水文计算已知小桥设计流量 $q_v = 30 \text{ m}^3/\text{s}$。根据下游河段流量-水位关系曲线，求得该流量时下游水深 $h = 1.0 \text{ m}$。桥前允许壅水水深 $H' = 2 \text{ m}$，桥下允许流速 $v' = 3.5 \text{ m/s}$。由小桥进口形式，查得各项系数：$\varphi = 0.9$，$\varepsilon = 0.85$，$\psi = 0.8$。试设计此小桥孔径。

**解：**（1）计算临界水深

$$h_c = \frac{\alpha \psi^2 v'^2}{g} = \frac{1.0 \times 0.8^2 \times 3.5^2}{9.8} = 0.8 \text{ m}$$

故此小桥过流为自由出流。

（2）计算小桥孔径

$$b = \frac{q_v}{\varepsilon v' \psi h_c} = \frac{30}{0.85 \times 0.8 \times 0.8 \times 3.5} = 15.8 \text{ m}$$

取标准孔径 $B = 16 \text{ m}$。

（3）重新计算临界水深

$$h_c = \sqrt[3]{\frac{\alpha q_v^2}{g(\varepsilon b)^2}} = \sqrt[3]{\frac{1 \times 30^2}{9.8 \times (0.85 \times 16)^2}} = 0.792 \text{ m}$$

$$1.3 h_c = 1.3 \times 0.792 = 1.03 \text{ m} > h$$

故小桥过流仍为自由出流。桥孔的实际流速为

$$v = \frac{q_v}{\varepsilon B \psi h_c} = \frac{30}{0.85 \times 16 \times 0.8 \times 0.792} = 3.48 \text{ m/s}$$

$v < v'$，不会发生冲刷。

（4）验算桥前壅水水深

$$H \approx H_0 = \frac{v^2}{2g\varphi^2} + \psi h_c = \frac{3.48^2}{19.6 \times 0.9^2} + 0.8 \times 0.792 = 1.396 \text{ m} \leqslant H'$$

$H < H'$，满足设计要求。

## 习 题

1. 在一矩形断面水槽末端设置一矩形薄壁堰，水槽宽 $B = 2.0 \text{ m}$，堰宽 $b = 1.2 \text{ m}$，堰高 $P = P' = 0.5 \text{ m}$。求水头 $H = 0.25 \text{ m}$ 时，非淹没堰流的流量。

2. 有一宽顶堰，堰顶厚度 $\delta = 16 \text{ m}$，堰顶水头 $H = 2 \text{ m}$，如上、下游水位及堰高均不变，当 $\delta$ 分别减小至 $8 \text{ m}$ 及 $4 \text{ m}$ 时，该堰是否还属于宽顶堰？

3. 已知设计流量 $q_v = 15 \text{ m}^3/\text{s}$,允许流速 $v' = 3.5 \text{ m/s}$。桥下游水深 $h = 1.3 \text{ m}$。取 $\varepsilon = 0.9$,$\psi = 1.0$,$\varphi = 0.9$,允许壅水高度 $H' = 2.2 \text{ m}$,试设计小桥孔径 $B$。

4. 一直角进口无侧收缩宽顶堰,堰宽 $b = 4.0 \text{ m}$,堰高 $P_1 = P_2 = 0.6 \text{ m}$,堰前水头 $H = 1.2 \text{ m}$,当堰下游水深 $h = 0.8 \text{ m}$ 和 $1.7 \text{ m}$ 时,其通过的流量 $q_v$ 分别是多少?

5. 某矩形河渠中建造曲线型实用溢流堰,下游堰高 $h_1 = 6 \text{ m}$,溢流宽度 $B = 60 \text{ m}$,过堰流量 $q_v = 480 \text{ m}^3/\text{s}$,堰的流量因数 $m = 0.45$,流速因数 $\varphi = 0.95$。求堰下游收缩断面水深 $h_c$。